Soil Erosion

LANDSCAPE SYSTEMS

A series in Geomorphology

Editor

M. J. Kirkby, *School of Geography,*
University of Leeds

Hillslope Hydrology:
Edited by
M. J. Kirkby, *School of Geography,*
University of Leeds

Soil Erosion:
Edited by
M. J. Kirkby, *School of Geography,*
University of Leeds

and

R. P. C. Morgan, *National College of Agricultural Engineering,*
Bedford

A publication of the British Geomorphological Research Group

Soil Erosion

Edited by

M. J. Kirkby
School of Geography
University of Leeds

and

R. P. C. Morgan
National College of Agricultural Engineering
Bedford

A Wiley–Interscience Publication

JOHN WILEY & SONS

CHICHESTER · NEW YORK · BRISBANE · TORONTO

British Library Cataloguing in Publication Data:

Soil erosion.—(Landscape systems).
  1. Soil erosion
  I. Kirkby, Michael John
 II. Morgan, R. P. C.
III. British Geomorphological Research Group
IV. Series
631.4′5   S623   80-40282

ISBN 0 471 27802 5

Set in Monophoto Times and printed in Great Britain by
Page Bros (Norwich) Ltd, Norwich.

Contributors

GARY D. BUBENZER, *Department of Agricultural Engineering, University of Wisconsin-Madison, 460 Henry Mall, Madison, Wisconsin 53706, USA*

R. U. COOKE, *Department of Geography, Bedford College, Regents Park, London NW1 4NS*

JAN DE PLOEY, *Instituut voor Aardwetenschappen, Laboratorium voor Experimentele Geomorfologie, Redingenstraat 16bis, 3000 Leuven, Belgium*

R. EVANS, *Soil Survey of England and Wales, Government Buildings, Block B, Brooklands Avenue, Cambridge CB2 2DR*

D. GABRIELS, *Leerstoel voor Bodemfysika, Rijksuniversiteit-Gent, B 9000 Gent, Belgium*

M. J. KIRKBY, *School of Geography, University of Leeds, Leeds LS2 9JT*

J. KENT MITCHELL, *Department of Agricultural Engineering, University of Illinois at Urbana-Champaign, 245 Agricultural Engineering Building, Urbana, Illinois 61801, USA*

R. P. C. MORGAN, *National College of Agricultural Engineering, Silsoe, Beds MK45 4DT*

J. B. THORNES, *Department of Geography, London School of Economics, Houghton Street, London WC2A 2AE*

S. J. WILSON, *Department of Geography, Bedford College, Regents Park, London NW1 4NS*

Contents

Series preface

Knowledge of our immediate physical environment has been gathered equally from *descriptive* and from *experimental* studies. Within descriptive studies I would include the analysis of geographical distributions, the construction of biological taxonomies, and the elucidation of the stratigraphic column. Experimental studies are mainly concerned with the study of processes at all scales. A science may develop from a purely descriptive to an experimental stage, and, in principle, the physical environment responds in a deterministic manner to external changes, and could therefore be studied as purely experimental science. In detail, however, our environment shows great variability even within a small area, so that accurate prediction often appears to be an impracticable goal. At the same time, there are, for instance, only a limited number of major climatic types, so that theories of, say, landscape formation can never be more than partially tested. With such difficulties, environmental science still seems to be a long way from a purely experimental stance, and consequently the need for collaboration between descriptive and experimental scientists will be with us for the foreseeable future.

Although it was, of course, backed by a considerable body of previous research, I believe that the present phase of quantitative geomorphology was signalled most significantly by the publication of Leopold, Wolman and Miller's *Fluvial Processes in Geomorphology* in 1964. Since that date, newer books have made this recent work accessible to schools, while others are more specialist and advanced texts. The present series, on *Landscape Systems*, is an attempt to bring work in active research fields to a wider audience at postgraduate level, and to draw together concurrent work in the several disciplines relevant to each of the fields covered.

Each volume of the series is being written by a number of authors, each of whom is contributing within his own field of expertise, but has been asked to collaborate within an overall framework. In this way, important topics in geomorphology are covered comprehensively enough to provide a coherent advanced text without masking differences of approach between the contributors. The series covers topics in geomorphology, but is highly relevant to other disciplines. For example, the volume on *Hillslope Hydrology* is very directly applicable to workers and students in Hydrology, Forestry, Agriculture or Soil Science; and a similar breadth of relevance is intended for the second (*Soil Erosion*) and subsequent volumes.   PROFESSOR M. J. KIRKBY

Preface

Research on soil erosion has been focused on the work of the US Soil Conservation Service since erosion was recognized as a serious agricultural problem in the late 1920s, and the SCS was established to combat it. Current work by Agricultural Engineers in the USA has continued to build on the very considerable experience of the SCS which has been systematized as the Universal Soil Loss Equation.

Since 1960 substantially new approaches to soil erosion research have been developed, both in the USA and in Europe. The most significant new development is that the geomorphological and hydrological processes of erosion are being re-examined in detail. The basic assumptions of the Universal Soil Loss Equation are being questioned, and workers in separate fields are beginning to make independent contributions to aspects of overland flow, infiltration, soil crusting, and soil removal. This book is an attempt to bring this work together for the first time in a research level text. The authors who have contributed are all active in soil erosion research, and bring a range of current viewpoints together. Each chapter may be read in isolation, but chapter topics have been designed to provide a sequential text which covers the whole of soil erosion as we see it.

Chapter 1 is an introductory survey of the field, showing some of the important areas for research development. Chapter 2 sets out the Universal Soil Loss Equation and how it may be used and developed further. Chapter 3 reviews established and new methods for measuring soil erosion. Chapters 4, 5, and 6 examine the processes of water erosion in detail, from empirical, theoretical, and modelling viewpoints respectively. Chapter 7 examines wind erosion similarly. Chapter 8 looks at the implications of new research for soil management, and Chapter 9 concludes briefly by looking to the future of erosion research.

In editing this book, we would like to acknowledge all the help we have received from all our fellow authors who have patiently revised and updated their chapters to bring the book together as a meaningful whole. Their encouragement in getting this book together from conception through to production has been a stimulating and essential ingredient of what we have always seen as a fully cooperative venture. Our thanks also go to colleagues at Leeds and Silsoe who have read and listened to drafts and ideas, and to Marjie Salisbury who has kept all the pieces together and typed or retyped quite a few of them.

M. J. KIRKBY
R. P. C. MORGAN

Soil Erosion
Edited by M. J. Kirkby and R. P. C. Morgan
© 1980 John Wiley and Sons Ltd.

CHAPTER 1

The problem

M. J. Kirkby
*Professor of Physical Geography,
University of Leeds, UK*

1.1 EROSION IN CONTEXT

Soil erosion is the removal of surface material by wind or water. Later chapters examine the processes acting, the measurements or prediction of their effects and ways to control harmful effects. Before getting involved in all this detail, it is helpful to stand back and survey soil erosion as a whole. The broadest view of soil erosion is to compare it with other processes of landscape denudation, because soil erosion should be recognized as the dominant problem only when and where it is the most rapid process. A broad view also helps to show the significance of rates of soil erosion within a geological time-scale, and asks what rates can be tolerated in the long-term. A second, narrower overview examines soil erosion with its immediate climatic and vegetational controls, and questions how well the processes involved in raindrop impact, flow generation, and sediment resistance are understood at this level. A third way of looking at soil erosion is through its broad patterns in time and space. At the present, the distributions of soil erosion through the year and around the world can only partially be explained, but the explanation helps to define the most urgent problems facing soil conservation in various environments. This chapter attempts to review soil erosion in these three ways, in order to stimulate new lines of approach and question traditional work. It is therefore a personal statement and not an annotated bibliography.

Soil erosion is one normal aspect of landscape development, but only in some parts of the world does it dominate other processes of denudation. The other major processes of sediment removal are *mass movement* and *solution*, and each of these is also dominant in suitable environments. A brief survey of these three major groups of processes helps to put soil erosion and soil conservation problems in perspective on a geological time-scale in a world-wide context.

An uplifted mountain mass is commonly dissected rapidly by rivers into steep-sided valleys where landslides, rock-fall, and other types of mass movement are

1

highly effective in widening the valleys and lowering their slope gradients. Rapid mass movements are the dominant processes at this stage, but they are characterized by a threshold angle, below which the slopes become stable. The stable angle is broadly related to the properties of the bedrock and soils of the area, and to a lesser extent on rock structures, climate, and groundwater conditions. Vegetation cover and type also play a minor part, so that land use changes can influence slope stability. Thus coarse sandstones typically stabilize at around 30°, whereas shales stabilize at 8–15° in humid climates, but at somewhat steeper slopes in arid areas. During the mass movement stage, soil erosion and solution, although active, usually play such a subsidiary role to the mass movements that they can be ignored in comparison.

Once slopes are at gradients which are stable with respect to rapid mass movements, the slower processes of soil erosion and solution become dominant. Where rainfall generally percolates into the soil, it flows slowly within soil and rock pores. As a result it can carry little or no material along in its flow, but is able to come to a near-equilibrium with the finest material in the soil. It is therefore highly effective in dissolving material from the soil and bedrock. This not only lowers the land surface, but also progressively converts bedrock into new soil. In humid temperate areas, where rainfall intensities are only moderate and soils are permeable, most rainfall follows this subsurface route, and solution may be the dominant process of landscape evolution. A climate like that of Britain fulfils these conditions, and solution dominates landscape lowering, at a rate of 10–100 μm/year. These rates are low, and may be exceeded by up to five times this amount under extremely humid climates, but it is inherent in the solution process that it can never be catastrophic in its effects.

Soil erosion by water is most active where solution is least active. Where rainfall cannot infiltrate the soil, but flows over the surface, it travels relatively fast, and is able to carry soil materials away through the hydraulic force of its flow. At the same time, it is only in contact with the surface soil for an hour or two, and not for the several days needed to pick up an appreciable amount of dissolved material. Thus where overland flow is dominant, soil erosion by water is likely to be the main process of denudation, and solution to be slight. As large depths of water can flow over the surface, and exert correspondingly large hydraulic forces, it follows that soil erosion sometimes acts catastrophically, even on low gradients. These conditions are most commonly met in semi-arid areas, but fields cleared for cultivation are liable to erosion in almost any climate. Severe soil erosion associated with gully incision can initiate mass movements from steepened slopes around the gullies. Such mass movements play what may be an important part in the total sediment removal.

Soil erosion by wind, like erosion by water, relies on the force which the fluid (in this case air) can exert on soil particles. For any fluid, this force depends to some extent on the roughness of the surface, but for wind roughness plays a particularly critical role because of the low density and hence transporting capacity of air. Where the surface is very rough, with for example plants or large stones which cannot be lifted by the wind, then the wind speed near the surface is low, and little erosion takes place. Any relatively smooth surface, like a bare field, is however liable to wind erosion, and the risk is increased where the soil contains appreciable amounts of silt-sized material, which settle out of the air only slowly once they have been picked up. These conditions are naturally most common in hot and cold deserts and along broad flood plains and coast-lines, but removal of vegetation and hedgerows for cultivation can allow wind erosion elsewhere.

There are few reliable measurements of wind erosion under natural conditions, but losses of up to 10 mm/year were recorded from Kansas 'Dustbowl' sites over the 20 year period before conservation measures were introduced. Water erosion rates do not normally match this extreme, although many undisturbed semi-arid areas erode at up to 1000 μm (1 mm)/year. In contrast, water erosion in lowland Britain under a vegetation cover rarely exceeds 1 μm/year. These rates are substantially exceeded where the natural vegetation is disturbed, and the effects of extreme water erosion are visually emphasized by the development of gully systems along which the erosion is concentrated.

In any study of soil erosion, it quickly becomes clear that effective ground cover, usually in the form of vegetation, is the key to any understanding of the processes involved. Under uncultivated conditions, a journey from desert to forest, through areas of progressively greater annual rainfall shows two effects on soil erosion. More rain leads to increased overland or subsurface runoff and to more vegetation. At the desert end of the journey, there is, to begin with, so little vegetation that the increasing runoff leads to increasing erosion. Wind erosion is marginally greatest in deserts in absolute terms, and appears relatively very dominant because water erosion rates are low. As soon as the rainfall is high enough to support semi-arid vegetation, the increasing vegetation cover does more to limit water erosion than the greater runoff does to increase it, so that net erosion actually decreases as rainfall increases. This reversal con-tinues until a complete forest cover is obtained. At still higher rainfalls, the vegeta-tion can offer little further protection and erosion may begin to increase again, though it never reaches the semi-arid peak values (Figure 1.1). This argument is supported by US data from Langbein and Schumm (1958) and some tropical data (Douglas, 1967), and seems plausible, although it is not universally accepted.

In general terms therefore, there seems to be an increase of erosion with rain-fall, and a decrease with vegetation cover. 'Natural' vegetation cover depends strongly on rainfall, producing the rather complex variation of erosion with rainfall described above, but agriculture, particularly in association with irriga-tion, can make rainfall and vegetation cover partially independent of one another. Without looking too closely, at this stage, into the mechanisms of erosion, it is clear that where farming produces a radical change in the vegetation cover, a corresponding change in erosion will result. Thus seasonal clearing of

M. J. Kirkby

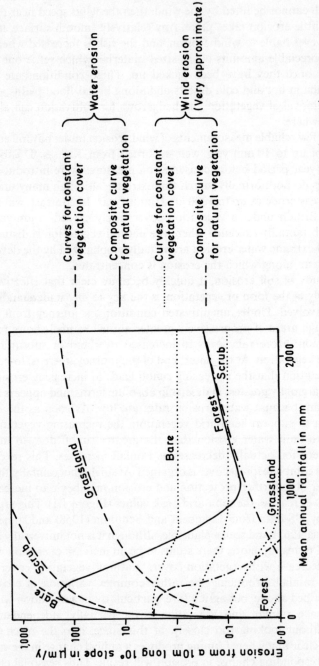

Figure 1.1. Estimated rates for soil erosion by wind and water as a function of rainfall and vegetation cover

the ground for sowing produces an increased risk of erosion for a month or so, especially if it occurs during a season of high intensity rain. Similarly heavy grazing may reduce the vegetation cover year-round and lead to an erosion rate which is always higher than the natural value. Sensitivity to increased erosion resulting from farming is seen to depend on the extent of the change in vegetation cover, the total rainfall at periods of low cover, and the intensity of the rains.

In addition to these man-induced changes, semi-arid areas under natural vegetation show a greater proportionate change in erosion rates with rainfall than other environments (see Figure 1.1), though in the reverse direction. Semi-arid areas are therefore very sensitive to small natural changes in climate, and in such areas it is difficult to separate natural from man-induced changes in erosion rates. The sensitivity of semi-arid areas to changes in erosion rate is therefore different in kind from that shown in stripping the vegetation of a humid area. The changes produced by man on semi-arid erosion may in fact be no greater than natural changes, and his role in accelerating erosion tends to be overstressed.

Although erosional lowering of the land surface at one millimetre per year has little apparent effect in a human lifetime, its long-term effect is considerable, as it represents a lowering of 1000 m over a million years. Soil conservation is not concerned with such time-spans, but erosion removes the topsoil, which contains a high proportion of the soil organic matter and the finer mineral fractions which provide water and nutrient supplies for plant growth. Even in the medium-term implicit in passing good farmland to the next generation, it is sound conservation practice not to allow erosion at a rate which depletes soil resources faster than they can be renewed.

In current soil conservation practice, it is normal to plan for acceptable rates of erosion of 2–10 tons/acre year, equivalent to a loss of about 0.2–1 mm/year from the surface. It is argued that this rate will keep pace with the rate of chemical weathering forming new soil. Although it has been stated above that natural rates of erosion indeed match those 'acceptable' rates in some semi-arid areas these are too high for natural rates in most humid areas. In the long-term perspective which is needed to support permanent agriculture, the balance between mechanical soil erosion and chemical soil formation needs to be examined more carefully. One complicating factor is that any increase in soil erosion tends to be associated with an increase in overland flow runoff. Consequently the subsurface runoff is slightly reduced, so that the amount of material removed in solution, which is a measure of the rate of bedrock weathering, is also slightly reduced. Thus any increase in erosion should reduce the 'acceptable' level set by the rate of new soil formation.

This approach is formalized in Figure 1.2. Bedrock is weathered to soil at rate W (in μm/year), and the soil is progressively weathered until, at the surface it is reduced so that only a proportion p_s ('by volume' analysis) of its original substance remains. This surface is lowered mechanically by soil erosion at rate T, and the soluble materials are removed at rate D, both measured in units equivalent to μm of bedrock removed per years. If the soil is to be treated as a renewable resource it must reach an equilibrium in which

$$W = T + D \tag{1.1}$$

and

$$T = W \cdot p_s \tag{1.2}$$

The first of these equations states that total bedrock lowering must be matched by total removal. The second states that the surface is being lowered at the same rate as the soil/bedrock interface: the condition for complete soil renewal.

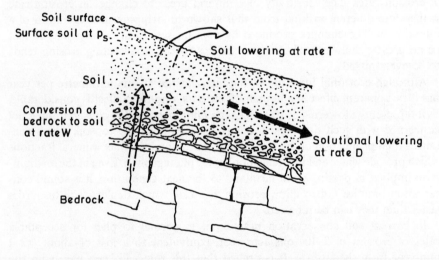

Figure 1.2. Equilibrium between erosion and soil formation by weathering

For an unweathered bedrock, $p_s = 1$ and lower values indicate progressively more weathered soil. What is important here is not perhaps the content of available inorganic nutrients so much as the degree of mechanical comminution of the rock material to form an adequate growth medium for plants. This can be specified as a maximum value of p_s, usually around 0.8. Given an appropriate maximum value of p_s, the maximum acceptable value for lowering by soil erosion, T, is given by:

$$T = D\left(\frac{p_s}{1 - p_s}\right) \tag{1.3}$$

Thus for Britain or N.E. United States with a solution rate of say 25 μm/year on non-calcereous rocks, the maximum acceptable erosion may be

$$T = 25 \times \frac{0.8}{1 - 0.8} = 100 \,\mu\text{m/year}$$

In the loess soils of the Great Plains, where solution is at about $50 \,\mu\text{m/year}$, the maximum erosion acceptable rises to $200 \,\mu\text{m/year}$; and in the south-west USA ($D \simeq 5 \,\mu\text{m/year}$), acceptable erosion may drop to $20 \,\mu\text{m/year}$. The acceptance of erosion rates as high as those currently used in conservation practice is tantamount in the short-term to treating the soil as a resource to be used up, and in the long-term to accepting a growth medium for which

$$p_s = \frac{T}{T + D} \tag{1.4}$$

Thus for example, acceptance of a loss at $1000 \,\mu\text{m/year}$ ($10 \,\text{T/acre year}$) for Britain would lead to a stabilized soil for which

$$p_s = \frac{1000}{1000 + 25} = 0.975$$

which is more like a gravel than a farming soil. Current values for acceptable erosion losses thus appear highly questionable in any long-term view of farmland as a resource which man cannot afford to squander, even over a period of centuries.

A process which is closely allied to soil erosion is the physical exhaustion of soil organic matter and soil nutrients used by plants. This process is important everywhere, although most dramatic in arid and semi-arid environments. Use through grazing or by some form of plant harvesting, for example as firewood, can rapidly deplete the soil and convert marginal lands to desert. This is an important process in its own right on low slopes, and on steeper slopes also leads to an increase in soil erosion by both wind and water. Loss of organic material can be reversed by application of fertilizers, but this solution is rarely economic for arid or semi-arid lands where yields are very low at best.

Soil erosion has been described above as the removal of inorganic soil grains, and the problem of soil nutrients has been largely ignored. Soils which are little weathered and consequently coarse-textured contain a high proportion of the original inorganic nutrients from the parent rock, but they are not readily available to plants, because they offer a low contact area with soil water and plant roots. Sufficient weathering to provide an adequate clay content of 5–10 per cent is therefore desirable to make rock minerals available to plants, and to chemically bind nutrients in fertilizers and prevent their rapid leaching in drainage waters. A moderate amount of weathering ($p_s \leqslant 0.8$) is therefore desirable to maintain an adequate level of clay minerals in the soil.

Organic matter in the soil fulfils a similar role to clays in holding water and inorganic nutrients, as well as organic nutrients, most critically nitrogen.

Organic matter is also very important in maintaining soil aggregates, and so providing a moist soil which remains highly permeable. As organic matter is concentrated near the surface, it appears liable to be eroded first. This is not usually a problem however because its high permeability and aggregate strength minimize the risk of overland flow. Loss of organic matter depends much more on the vegetation cover and its management. Partial removal of vegetation, for example by sending crops to market or through grazing, or wholesale clearance, prevents the addition of dead plant material to the soil as the source of new organic matter for the soil. Over a period of a few years, this loss of organic matter results in a proportionate loss of plant nutrients, and, in a dry climate, a significant reduction in soil moisture. A vicious circle can readily begin, in which less plants can grow, and the attempt to sustain some sort of yield accelerates the decline towards a completely barren surface. This process can remove the vegetation cover in a semi-arid area within less than a decade, even without the physical soil erosion which it also increases as vegetation cover and humus decline. Figure 1.3 shows, in qualitative terms, how unirrigated dryer areas can only support a decreasing sustained yield. In this figure, the effects of organic matter are somewhat exaggerated by ignoring mineral soil as a store for water or nutrients, but the general trend remains valid except near the dryest extremes. The figure shows how a sustained yield can be maintained, although with a thinner cover than if there were no removal of the plants. The *proportion* of the plants used, however, as well as the *absolute* yields, must fall in dryer areas if plant use is to be sustained. Thus, for example, if the land is being used for grazing, it is argued that not only do the animals need to graze over larger areas

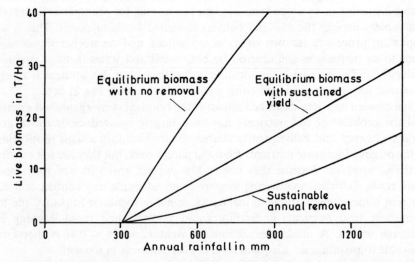

Figure 1.3. Substantial yields in a tropical climate. Values are estimated on assumption that organic matter is the only store for water and nutrients

each day in dryer areas, but that each area should be given longer to recover. This cycle of organic depletion leads to lower infiltration capacity and increased overland flow. On slopes of more than a few degrees, erosion therefore follows close behind it, and may receive more than its share of the blame for the loss of nutrients.

1.2 APPROACHES TO UNDERSTANDING THE MECHANISMS OF EROSION

Most of our present knowledge of soil erosion mechanisms and rates originates in the work of the US Soil Conservation Service, which was established in the 1930s to combat the severe soil losses occurring in the South Eastern Piedmont and Great Plains. The emphasis of the SCS has always been pragmatic, and its predictions of erosion rates have centred round the development and extension of the Universal Soil Loss Equation (USLE) which is fully described in Chapter 2. Both the strength and weakness of the USLE lie in its estimation of erosion as the product of a series of terms for rainfall, slope gradient, slope length, soil, and cropping factors. This allows extensive tabulations of individual factors, incorporating the results of a vast experience, but it does not allow for any sort of non-linear interactions between the factors, and this may be a fatal flaw.

Looking at mechanisms in slightly more detail, there have been major shifts of opinion concerning the role of raindrop impact compared to the role of overland flow, and this balance is still far from clear. Today nobody disputes the role of raindrops in sealing the soil surface and so reducing infiltration and increasing overland flow; and the effect of a vegetative or any other canopy in reducing the momentum or energy of raindrops and preventing this surface sealing. Similarly the transporting capacity of overland flows on a slope is not disputed. However, the importance of direct movement of material by raindrop splash relative to movement in a flow remains open to discussion; and the relative balance between transporting capacities and the amounts of material available for transport appears highly controversial.

The kind of non-linearity which can arise in erosion is most simply illustrated in the context of overland flow estimation. Even though transport by overland flow is only one part of the total movement of soil, it is clear that changes in overland flow are responsible for changes in soil transport rates. Ignoring evapotranspiration as an irrelevance for present purposes, the overland flow generated in a small quadrat (and added to flow accumulated from upslope) is equal to rainfall intensity *minus* infiltration rate. Since maximum infiltration rate depends on soil rather than rainfall properties, it is plain that neither overland flow runoff, nor the soil transported by it can be expressed as a rainfall factor *multiplied* by a soil factor, as is demanded by the USLE. Instead there is a threshold effect, with a threshold level determined by the soil properties and initial moisture (where rainfall intensity is high) or by total rainfall amount (where intensities are low). Alternative predictive models are explored more fully in Chapter 6.

Direct transport of soil particles by raindrop splash has been well documented and measured, mainly since Ellison's (1945) classic work. The formation of pedestals several centimetres high beneath protecting stems, shells, roots, etc., demonstrates clearly that appreciable surface lowering of an unvegetated surface can take place by splash with no overland flow. Bare surfaces can give transport rates of up to 100 cm^3/cm year by splash alone, though the totally inhibiting effect of ground cover leads to a natural vegetation maximum of 2 to 5 cm^3/cm year, which occurs under semi-arid conditions. Ignoring, for a moment, the interactions between raindrops and overland flow, then data on rates of total wash and rates of splash transport (summarized in Carson and Kirkby, 1972, Equations (8.1) and (8.10)) indicate that overland flow transports considerably more material than splash on any unvegetated slope greater than 10 metres long. What this implies is that rainsplash can generally be ignored as a major transporting agent except near the top of a plot or hillside. This general statement however ignores soils with exceptionally high infiltration rates (particularly well-sorted sands) and intense rainstorms which are too short to produce overland flow: it also ignores the spatial pattern of flow, which invariably forms into a braided pattern with threads of appreciable flow separated by areas with only a film of water. Overland flow moves soil material most rapidly along these threads where the flow is deepest, and splash plays an important role in supplying additional material to the threads. Grains of soil may thus splash into a flow thread and be carried by the flow, so that much of the transporting capacity of the flow is supplied by the splash, and erosional lowering is not limited to the flow threads, but distributed over the whole soil surface. This is perhaps another way of describing the role of splash mainly as that of soil detachment, and of wash mainly as transportation.

As well as the division of labour between splash and wash just described, there is a strong interaction between the two because a layer of flowing water protects the surface from direct raindrop impact. The raindrops are however still able to disturb the flow, and increase the concentration of material suspended in it. The overall effect on sediment transport is thought to be as shown qualitatively in Figure 1.4. At high discharge, the effect of raindrops is clearly minimal, as the flow disturbance caused by the drops is well above the soil surface. At low discharges, the two effects of rainsplash impact and flow disturbance work in opposite directions, so that the curve for sediment transport depends mainly on the rainfall intensity. It is again apparent that there is a non-linear interaction, this time between raindrops and flow transport. The details required to quantify Figure 1.4 can be seen to depend not only on soil aggregate properties, but also on surface roughness characteristics which relate an average flow condition to an actual distribution of flow depths across the slope.

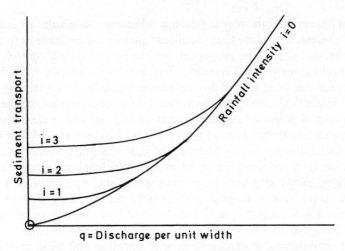

Figure 1.4. Qualitative relationship between sediment transport and intensities of rainfall and flow (All units arbitrary; scales arithmetic)

Another aspect of soil erosion mechanics which may be strongly seen in most experimental runs is the phenomenon of available sediment and its exhaustion. It is plain that each new flow and/or rainfall condition causes a peak in the sediment transport, with a rapid decline thereafter while constant hydraulic conditions are maintained. This appears to suggest the build-up of a reservoir of readily eroded material, mainly between storms, and its rapid depletion during an eroding storm. Significant additions to this reservoir may be due to drying of surface aggregates, break-up by frost, and the action of insects in bringing casts and other loose material to the surface. Depletion of this 'available soil' reservoir is likely to depend not only on direct exhaustion by removal, but also through re-attachment due to wetting. Even where no soil is available in this way, it is also clear that sufficiently rapid flows can erode material for transport, although to a lesser extent. Little research appears to have been done so far to predict variations in erosion and transport rates occurring in this way.

The mechanics of soil erosion by water are therefore seen as rather poorly understood, with considerable scope for re-appraisal. A fuller understanding can best be based on a clear separation between hydrology and hydraulics, and the soil properties which are significant for each. Hydrological soil properties determine the rate of infiltration, and so decide how much of the rainfall contributes to overland flow. Hydraulic soil properties determine the resistance of the soil to transportation by flow or raindrops. Other sets of soil properties determine the way in which aggregates break down to seal the surface under raindrop impact; and the way in which soil becomes 'available' for easy transport. At present, little more than intelligent guesses can be made as to the relative importance of all these properties, and the processes each controls.

The mechanisms of wind erosion are in some ways simpler to understand, but research has been limited by the greater difficulty in measuring rates of erosion. Water consistently carries material downslope, so that sediment traps in experimental plots and catchments can be laid out with a clear knowledge of the source area from which trapped sediment has been eroded. For wind no such simple assumptions can be made, so that it is much harder to interpret measurements of sediment transport in terms of localized erosion and vice versa. Since, except at a very large scale, winds are imposed upon a surface, there are no problems analogous to those of predicting runoff. The mechanism of wind erosion is therefore one of aerodynamic flow over a more or less rough surface. Complications arise in discussing soil erodibility and the interaction of direct entrainment in combination with abrasion and wind-drying of the surface. Other difficulties occur at high sediment transport rates, when the transported sediment forms its own roughness elements as ripples and/or dunes. These issues are discussed fully in Chapter 7 (below).

1.3   THE DISTRIBUTION OF EROSION IN TIME AND SPACE

At a given site, the growth of vegetation closely follows the monthly pattern of actual evapotranspiration, which is considered to follow potential evapotranspiration (strongly dependent on temperature) and available moisture (dependent on rainfall). When combined with the seasonal pattern of leaf-fall, the amount of live natural vegetation cover follows an annual cycle which normally lags 1 to 3 months behind the evapotranspiration cycle. In temperate Britain, this means that natural vegetation cover is greatest at the end of the summer, and declines during the rainy winter but is never absent. Annual crops follow a similar cycle in exaggerated form, often with completely bare ground in late autumn and winter. In a semi-arid climate the actual evapotranspiration closely follows the rainfall, so that the highest rainfalls occur before the annual vegetation cover is established.

Rainfall intensity is at least as important for erosion as total rainfall, and a rough estimate of the overland flow produced can be obtained by (i) considering the distribution of daily rainfalls, which is characterized by the variable, $r_0 =$ 'mean rain per rainday', and (ii) assuming that overland flow occurs whenever the daily rain exceeds a notational $r_c =$ 'soil storage capacity', which is dependent on the vegetation cover (which prevents surface sealing and so increases effective storage capacity) and the soil organic matter. High storage capacity (100 mm say) is thus associated with a dense vegetation cover, and low capacities (10 mm) with bare ground. This simple model (Carson and Kirkby, 1972, pp. 216–218) is discussed more fully in Chapter 6 below. For the present, it provides a qualitative example of the differences in overland flow over a range of rainfall and vegetation regimes (Figure 1.5). The figure shows how overland flow and hence erosion increase with both rainfall intensity and bare surface exposure. The effect of seasonal differences may be seen by combining seasonal differences of rainfall intensity

    (i) under a uniform vegetation cover: A *vs.* B in Figure 1.5
   (ii) where vegetation cover is less at times of intense rainfalls: A *vs.* C
  (iii) where vegetation cover is more at times of intense rainfalls: B *vs.* D

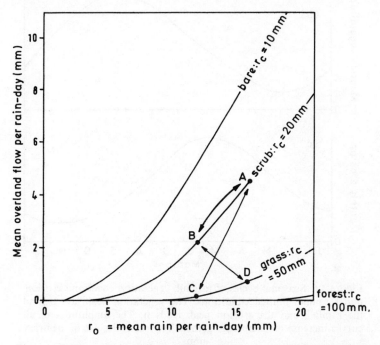

Figure 1.5. The relationship between estimated overland flow as an indicator of erosion, and (i) rainfall intensity given by r_0 = mean rain per day; (ii) vegetation cover indicated by r_c = storage capacity for each day's rainfall

It is clear that erosion is greatest where intense rain and vegetation cover are out of phase, as is normally the case to a significant extent in semi-arid and mediterranean climates. Following these contrasts through the year in Figure 1.6, it can be seen that erosion is greatest when the rains increase but vegetation is not yet established, so that the erosional peak leads the rainfall peak. A wide range of possible combinations of timings for rainfall and vegetation can occur, but that shown qualitatively in Figure 1.6 is common and indicates a severe

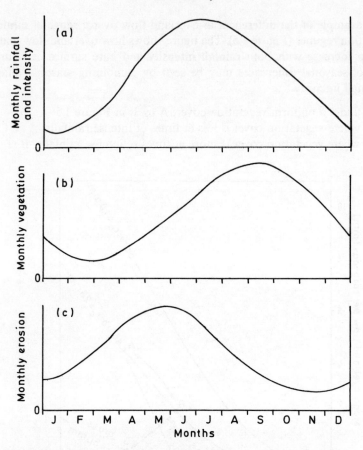

Figure 1.6. Seasonal cycles of rainfall, vegetation cover and erosion, under rainfall limited conditions. The vegetation peak lags behind the rainfall, and the erosion peak leads it. The amplitudes of all curves increase or decrease together. Vertical scales are in arbitrary units

erosion hazard. It can be seen that as the same annual rainfall is spread in a more strongly seasonal fashion, the curves in Figure 1.6 all show a greater amplitude, and the total erosion over the year is increased. This phenomenon is clearly shown in Fournier's (1960) empirical study in which the most significant predictor of erosion is the variable, $p^2/P =$ (rainfall of wettest month)^2/(annual rainfall). It can however be seen that any such simple variable must suppress a good deal of the relevant detail.

A discussion of the seasonal distribution of erosion leads naturally to discussion of where soil erosion is most severe, and where the human hazards need to be reduced most urgently. Two distinct types of area appear to be at great risk. The first to be recognized, and the one on which conservation work has concentrated in the past, is the humid areas which have been seasonally stripped of vegetation for crop cultivation. The second type of area at great risk is the arid and semi-arid zone, in which exhaustion of organic matter through total removal of vegetation either as a crop or through grazing may play as important a part in the destruction of usable land as physical erosion by wind or water.

The cultivation of humid areas produces erosion which has been combatted by contour ploughing, terracing, strip cropping, and mulch spreading with great success. Nevertheless it has been argued above that current conservation practice aims at controlling erosion at a level which is still too high for sustained farming over long periods. One result of this high threshold of acceptance is that temperate areas with modest erosion are not recognized as potentially at risk at all, even though light soils on ploughed slopes of more than 5° are frequently rilled after storms, even under the low intensity rains of Britain. Now that most areas of spectacular erosion are at least partially protected, there is a great danger of a complacency which will allow the land to be destroyed during the next century. Figure 1.7 gives a rough estimate of the expected erosion rates for bare soil on a range of slope gradients, related to the rainfall intensity

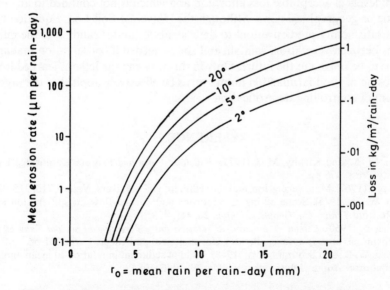

Figure 1.7. Estimated erosion per rain day on bare surfaces, as a function of rainfall intensity (r_0) in the bare-soil period, and slope gradient. Values are orders of magnitude only. Total erosion should be estimated by multiplying by number of rain days during bare-soil period

at seasons when the soil is exposed. The estimates refer to a 10 m long slope, and show the trade-offs between the duration of bare soil, rainfall intensity, and slope gradient.

The sustained use of arid and semi-arid lands poses a different set of problems, many of which are economic or social. Possible yields are generally so low near desert margins that it is rarely worth spending large amounts of money on conservation, and good management must work with natural processes as far as possible. On low slopes, the direct loss of organic material leads to depletion of moisture and nutrient storage, and physical erosion is a secondary problem. On slopes of more than 1° or 2°, water erosion becomes dominant, but both processes are part of the wider syndrome of desertification. Rational management on low slopes consists of controlling yields at a low and sustainable level, and varying the level sensitively with climatic zones, and with wet and dry runs of years. On steep slopes, or where there are mobile dunes, vegetation needs to be encouraged at all costs, and any grazing or burning discouraged. Such management may be possible with nomadic herding in a thinly populated continent, but raises severe social problems as populations grow and settle. It is perhaps these arid and semi-arid areas which present the greatest challenge to soil conservationists today, but one which they have only recently begun to face as a problem separate from that of erosion on humid farmlands.

Soil erosion is therefore seen as a problem which is more widespread than current levels of acceptable loss allow for, and which is not confined to any one climatic or geographical zone. Although the Universal Soil Loss Equation has been a satisfactory predictive tool to date, a subtler understanding of the interactions between vegetation, rainfall, and soil is needed if conservation management is to be refined in the future. Only in this way can the lessons learned from the erosion of good farmland in humid areas be effectively applied to the urgent problem of controlling desertification.

REFERENCES

Carson, M. A., and Kirkby, M. J. (1972). *Hillslope Form and Process*. Cambridge University Press. 475 pp.

Douglas, I. (1967). Man, vegetation, and the sediment yield of rivers. *Nature*, **215**, 925–928.

Ellison, W. D. (1945). Some effects of raindrops and surface-flow on soil erosion and infiltration. *Trans. Am. Geophys. Union*, **26**, 415–429.

Fournier, F. (1960). *Climat et érosion: la relation entre l'érosion du sol par l'eau et les precipitations atmospheriques*, Presses Universitaires de France, Paris. 201 pp.

Langbein, W. B., and Schumm, S. A. (1958). Yield of sediment in relation to mean annual precipitation. *Trans. Am. Geophys. Union*, **39**, 1076–1084.

Soil Erosion
Edited by M. J. Kirkby and R. P. C. Morgan
© 1980 John Wiley and Sons Ltd.

CHAPTER 2

Soil loss estimation

J. Kent Mitchell

*Associate Professor, Department of Agricultural Engineering,
University of Illinois at Urbana-Champaign, USA*

and

Gary D. Bubenzer

*Professor, Department of Agricultural Engineering,
University of Wisconsin-Madison, USA*

Soil conservation specialists have for many years attempted to estimate soil loss from individual fields or slopes to determine land use practices which will ensure long-term productivity of the soil. Recent environmental concerns require that soil loss and sediment yield predictions be made to evaluate the extent of non-point pollution sources. Researchers include current soil loss estimation procedures in extensive hydrologic models.

Soil erosion, soil loss, and sediment yield are terms with distinct meanings in soil erosion technology. Soil erosion is the gross amount of soil moved by drop detachment or runoff. Soil loss is the soil moved off a particular slope or field. Sediment yield is the soil loss delivered to a point under evaluation. For example, most slopes have topographic and cultural irregularities that cause both erosion and deposition to occur. Thus, the erosion at selected points on the slope often differs from the soil loss at the base of the slope. Further deposition often occurs in field boundaries and watercourse borders, thereby reducing the sediment yield of the watershed.

Soil loss prediction techniques have developed over many years as understanding of the erosion process expanded and increasingly more erosion research was conducted. Early estimates were primarily qualitative in nature and illustrated that some cultural practices differed in their ability to control soil erosion. Initially, equations were developed to describe soil loss using a single independent variable. These single factor equations were for local situations where other contributing factors were nearly constant. Multiple factor equations were developed as more data became available and researchers were better able to describe contributing factors. These analyses culminated in the equation most widely used today for soil loss prediction—the Universal Soil Loss Equation (USLE). Recent investigations have focused on defining the parameters of the USLE for a greater range of conditions. Considerable work continues to define the soil erosion process and, hence, eventually predict soil loss, from a physical basis as contrasted with the predominantly empirical soil loss predictions of the past and present.

Sediment yield prediction methods are quite varied in form and extent. Empirical methods have been developed which relate sediment concentrations to flow stage or which relate sediment yield to watershed and hydrologic parameters. In other cases, sediment yield is estimated from gross soil loss estimates using the sediment delivery ratio concept. Others have attempted to modify the USLE to estimate sediment yields. Attempts are currently under way to combine soil loss and hydrologic models so that sediment movement within and from the watershed may be estimated. Because the erosion-sediment transport processes are not fully described, a physically based sediment yield estimation procedure is likely to not be developed for some time.

2.1 UPLAND EROSION PREDICTION EVOLUTION

The first scientific study of erosion effects is thought to have been done by Wollny in the late nineteenth century (Hudson. 1971), The first quantitative experiments in America were begun by the Forest Service in 1915. In 1917, M. F. Miller began a plot study of the effect of crops and rotations on runoff and erosion. The wide spread concern of the dangers of soil erosion in the 1920s and early 1930s resulted in an increase in scientific erosion research. The results of this early work were, of necessity, qualitative in nature. However, a basic understanding of most of the factors affecting erosion was developed during this period (Ayres, 1936). The importance of raindrop impact in the erosion process was not fully appreciated until the natural rainfall studies of Laws (1940) and the analysis of the mechanical action of raindrops by Ellison (1947).

Several scientists began to develop empirical equations for soil erosion prediction as data were accumulated and exchanged. The first of these related soil loss to steepness and length of slope (Zingg, 1940). Using plots under simulated rainfall and field conditions, Zingg demonstrated that doubling the degree of slope increased the soil loss 2.61 to 2.80 times and doubling the horizontal length of slope increased the soil loss in runoff 3.03 times. This relationship was expressed by:

$$A = CS^m L^{n-1} \tag{2.1}$$

where:

A = average soil loss per unit area from a land slope of unit width,
C = a constant of variation,

S = degree of land slope,
L = horizontal length of land slope,
m, n = exponents of degree and horizontal length of land slope, respectively.

Values of 1.4 and 1.6 were proposed by Zingg for m and n, respectively. As can be observed by comparing this equation with later developments, the constant of variation, C, combines the effects of rainfall, soil, crop, and management.

Smith (1941) evaluated the effects of mechanical conservation practices for four combinations of crop rotation and soil treatment on one soil. He determined that:

(a) the soil loss from contouring is 57 per cent of that from up-and-down hill operation,
(b) the soil loss from rotation strip cropping is 25 per cent of that from up-and-down hill operation, and
(c) the soil loss from terracing is 3 per cent of that from up-and-down hill operation.

Curves were presented to enable the field conservationist to evaluate the need for mechanical practices for each of the four crop-treatment combinations on Shelby soil with an allowable soil loss of $0.9 \, kg/m^2$ per year. Browning *et al.* (1947) expanded the procedure proposed by Smith (1941) for several crop rotations and for most soils of Iowa. Additionally, an estimate was made of the effect of soil treatment practices on soil loss.

The relationship of rainfall characteristics to the amount of soil eroded was introduced by Musgrave (1947). Using data from several stations, erosion, E, was determined to be proportional to $P_{30}^{1.75}$, where P_{30} is the maximum 30-minute rainfall. The larger data base analysed at that time indicated that the slope steepness and length factor exponents should be 1.35 and 0.35, respectively. A procedure was presented to estimate soil loss, in mm for several vegetal covers and for soils of many portions of the eastern and central United States. Although not stated explicitly, the equation proposed by Musgrave was:

$$E = (0.00527)IRS^{1.35}L^{0.35}P_{30}^{1.75} \qquad (2.2)$$

where:

E = the soil loss, mm per year,
I = the inherent erodibility of a soil at 10 per cent slope and 22 m slope length, mm per year,
R = a vegetal cover factor,
S = degree of slope, per cent,
L = length of slope, metres, and
P_{30} = the maximum rainfall, mm.

The Musgrave equation was used extensively for estimating gross erosion from watersheds. Lloyd and Eley (1952) provided a graphical solution of the Musgrave equation for use in the Northeastern United States.

A disadvantage to the steepness of slope evaluation by Zingg (1940) was that the soil loss from slopes less than four per cent was underpredicted and zero soil loss was computed for a zero per cent slope. Smith and Whitt (1947, 1948) proposed an equation of the form:

$$A = a + bS^n \qquad (2.3)$$

to describe the effect of slope on soil loss. They also presented a method of estimating soil loss for the claypan soils of Missouri using the effects of slope steepness, length of slope, crop rotations, conservation practices, and soil groups. The equation presented was:

$$A = CSLKP \qquad (2.4)$$

where:

A = the average annual soil loss,
C = the average annual rotation soil loss from plots, and
S, L, K and P = are multipliers to adjust the plot soil loss C for slope steepness, length, soil group, and supporting conservation practice, respectively.

It is interesting to note that this equation is similar in form to the Universal Soil Loss Equation introduced 11 years later. However, Equation (2.4) does not contain a separate rainfall factor. Van Doren and Bartelli (1956) evaluated the factors affecting soil loss for Illinois soils and conditions. Their soil loss estimation procedure provided two factor tables; one that included the effects of soil type, slope steepness, slope length, and mechanical practices, and a second that included soil erodibility effects, previous erosion, crop rotation, management, and intensity and frequency of 30-minute rainfall.

In 1954 soil erosion prediction research was consolidated in a cooperative effort aimed at overcoming many of the disadvantages inherent in local or regionalized research projects. More than 8000 plot-years of erosion research data were compiled from 36 locations in 21 states. A re-evaluation of the various factors affecting soil loss (Smith and Wischmeier, 1957; Wischmeier and Smith, 1958; and Wischmeier et al., 1958) was made which led to development of the widely used soil loss prediction method called the Universal Soil Loss Equation.

Concurrently, Hudson (1961) presented an erosion equation:

$$E = TSLPMR \qquad (2.5)$$

where E is erosion and the remaining factors are functions of soil type, slope gradient, and length, agronomic or agricultural practice, mechanical protection, and rainfall, respectively. The problems of adequately evaluating each of these factors were discussed by Hudson and he reported on extensive research on the erosivity of rainfall in the sub-tropics of Africa. Equation (2.5) is identical in concept to the Universal Soil Loss Equation.

Elwell (1977) developed a Soil Loss Estimation System for Southern Africa:

$$Z = KCX \tag{2.6}$$

where:

Z = predicted mean annual soil loss,

K = mean annual soil loss, from a standard field plot 30 m × 10 m at a 4.5 per cent slope for a soil of known erodibility under bare fallow,

C = the ratio of soil lost from a cropped plot to that lost from the standard plot, and

X = the ratio of soil lost from a plot of length L and slope S to that lost from the standard plot.

The K factor is dependent on rainfall kinetic energy and soil erodibility. The The soil erodibility index is defined by basic soil type and may be adjusted for permeability, structure, and conservation practices. The crop cover factor, C, is a function of the percentage rainfall energy intercepted which is determined from a crop cover distribution curve for the assumed crop and the distribution of rainfall energy. The X factor is the same as the LS factor of the Universal Soil Loss Equation.

The development of prediction equations was of necessity evolutionary in nature. The predominantly qualitative descriptions led to the evaluation of one or two factors that could be quantified with the data from a locale or region. The evolutionary process continued with additional factors being quantified as more data were obtained. The consolidation of data from many stations enabled researchers to develop prediction equations applicable to a region or a number of regions. Each of the predictive techniques was limited in its applicability by the limits of the data from which it was developed. Hence, it was usually useful only for a local area, specific soil type, or perhaps a region. As more data sources became available, more conditions could be estimated and the area of applicability expanded. The centralization of data led to an equation widely accepted in the United States and which has been adopted world-wide.

2.2 THE UNIVERSAL SOIL LOSS EQUATION

The most widely used method of soil loss prediction by conservationists in the United States is the Universal Soil Loss Equation (USLE):

$$A = (0.224)RKLSCP \tag{2.7}$$

where:

A = the soil loss, kg/m^{2}s,

R = the rainfall erosivity factor,

K = the soil erodibility factor,
L = the slope length factor,
S = the slope gradient factor,
C = the cropping management factor, and
P = the erosion control practice factor.

The USLE is also being adapted for soil loss estimation in other countries as will be discussed throughout the description of the elements of the equation.

Equation (2.7) was designated 'universal' because it was free of some of the generalizations and geographic and climatic restrictions inherent in earlier models. It has been criticized as not being universal because parameter values were presented for conditions of the eastern two-thirds of the United States. However, as data are accumulated, parameters are being identified for use in more areas, including regions in other continents.

The USLE was developed as a method to predict average annual soil loss from interrill and rill erosion. With the parameter values available, cropping and management alternatives can be determined to reduce the estimated soil loss to suggested tolerance values for the soil type. As detailed by Wischmeier (1976) the USLE may properly be used to:

(1) predict average annual soil loss from a field slope with specific land use conditions,
(2) guide the selection of cropping and management systems, and conservation practices for specific soils and slopes,
(3) predict the change in soil loss that would result from a change in cropping or conservation practices on a specific field,
(4) determine how conservation practices may be applied or altered to allow more intensive cultivation,
(5) estimate soil losses from land use areas other than agricultural, and
(6) provide soil loss estimates for conservationists to use for determining conservation needs.

Although the USLE was primarily designed as a convenient aid for conservation planning it may also be a useful research tool provided one is more precise in evaluating the equation factors than for conservation planning. It can be used to generate soil loss data for correlation with other parameters. The equation was developed to estimate long-term average annual soil loss. Therefore, its application to a specific year or storm may not be appropriate. When used with a specific storm it will estimate the average soil loss for numerous recurrences of that event and, as with any average, the soil loss from any one of these events may vary considerably. Applying the equation to situations for which factor values are not yet determined is especially dangerous. Although expedient and often necessary for conservation planning purposes, extrapolation is always hazardous.

The factors of the USLE were developed using an evaluation unit called the standard plot. A standard plot is 22.13 metres long on a uniform lengthwise slope of 9 per cent. The plot was tilled up and down slope and was in continuous fallow for at least two years. The standard plot is simply a result of the historical development of the USLE. Early basic data were often obtained from plots 40.5 m² (0.01 acre) in size. For a convenient width of 1.83 metres (6 ft) this gave a plot length of 22.13 metres (72.6 ft). Much of the data were taken from surfaces that deviated from the standard plot and provided the ranges for the statistical analysis that lead to the development of the USLE. However, the standard plot was used as the basis for defining the variation in *L*, *S*, *C*, and *P*.

Aside from the general considerations of the USLE, specific uses and limitations are presented with the discussion of each factor. The development and evaluation of each factor of the equation are presented in the following subsections.

2.2.1 The rainfall erosivity factor, *R*

The rainfall erosivity factor in the USLE is the rainfall erosion index as presented by Wischmeier (1959). The *R* factor is a definition of the erosivity of rainfall events and is defined as the product of two rainstorm characteristics: kinetic energy and the maximum 30 minute intensity.

Wischmeier and Smith (1958) analysed drop size distribution data published by Laws and Parson (1943) and rain drop terminal velocity reports by several researchers. A regression equation describing the kinetic energy of a rainstorm or portion of a rainfall event was developed:

$$E = 1.213 + 0.890 \log_{10} I \qquad (2.8)$$

where:

E = the kinetic energy, kg m/m² mm, and ...
I = rainfall intensity, mm per hour.

The kinetic energy for an intensity increment is obtained by multiplying the kinetic energy from Equation (2.8) by the rainfall amount for that intensity increment. The total energy, in kg m/m², for a rainfall event can be computed by accumulating the kinetic energy for each distinct intensity increment of the event.

Regression analyses showed that storm soil losses from cultivated continuous fallow plots were highly correlated to the cross-product of the total kinetic energy and the maximum 30-minute rainfall intensity (Wischmeier, 1959). This product, designated by *EI*, is a measure of the manner in which energy and intensity are combined in a storm and defines the combined effects of raindrop impact and turbulence of runoff to transport soil particles from a field. The rainfall erosivity factor, *R*, is obtained by dividing the *EI* product by 173.6

The computation of the rainfall erosivity factor, R, for a storm is defined by:

$$R = \frac{\left[\sum_{j=1}^{n} (1.213 + 0.890 \log_{10} I_j)(I_j T_j) \right] I_{30}}{(173.6)} \qquad (2.9)$$

where:

R = the rainfall erosivity index,
I_j = the rainfall intensity for a specific storm increment, mm/hr,
T_j = the time period of the specific storm increment, hr,
I_{30} = the maximum 30-minute rainfall intensity for the storm, mm/hr,
j = the specific storm increment, and
n = the number of storm increments.

Although there are units associated with the R factor, the USLE is not dimensionally correct without assuming units for the K factor. For this presentation the conversion from English units to SI units has been done simply by converting the original English unit equations. Although the units for intensity in Equation (2.9) are SI the R factor value is identical to the value one would obtain if the equivalent English units were used with the traditional equation (Wischmeier and Smith, 1965). This procedure has the advantage of not requiring conversion of existing R and K factor tables and maps.

The rainfall erosivity indexes can be summed for any time period to provide a numerical measure of the erosivity of the rainfall during that period. The

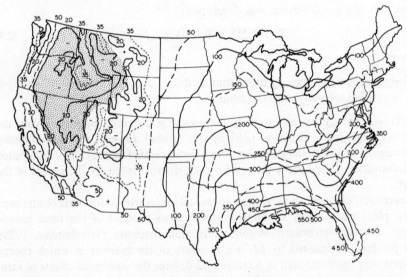

Figure 2.1. Average annual values of the rainfall-erosivity factor, R, for the
USA

reduction of long-term rainfall records provides average annual values of the rainfall erosivity index or rainfall factor, R. These rainfall factor values for a large area can be presented as curves of equal erosivity (iso-erodents) on a map of the area of interest.

Studies have shown that the median dropsize of rainfall does not continue to increase with rainfall intensities above 76.2 mm/hr (Carter *et al.*, 1974; Hudson, 1971). Therefore, current recommendations (Wischmeier and Smith, 1978) are that a limit of 76.2 mm/hr be used in the computation of energy per unit of rainfall by Equation (2.8). Also, it is recommended that a limit of 63.5 mm/hr be placed on the I_{30} component of Equation (2.9).

Average annual R values for the contiguous United States are shown in Figure 2.1. A more detailed map of R values has recently been presented by Wischmeier and Smith (1978). It is recommended that the Soil Conservation Service Technical Center, Portland, Oregon, be contacted for specific recommendations for locations in the shaded region of the west and northwest. Average annual values of the rainfall-erosivity factor, R, for Hawaii are shown in Figure 2.2.

For any hydrologic evaluation there is an associated annual distribution of the phenomena being investigated. Because the rainfall factor, R, does not completely describe the differences in the effect of rainfall pattern on soil erosion due to location, the area east of the 104° meridian in the United States was

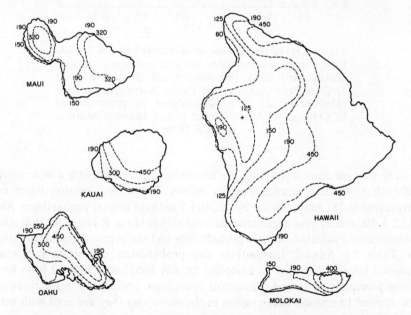

Figure 2.2. Estimated average annual values of the rainfall-erosivity factor, R, in Hawaii (ARS, 1975)

divided into 33 geographic areas and a distribution curve presented for each area (Wischmeier and Smith, 1965). The monthly and annual distribution for each area are different for some portion of the distribution curve. Typical distribution curves are shown in Figure 2.3 for three areas. These distributions are used with detailed cropping management data to obtain seasonal cropping factors or to combine crop rotation information as described in Section 2.2.4.

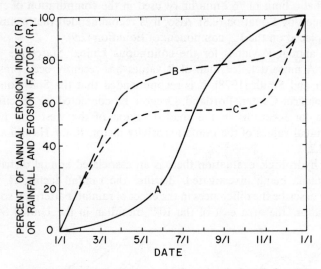

Figure 2.3. Distribution curves of erosion-index, *R*, for: (A) Central corn belt (northern Missouri and central Illinois, Indiana, and Ohio) (Wischmeier and Smith, 1965); (B) Dryland grain region of the Pacific Northwest, tentative (McCool *et al.*, 1977) (Reproduced by permission of SCSA); and (C) South Central part of Island of Molakai (USSCS, 1976)

The average annual *R* values are determined from data with a wide scatter of both storm and annual erosivity values. The annual erosion index was evaluated at 181 locations for 50, 20, and 5 per cent annual probabilities. Also, 1, 2, 5, 10, and 20 year recurrence interval values of the *R* factor for individual storms were evaluated. Sample probabilities and recurrence intervals are listed in Table 2.1. Annual distributions and probabilities and storm recurrence interval information may be compiled for any location of interest from long-term precipitation records. Standard hydrologic probability evaluations can be applied to erosion index values in the same way they are used with other hydrologic phenomenon.

Additional methods of defining an erosion factor have been proposed to

Table 2.1. Probability values of erosion index R, for central corn belt, USA[a]

Location	Probability values of annual erosion index (EI)			Single event index values normally exceeded once in				
	50 per cent	20 per cent	5 per cent	1 year	2 years	5 years	10 years	20 years
Illinois:								
Moline	158	221	303	39	59	89	116	145
Rantoul	152	201	263	27	39	56	69	82
Springfield	154	210	283	36	52	75	94	117
Indiana:								
Fort Wayne	127	183	259	24	33	45	56	65
Indianapolis	166	225	302	29	41	60	75	90
Terre Haute	190	273	389	42	57	78	96	113
Missouri:								
Columbia	214	297	406	43	58	77	93	107
Kansas City	170	248	356	30	43	63	78	93
Rolla	209	287	387	43	63	91	115	140
St. Louis	168	290	488	45	62	86	106	126
Ohio:								
Cincinnati	146	211	299	27	36	48	59	69
Columbus	113	158	216	27	40	60	77	94
Coshocton	158	235	343	27	45	77	108	143
Dayton	125	175	240	21	30	44	57	70

[a] From Wischmeier and Smith, 1965.

simplify the original method and to obtain R values more applicable to regional conditions. Studies by the United States Agricultural Research Service (Wischmeier, 1974) and the United States Environmental Protection Agency (Ateshian, 1974; EPA, 1973) were directed toward correlating the rainfall erosion index with various rainfall intensity–duration–frequency information. Equations were developed whereby the erosion index can be obtained from the 2-year, 6-hour duration rainfall for the two rainfall distribution types defined by the United States Soil Conservation Service (1973) in their runoff estimation procedures. The United States Soil Conservation Service (1975) also defined a third rainfall distribution type and Woodward (1975) presented the erosion index equation for that condition. These equations for the average annual erosion index are:

$$\text{Type IA}: R = 0.00829 \, P_{2,6}^{2.2} \tag{2.10}$$

$$\text{Type I}: \quad R = 0.0134 \, P_{2,6}^{2.2} \tag{2.11}$$

$$\text{Type II}: \quad R = 0.0245 \, P_{2,6}^{2.17} \tag{2.12}$$

where:

R = the rainfall erosion index, and

$P_{2,6}$ = the 2-year recurrence interval, 6-hour duration rainfall, mm.

Type IA storm distribution is characteristic of the coastal side of the Cascade and Sierra Nevada Mountains in Oregon, Washington, and Northern California, and the coastal regions of Alaska. Type I is applicable to the coastal side of the Sierra Nevada Mountains in Southern California, to Hawaii, and to the interior regions of Alaska. Type IA and I storm distributions are representative of maritime climates with wet winters and dry summers. Type II storm distribution is characteristic of the intense storms that occur over the remainder of the United States, Puerto Rico, and the Virgin Islands. Wischmeier (1974) cautions against using Equations (2.10)–(2.12) if more detailed data are available. The equations and the 2-year, 6-hour rainfall amounts are both subject to estimation errors. Values of EI computed directly or obtained from iso-erodent maps developed from direct computation are preferable to those derived from empirical equations.

Ateshian (1974) also developed equations for Type I and II storms for 2-year recurrence interval storms of duration to 24-hours. The equations describing the individual storm erosion index are:

$$\text{Type I}: R_i = 0.0122 \, P^{2.2}/H^{0.6065} \tag{2.13}$$

$$\text{Type II}: R_i = 0.0156 \, P^{2.2}/H^{0.4672} \tag{2.14}$$

where:

R_i = individual storm erosion index,

P = storm rainfall depth, mm, for 2-year recurrence interval and duration as specified by H, and

H = duration of storm, hours.

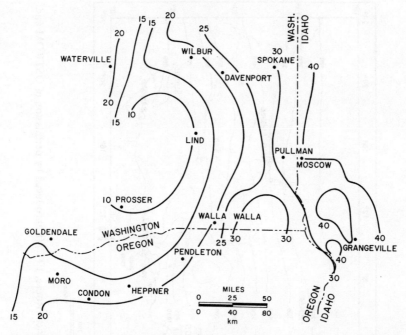

Figure 2.4. Tentative rainfall and runoff erosivity factor for the dryland grain region of the United States Pacific Northwest (McCool *et al.*, 1977) (Reproduced by permission of SCSA)

The special conditions of the dryland grain region of the United States Pacific Northwest were examined by McCool *et al.* (1974, 1976). For this region, defined by Figure 2.4, soil loss was found to be correlated with the December through March precipitation. This discovery lead to a compromise relationship of:

$$R_T = 0.0245\, P_{2,6}^{2.17} + 0.0591\, P_{(D-M)} \tag{2.15}$$

where:

R_T = the 'rainfall and runoff' erosivity factor,
$P_{2,6}$ = the 2-year recurrence interval, 6-hour duration rainfall, mm, and
$P_{(D-M)}$ = the December through March precipitation, mm.

A comparison of R values for the region defined by Figure 2.4 and the same portion of Figure 2.1 indicates as much as a two-fold increase in the R factor using Equation (2.15). The distribution curve for this region (Figure 2.3) shows that the major erosion potential is obviously in the winter months and occurs in the form of low intensity rainfall or snow. This low intensity precipitation does not increase the R value defined by Equation (2.9) sufficiently but is a significant erosion agent in the region.

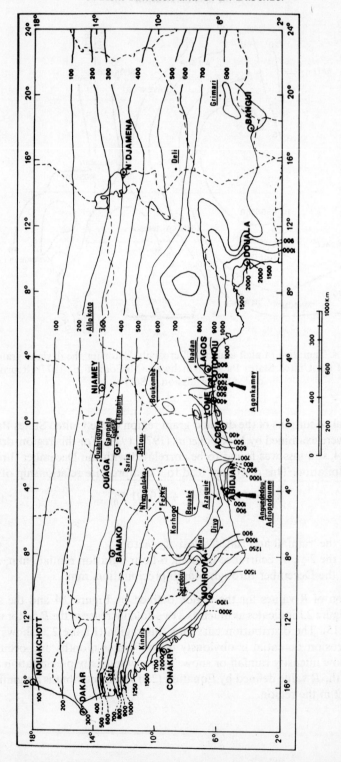

Figure 2.5. Average annual values of the rainfall-erosivity factor, *R*, in West Africa (Roose, 1977) (Reproduced by permission of OSTROM and SCSA)

Recent studies have shown that thaw and snowmelt runoff erosion are significant. In areas of the United States where this is the case, current recommendations (Wischmeier and Smith, 1978) are that an R_s factor equal to 0.0591 times the December-through-March precipitation (mm) be added to the R factor value computed by Equation (2.9).

Average annual R values for a large part of West Africa were developed by Roose (1977). A detailed study of the correlation between daily rainfall and the rainfall erosivity index was conducted that resulted in a single empirical relationship:

$$R_{an}/H_{an} = 0.50 \pm 0.05 \qquad (2.16)$$

where:

R_{an} = average annual erosivity index, and
H_{an} = average annual rainfall amount, mm.

This relationship was found appropriate to several recording stations in Ivory Coast, Upper Volta, Senegal, Niger, and Tchad. Roose stated that the iso-erodent map (Figure 2.5) has sufficient precision (5 per cent), outside the mountainous and coastal zones, to permit use of the USLE in West Africa.

Iso-erodent maps for Tunisia (Masson, 1972) and the Tet River basin in southern France (Masson, 1976) have been developed as well as erosion index distribution curves for those areas. In an earlier study, Cormary and Masson (1964) described the USLE factor evaluations for Tunisia. Equations relating the USLE erosivity factor, R, to local precipitation amount and intensity were developed by Delwaulle (1973) for Niger. Bailly *et al.* (1976) divided Madagascar into regions and presented equations to determine rainfall energy from precipitation amount for each region.

Hudson (1971) conducted extensive soil erosion studies on soils in subtropical Africa. He found that soil loss was most closely correlated to the kinetic energy of individual storms of 25.4 mm (1 in) per hour and greater intensity. He described this parameter as the KE > 25 index.

Several authors have discussed the problems of determining an adequate rainfall erosivity index for areas outside of those for which the USLE was developed. Renard and Simanton (1975) describe the large variability in computed EI values between locales relatively close together affected primarily by high intensity, short duration orographic storms. Preliminary studies by Singer *et al.* (1977) indicate that the soil loss estimate using EI values is too large for low intensity rainfall of California rangelands. They point out that more data are needed to verify this indication.

2.2.2 The soil erodibility factor, K

The soil erodibility factor, K, in the USLE is a quantitative description of the inherent erodibility of a particular soil. This factor reflects the fact that differ-

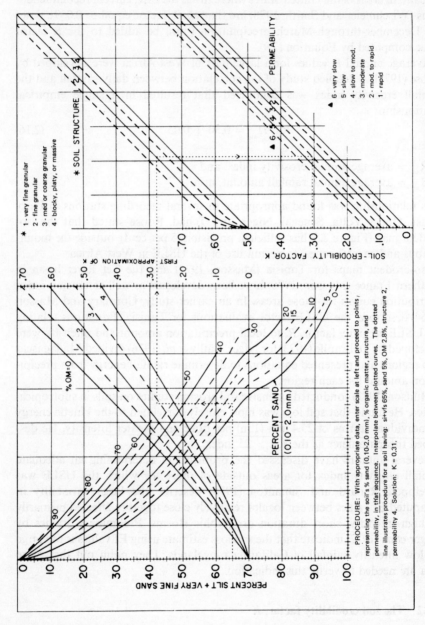

Figure 2.6. Nomograph for determining the soil-erodibility factor, *K*, for United States mainland soils (ARS, 1975)

ent soils erode at different rates when the other factors that affect erosion are the same. Soil properties that affect infiltration rate, permeability, total water capacity, dispersion, splash, abrasion, and transporting forces also affect erodibility. For a particular soil, the soil erodibility factor, K, is the rate of erosion per unit of erosion index from a standard plot.

Direct measurements of K were made for 23 major soils in the United States (Olson and Wischmeier, 1963; Wischmeier and Smith, 1965). Following these initial measurements, the K values for numerous other soils were approximated by considering the characteristics of the soil in comparison to measured soils. Additional direct measurements have been made at many locations world-wide.

Direct measurement of the K factor requires considerable time and equipment and is costly to perform. In an effort to eliminate this procedure, a study was conducted to describe the K factor using 15 soil properties and their interactions (Wischmeier and Mannering, 1969). A multiple-regression equation of 24 terms was developed and was considered valid for a broad range of medium textured

Table 2.2. Indications of the general magnitude of the soil-erodibility factor, K^a

Texture class	Organic matter content		
	< 0.5 per cent	2 per cent	4 per cent
	K	K	K
Sand	0.05	0.03	0.02
Find sand	0.16	0.14	0.10
Very fine sand	0.42	0.36	0.28
Loamy sand	0.12	0.10	0.08
Loamy fine sand	0.24	0.20	0.16
Loamy very fine sand	0.44	0.38	0.30
Sandy loam	0.27	0.24	0.19
Fine sandy loam	0.35	0.30	0.24
Very fine sandy loam	0.47	0.41	0.33
Loam	0.38	0.34	0.29
Silt loam	0.48	0.42	0.33
Silt	0.60	0.52	0.42
Sandy clay loam	0.27	0.25	0.21
Clay loam	0.28	0.25	0.21
Silty clay loam	0.37	0.32	0.26
Sandy clay	0.14	0.13	0.12
Silty clay	0.25	0.23	0.19
Clay		0.13–0.29	

[a] The values shown are estimated averages of broad ranges of specific-soil values. When a texture is near the borderline of two texture classes, use the average of the two K values. For specific soils, use of Figure 2.6 or Soil Conservation Service K-value tables will provide much greater accuracy. From ARS, 1975.

soils. However, it proved too cumbersome for general use. This deficiency, plus additional investigation, prompted the development of the soil erodibility nomograph, Figure 2.6 (Wischmeier *et al.*, 1971). The soil erodibility nomograph can be used to obtain the soil erodibility factor, K, for soils for which the K value has not previously been determined. It is particularly helpful for use in construction areas where the K values for subsoils are not known. Two important findings were made in the analysis leading to the development of this nomograph. The silt-size fraction was expanded to include the very fine sand classification and improved the prediction value of both sand and silt. The product of per cent silt and per cent sand-and-silt accounted for 85 per cent of the variance in observed K values for the soils analysed.

Five soil parameters are needed to use the nomograph: per cent silt (0.002–0.05 mm) plus very fine sand (0.05–0.10 mm), per cent sand (0.10–2.0 mm), organic matter content, structure, and permeability. Structure and permeability values are given in Figure 2.6. Additional guides for permeability described by Wischmeier *et al.* (1971), are: code fragipan soils as 6, code more permeable surface soils underlain by massive clay or silty clay as 5, code moderately permeable surface soils underlain by a silty clay or silty clay loam having a weak subangular or angular blocky structure as 4, and code as 3 if the subsoil structure grade is moderate or strong or the texture is coarser than silty clay loam.

If the measured soil erodibility value is available for a soil of interest it should be used as first choice in the USLE. The United States Soil Conservation Service has evaluated K for most soils in the United States. The soil erodibility nomograph is valuable for those soils where K values are not available from the sources mentioned above. Table 2.2 is available for a general estimation of K if the above data are not available.

2.2.3 The slope length factor, L, and the slope gradient factor, S

The effects of slope length and gradient are represented in the USLE as L and S, respectively; however, they are often evaluated as a single topographic factor, LS. Slope length is defined as the distance from the point of origin of overland flow to the point where the slope decreases sufficiently for deposition to occur or to the point where runoff enters a defined channel. The channel may be part of a drainage network or a constructed channel. Slope gradient is the field or segment slope, usually expressed as a percentage. The development of the USLE was based on a standard plot length of 22.13 metres (Wischmeier and Smith, 1965); therefore, the slope-length factor was defined as:

$$L = \left(\frac{x}{22.13}\right)^m \tag{2.17}$$

where:

L = slope length factor,

x = slope length, metres, and
m = an exponent.

Current recommendations (Wischmeier and Smith, 1978) for the exponent m are:

m = 0.5 if slope $\geqslant 5$ per cent,
m = 0.4 if slope < 5 per cent and > 3 per cent,
m = 0.3 if slope $\leqslant 3$ per cent and $\geqslant 1$ per cent, and
m = 0.2 if slope < 1 per cent.

These recommendations are reflected in the construction of the slope effect chart (Figure 2.7).

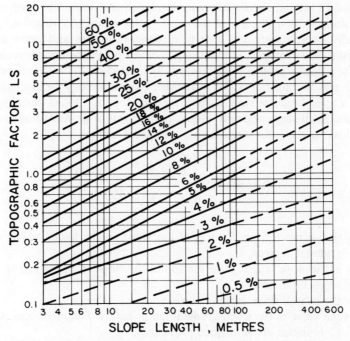

Figure 2.7. Slope length and gradient factor, *LS*, for use with the Universal Soil Loss Equation

Smith and Wischmeier (1957) also determined that soil loss was correlated with a parabolic description of the effect of slope steepness or gradient. Normalizing this equation to a standard plot slope of 9 per cent resulted in a description of the slope-gradient factor:

$$S = \frac{0.43 + 0.30s + 0.043s^2}{6.613} \tag{2.18}$$

where:

   S = the slope gradient factor, and
   s = the gradient, per cent.

Equation (2.18) is recommended for the slope gradient factor and used in the development of the slope effect chart (Figure 2.7). Values of LS may be computed from:

$$LS = \left(\frac{x}{22.13}\right)^m (0.065 + 0.045s + 0.0065s^2) \qquad (2.19)$$

where all terms have been previously defined. Figure 2.7 is a convenient aid in obtaining the topographic factor, LS, for use in the USLE.

Equation (2.19) and Figure 2.7 were developed and should be used only for single uniform slopes. The use of the topographic factor, LS, described above, will usually overestimate soil loss from concave slopes and underestimate the loss from convex slopes. The first step in developing a method for computing soil loss for complex profiles was presented by Onstad *et al.* (1967) and continued by Foster and Wischmeier (1974) to provide a methodology for evaluating the effects of LS for irregular slopes. The irregular slope is divided into a series of n segments. Each slope segment should be uniform in gradient and soil type. The soil loss for the entire slope is then computed using:

$$A = (0.224)RKCP \left[\frac{\sum_{j=1}^{n} (S_j x_j^{m+1} - S_j x_{j-1}^{m+1})}{x_e (22.13)^m} \right] \qquad (2.20)$$

where:

   x_j = the distance from the top of the slope to the lower end of the jth segment, metres,
   x_{j-1} = is the slope length from the top of the slope to upper end of the jth segment, metres,
   x_e = the overall slope length, metres,
   S_j = the value of the slope-gradient factor for the j segment, and A, R, K, C, P, and m are as defined previously.

Wischmeier (1974) presented a simplified method for solving Equation (2.20) when the complex slope may be divided into two to five segments of equal length.

   To illustrate the use of Equation (2.20), assume a convex slope of the following segments: 50 metres at 5 per cent slope, 40 metres at 8 per cent slope and 30 metres at 11 per cent slope. Table 2.3 is a result of the step by step solution of Equation (2.20). A tabular solution of Equation (2.20) is made more expedient by defining:

$$u_j = \frac{S_j x_j^{m+1}}{(22.13)^m} \quad \text{and} \quad u_{j-1} = \frac{S_j x_{j-1}^{m+1}}{(22.13)^m} \qquad (2.21)$$

Table 2.3. Determination of LS factor for a convex slope

Segment No.	Slope per cent	S_j	Length, metres	x_j	x_{j-1}	u_j	u_{j-1}	$u_j - u_{j-1}$	Segment LS	Percent of total soil loss
1	5	0.454	50	50	0	34	0	34	0.68	13
2	8	0.844	40	90	50	153	63	90	2.25	35
3	11	1.351	30	120	90	378	245	133	4.43	52
Sum			120					257		100
Entire slope									2.14	

The entire slope LS is obtained by dividing the sum of the $(u_j - u_{j-1})$ by x_e. The segment LS is obtained by dividing the $(u_j - u_{j-1})$ for that segment by the length of the segment. The last column of Table 2.3 shows the portion of the total soil loss that comes from each slope segment and is obtained by dividing the $(u_j - u_{j-1})$ for a segment by the sum of the $(u_j - u_{j-1})$. A concave slope of 120 metres total length and elevation difference of 9 metres can be examined by reversing the order of the slope segments of Table 2.3. The LS factor for this concave slope is 1.61 and the LS factor for a uniform 7.5 per cent slope of 120 metres length is 1.79.

It would be reasonable to expect that while progressing down a slope different soil types, hence, different K values would be found. This change may be incorporated into the procedure as follows:

$$A = (0.224)\frac{RCP}{x_e} \sum_{j=1}^{n} K_j \left[\frac{S_j x_j^{m+1} - S_j x_{j-1}^{m+1}}{(22.13)^m} \right] \qquad (2.22)$$

where:

K_j is the soil erodibility factor for the jth slope segment.

It is also reasonable to treat a change in the cropping management factor, C in a similar manner if land use changes on the slope.

2.2.4   The cropping management factor, C

The cropping management factor represents the ratio of soil loss from a specific cropping or cover condition to the soil loss from a tilled, continuous fallow condition for the same soil and slope and for the same rainfall. This factor includes the interrelated effects of cover, crop sequence, productivity level, growing season length, cultural practices, residue management, and rainfall distribution. The evaluation of the C factor is often difficult because of the many cropping and management systems. Crops can be grown continuously or rotated with other crops. Rotations are of various lengths and sequences. Residues can be removed or left on the field or incorporated into the soil. The soil may be clean tilled or one of several conservation tillage systems may be used. Each segment of the cropping and management sequence must be evaluated in combination with the rainfall erosivity distribution for the region.

Several effects of cropping and management were presented by Wischmeier (1960) and Wischmeier and Smith (1965) in the form of extensive tables of soil loss ratios. Greatly expanded tables of C factor values were presented by Wischmeier and Smith (1978) that represent most cropping and management schemes in use in the United States. A small portion of these recommendations are presented in Table 2.4. Six crop-stage periods are identified in Table 2.4

Table 2.4. Soil loss ratios for selected crops[a]

Cover, Sequence, and Management	Productivity kg/m²	Soil Loss Ratio, Percentage For Crop-Stage Period[b]					
		F	SB	1	2	3	4
First-year corn after grain and legume hay, spring turn plough, conventional tillage, residue left	0.6+	8	22	19	17	10	14
Small grain in disked row-crop residue, after one year corn after meadow, residue left		—	12	12	11	7	2
Grass and legume meadow	0.7+						0.4

[a] From Wischmeier and Smith, 1978.
[b] Crop—stage periods:
F—Rough fallow
SB—Seedbed
1—Establishment
2—Development
3—Maturing Crop
4—Residue or stubble.

Table 2.5. Example of cropping-management (C) factor evaluation[a]

Crop Stage Period	Dates	Percent of Ann. Erosion Index[b]	Crop Stage Soil Loss, C Percentage	C Factor
Meadow	1/1–4/15	10	0.4	0.0004
Rough ploughed sod	4/15–5/5	5	8	0.0040
Disked and corn seedbed	5/5–6/1	10	22	0.0220
10–50 per cent canopy	6/1–6/20	13	19	0.0247
50–75 per cent canopy	6/20–7/10	14	17	0.0238
75 per cent canopy–harvest	7/10–10/15	40	10	0.0400
Residue	10/15–12/31	8	14	0.0112
	1/1–4/1	8	14	0.0112
Oat seedbed	4/1–4/15	2	12	0.0024
10–50 per cent canopy	4/15–5/1	4	12	0.0048
50–75 per cent canopy	5/1–6/1	11	11	0.0121
75 per cent–harvest	6/1–6/15	9	7	0.0063
New meadow in oat stubble	6/15–8/15	38	2	0.0076
Meadow (16.5 months)	8/15–1/1	128	0.4	0.0051
Total		300		0.1756
Average Annual C Factor				0.0585

[a] For assumed COM rotation in Central Corn belt, United States.
[b] From curve A, Figure 2.3.
[c] From Table 2.4.

Table 2.6. Generalized values of the cover and management factor, *C*, in the United States east of the 104° meridian[a]

Crop, rotation, and management[c]	Productivity level[b]	
	High	Mod.
	C value	
Base value: continuous fallow, tilled up and down slope	1.00	1.00
CORN		
C, Rdr, fall TP, conv(1)	0.54	0.62
C, RdR, spring TP, conv(1)	0.50	0.59
C, RdL, fall TP conv(1)	0.42	0.52
C, RdR, wc seeding, spring TP, conv(1)	0.40	0.49
C, RdL, standing, spring TP, conv(1)	0.38	0.48
C, fall shred stalks, spring TP, conv(1)	0.35	0.44
C(silage)–W(RdL, fall TP) (2)	0.31	0.35
C, RdL, fall chisel, spring disk, 40–30%rc(1)	0.24	0.30
C(silage), W wc seeding, no-till pl in c-k W(1)	0.20	0.24
C(RdL)–W(RdL, spring TP) (2)	0.20	0.28
C, fall shred stalks, chisel pl, 40–30%rc(1)	0.19	0.26
C-C-C-W-M,RdL, TP for C, disk for W(5)	0.17	0.23
C, RdL, strip till row zones, 55–40%rc(1)	0.16	0.24
C-C-C-W-M-M, RdL,TP for C, disk for W(6)	0.14	0.20
C-C-W-M, RdL, TP for C, disk for W(4)	0.12	0.17
C, fall shred, no-till pl, 70–50%rc(1)	0.11	0.18
C-C-W-M-M, RdL, TP for C, disk for W(5)	0.087	0.14
C-C-C-W-M, RdL, no-till pl 2d and 3rd C(5)	0.076	0.13
C-C-W-M, RdL, no-till pl 2d C(4)	0.068	0.11
C, no-till pl in c-k wheat, 90–70%rc(1)	0.062	0.14
C-C-C-W-M-M, no-till pl 2d and 3rd C(6)	0.061	0.11
C-W-M, RdL, TP for C, disk for W(3)	0.055	0.095
C-C-W-M-M, RdL, no-till pl 2d C(5)	0.051	0.094
C-W-M-M, RdL, TP for C, disk for W(4)	0.039	0.074
C-W-M-M-M, RdL, TP for C, disk for W(5)	0.032	0.061
C, no-till pl in c-k sod, 95–80%rc(1)	0.017	0.053
COTTON[d]		
Cot, conv(Western Plains) (1)	0.42	0.49
Cot, conv(South) (1)	0.34	0.40
MEADOW		
Grass and Legume mix	0.004	0.01
Alfalfa, lespedeza, or sericia	0.020	
Sweet clover	0.025	
SORGHUM, GRAIN (Western Plains)[d]		
RdL, spring TP, conv(1)	0.43	0.53
No-till pl in shredded 70–50%rc	0.11	0.18

Table 2.6. (*Continued*)

Crop, rotation, and management[c]	Productivity level[b]	
	High	Mod.
	C value	
SOYBEANS[d]		
B, RdL, spring TP, conv(1)	0.48	0.54
C-B, TP annually, conv(2)	0.43	0.51
B, no-till pl	0.22	0.28
C-B, no-till pl, fall shred C stalks (2)	0.18	0.22
WHEAT		
W-F, fall TP after W(2)	0.38	
W-F, stubble mulch, 0.06 kg rc(2)	0.32	
W-F, stubble mulch, 0.11 kg rc(2)	0.21	
Spring W, RdL, Sept TP, conv (N&S Dak) (1)	0.23	
Winter W, RdL, Aug TP, conv (Kans) (1)	0.19	
Spring W, stubble mulch, 0.08 kg rc(1)	0.15	
Spring W, stubble mulch, 0.14 kg rc(1)	0.12	
Winter W, stubble mulch, 0.08 kg rc(1)	0.11	
Winter W, stubble mulch, 0.14 kg rc(1)	0.10	
W-M, conv(2)	0.056	
W-M-M, conv(3)	0.026	
W-M-M-M, conv(4)	0.021	

[a] This table is for illustrative purposes only and is not a complete list of cropping systems or potential practices. Values of *C* differ with rainfall pattern and planting dates. These generalized values show approximately the relative erosion-reducing effectiveness of various crop systems, but locationally derived *C* values should be used for conservation planning at the field level. Tables of local values are available from the Soil Conservation Service. From ARS, 1975.

[b] High level is exemplified by long-term yield averages greater than $0.47 \ kg/m^2$ corn or $0.7 \ kg/m^2$ grass-and-legume hay; or cotton management that regularly provides good stands and growth.

[c] Numbers in parentheses indicate number of years in the rotation cycle. No. (1) designates a continuous one-crop system.

[d] Grain sorghum, soybeans, or cotton may be substituted for corn in rotation lines to estimate *C* values for sod-based rotations.

Abbreviations defined:

B—soybeans
C—corn
c-k—chemically killed
conv—conventional
cot—cotton

F—fallow
M—grass and legume hay
pl—plant
W—wheat
wc—winter cover

kg rc—kg of crop residue per m^2 remaining on surface after new crop seeding
%rc—percentage of soil surface covered by residue mulch after new crop seeding
70-50%rc—70% cover for *C* values in first column, 50% for second column
RdR—residues (corn stover, straw, etc.) removed or burned
RdL—all residues left on field (on surface or incorporated)
TP—turn plowed (upper 125 or more mm of soil inverted, covering residues).

and are defined as:

Period F—Rough fallow. Inversion ploughing to secondary tillage.

Period SB—Seedbed. Secondary tillage for seedbed to 10 per cent canopy cover.

Period 1—Establishment. 10–50 per cent canopy cover.

Period 2—Development. 50–75 per cent canopy cover.

Period 3—Maturing crop. 75 per cent canopy cover to crop harvest.

Period 4—Residue or stubble. Crop harvest to ploughing or new seeding.

The values in the body of Table 2.4 represent the soil loss expected for a specific condition during a specific crop-stage period as a percentage of the soil loss from continuous fallow. The crop stage tabular values were designed for use with erosion index distribution curves similar to Figure 2.3.

To derive a C value for a crop rotation, the year is divided into crop-stage periods as determined by the local ploughing, seeding, and harvest dates. The appropriate erosion index distribution curve is entered to obtain the percentage of annual erosion index expected within each crop stage period. The crop-stage C value multiplied by the corresponding value obtained from the distribution curve is the C value for that period. All the crop period C values are summed for the rotation and when divided by the number of years of the rotation the average annual C values for use in the USLE is obtained.

An example will serve to further explain the method for computing the average annual C factor. Assume a field in the central cornbelt of the United States (Region 16) that is in a 3 year rotation of corn–oats–grass and legume meadow. Corn yields are greater than 0.6 kg/m^2, meadow yields are 0.7–1.1 kg/m^2, and residues are left on the field. The crop stage dates are as listed in Table 2.5 which is a compilation of the method of evaluating the C factor for this situation. As a specific example consider the data for the corn seedbed period. From the erosion index distribution curve for Region 16 of Figure 2.3, 15 per cent and 25 per cent are obtained for May 5 and June 1, respectively. Thus, the portion of the annual erosion index for that period is 10 per cent. The soil loss ratio from Table 2.4 is 22 per cent for first year corn, after grass and legume hay, spring ploughed, conventional tillage, crop stage SB. The C factor for the corn seedbed crop period (0.022) is obtained by multiplying the percentage annual erosion index (0.10) by the crop stage soil loss ratio (0.22). The crop stage soil loss ratios for the other crop stages are determined in the same way and their sum (0.1756) divided by the number of years of the rotation (3) yields an annual average C factor of 0.0585 for this cropping management condition in the region. The effect of lower corn or meadow yields, of minimum tillage, or of different crop sequences in the rotation may be made by comparing C factor values.

This method of computation of local C values was not convenient for daily use by farm planners (Wischmeier, 1960). Therefore, it was recommended that the table of crop stage soil loss ratios (excerpted in Table 2.4) be used to develop handbook values of average annual C factors for the common situations of specific areas. A compilation similar to Table 2.6 can be developed for such a purpose.

Table 2.6 is a list of generalized values of the C factor for the area east of the Rocky Mountains in the United States. It is for illustrative purposes and special notice should be taken of footnote 1 of the table. Cropping management factor evaluations can be obtained from local Soil Conservation Service handbooks for areas of specific interest in the United States.

Table 2.7. C values, percentage for undisturbed land[a]

Vegetal canopy		Mulch or vegetation at ground surface[b] Percentage cover					
Type and Height	Percentage Cover[c]	0	20	40	60	80	95–100
None		45	24	15	9.1	4.3	1.1
Tall weeds or short brush,							
0.5 m effective height	25	36	20	13	8.3	4.1	1.1
	75	17	12	9	6.8	3.8	1.1
Brush or bushes,							
2 m effective height	25	40	22	14	8.7	4.2	1.1
	75	28	17	12	7.8	4.0	1.1
Trees, 4 m effective height	25	42	23	14	8.9	4.2	1.1
	75	36	20	13	8.4	4.1	1.1
Factor to obtain C values with grass or compacted duff ground cover[d]		1.0	0.83	0.67	0.46	0.30	0.27

[a] From Wischmeier, 1974. (Reproduced by permission of SCSA).
[b] C values in Table are for ground cover of weeds or undecayed residue.
[c] For canopy cover between 25–75 per cent, straight line interpolation is appropriate.
[d] To obtain C values for ground cover of grass or compacted duff, multiply the C value from the table by the factor.

Table 2.7 lists C values that have been determined for permanent pasture, rangeland, and idle lands (Wischmeier, 1974). Smith and Ports (1976) have compiled recommended values of the C factor for construction areas treated with mulches and seeded, Table 2.8.

Elwell and Stocking (1976) presented a crop cover and classification approach for conditions tested in Rhodesia. The results of this approach included a relationship of soil loss as a function of the mean seasonal vegetal cover. Roose (1977) has developed average annual C factor values, (Table 2.9), for vegetal cover and cultural techniques for the area of West Africa described in Figure 2.5.

Table 2.8. C values for mulched and seeded areas[a]

Treatment	Mulch Rate, kg/m²	Slope, Percentage	C Value, Percentage[b] Period 1	Period 2
No mulch or seeding		all	100	—
Grain or fast growing grass:	none	all	70	10
Straw mulch[c]	0.22	≤10	20	7
	0.34	≤10	12	5
	0.45	≤10	6	5
		11–15	7	5
		16–20	11	5
		21–25	14	5
		26–33	17	5
		34–50	20	5
Crushed stone	30.2	≤15	5	5
	53.8	≤20	2	2
Woodchips	1.6	≤15	8	5
	2.7	≤15	5	2
	5.6	≤15	2	2
Permanent seeding,		all	—	1
second year		all	1	1

[a] From Smith and Ports, 1976 (Reproduced by permission of SCSA).
[b] Period 1: Through first six weeks of growing period; Period 2: after six weeks of growing period.
[c] Straw or hay mulch, tied down by anchoring and tracking equipment used on slope.

Table 2.9. The vegetal cover factor and cultural techniques (C factor) in West Africa[a]

Practice	Annual Average C Factor
Bare soil	1
Forest or dense shrub, high mulch crops	0.001
Savannah, prairie in good condition	0.01
Over-grazed savannah or prairie	0.1
Crop cover of slow development or late planting: 1st year	0.3 to 0.8
Crop cover of rapid development or early planting: 1st year	0.01 to 0.1
Crop cover of slow development or late planting: 2nd year	0.01 to 0.1
Corn, sorghum, millet (as a function of yield)	0.4 to 0.9
Rice (intensive fertilization)	0.1 to 0.2
Cotton, tobacco (2nd cycle)	0.5 to 0.7
Peanuts (as a function of yield and the date of planting)	0.4 to 0.8
1st year cassava and yam (as a function of the date of planting)	0.2 to 0.8
Palm tree, coffee, cocoa with crop cover	0.1 to 0.3
Pineapple on contour (as a function of slope) { burned residue	0.2 to 0.5
buried residue	0.1 to 0.3
surface residue	0.01
Pineapple and tie-ridging (slope 7 per cent)	0.1

[a] From Roose, 1977 (Reproduced by permission of ORSTOM and SCSA).

2.2.5 The erosion control practice factor, *P*

The erosion control practice factor is the ratio of soil loss using the specific practice compared with the soil loss using up-and-down hill culture. The erosion control practices usually included in this factor are contouring, contour strip-cropping, and terracing. Conservation tillage, crop rotations, fertility treatments, and the retention of residues are important erosion control practices. However, these cultural practices are included in the cropping management factors described earlier.

Table 2.10. Erosion control practice factor, P^a

Land Slope, percentage	Contouring	Contour Strip cropping and Irrigated Furrows	Terracing^b
1–2	0.60	0.30	0.12
3–8	0.50	0.25	0.10
9–12	0.60	0.30	0.12
13–16	0.70	0.35	0.14
17–20	0.80	0.40	0.16
21–25	0.90	0.45	0.18

^{a} From Wischmeier and Smith, 1978.
^{b} For prediction of contribution to off-field sediment load.

The practice factors for the three major mechanical practices as recommended by Wischmeier and Smith (1978) are shown in Table 2.10. Within a practice type the *P* factor is most effective for the 3–8 per cent slope range and values increase as the slope increases. As slope decreases below 2 per cent the practice factor value increases due to the reduced effect of the practice when compared with up-and-down-hill cultivation. The factor for terracing in Table 2.10 is for the prediction of the total off-the-field soil loss. If within-terrace interval soil loss is desired, the terrace interval distance should be used for the slope length factor, *L*, and the contouring *P* value used for the practice factor.

2.2.6 Soil loss tolerance

Soil loss tolerance is the maximum rate of soil erosion that permits a high level of productivity to be sustained. Tolerances for specific soils in the United States have been the subject of several workshops where soil scientists have collectively evaluated the soils and then recommended tolerance values. In general, deep, medium textured, moderately permeable soils that have sub-soil characteristics favourable for plant growth were assigned tolerances of 1.1 kg/m^2/year. Soils with a shallow root zone or other detrimental characteristics were assigned lower tolerances. Recommended soil loss tolerance values for soils in the United States may be obtained from Soil Conservation Service handbooks.

The soil loss tolerance for a specific soil is used as a guide for soil conservation planning. The USLE is used to estimate the actual soil loss and to evaluate how changes in practices can be applied to reduce soil loss to below the tolerance level.

2.2.7 Application

Average annual soil loss can be estimated using the USLE. As an example of the complete procedure assume a field in Fulton County, Illinois, USA, on Fayette silt loam soil, having a 9 per cent slope approximately 88 metres long. The cropping system is a three year corn–oats–meadow rotation as described previously in Section 2.2.4. The value for each factor of the USLE is obtained from the charts and tables discussed in the preceding subsections. The value of the rainfall factor, R, is obtained from Figure 2.1; by linear interpolation, $R = 195$. The value of the soil erodibility factor, K, was obtained after having determined that this soil is 70 per cent silt plus very fine sand, 10 per cent sand, and 1.8 per cent organic matter and has a fine granular structure with moderate to rapid permeability. The K value from Figure 2.6 is 0.37. The slope effect chart (Figure 2.7) shows that $LS = 2.0$ for an 88 metre, 9 per cent slope. This value could also be obtained by substitution into Equation (2.19). The cropping management factor, C, for a corn–oats–meadow rotation was determined in Section 2.2.4 (Table 2.5) to be 0.0585. If this field were contour farmed, the erosion control practice factor, P, would be 0.6.

Solution of the USLE provides $A = (0.224) (195) (0.37) (2.0) (0.0585) (0.6) = 1.13 \text{ kg/m}^2/\text{year}$ soil loss. This value of the average annual soil loss is greater than the $0.9 \text{ kg/m}^2/\text{year}$ soil loss tolerance recommended for Fayette silt loam. Strip cropping would reduce the P value to 0.3 which reduces the estimated soil loss to $0.57 \text{ kg/m}^2/\text{year}$. However, other management alternatives may also provide the desired level of soil loss protection and should be investigated.

The soil loss probability may be evaluated using the data of Table 2.1. The R value for Fulton County is essentially the same as that for Springfield, Illinois. Therefore, the 5 per cent probability of occurrence of the R index for Fulton County is assumed to be 283. This is $283 \div 195$, or 1.45 times the average annual value and the soil loss from the corn year of the example would be expected to equal or exceed $1.45 \times 2.45 = 3.55 \text{ kg/m}^2$ in 5 per cent of the years.

Soil loss for a storm of a particular return interval can be estimated using the data of Table 2.1. Assume a storm that will occur on the average once in 20 years on the field of the example three weeks after corn planting. From Table 2.1, the R factor equalled or exceeded once in 20 year is 117 for Springfield, Illinois. The C factor for three weeks after corn planting is 0.022. The estimated average soil loss from such a storm on corn of the example is (0.224) (117) (0.37) (2.0)(0.022) (0.6) = 0.43 kg/m^2.

Specific year losses cannot, of course, be estimated in advance because single year deviations from the average for all factors are large and cannot be estimated in advance. However, average soil loss can be estimated and used to recommend adjustments in land use to enhance soil retention and reduce detrimental effects from soil loss.

The estimation of soil loss from areas of the world for which factor values have not been established may be difficult but not impossible. The difficulty may be particularly acute for countries where extensive research programmes are not possible. However, simple field experiments may be established to obtain first approximations of factor values for common situations. Extension of these values for other situations can be made by pooling the judgment of field researchers. These first steps provide a way for soil loss estimation, and hence, for erosion control practices to be applied and a programme started to save a country's valuable soil resources. Excellent suggestions of ways to accomplish first step approximations to the USLE were presented by Hudson (1971).

2.2.8 Modifications

Several modifications of the USLE have been proposed for various applications. Modifications to improve the estimation of R values for a region were described in the discussion of the rainfall erosivity factor. Most of the additional modifications are extensions that attempt to apply the USLE to sediment yield prediction.

The special conditions of semi-arid rangeland of the Southwest United States were discussed by Renard *et al.* (1974). They applied the USLE to sediment yield estimates from small watersheds by describing sediment yield as:

$$A = (0.224)(RKLSCP)\,E_c \tag{2.23}$$

where:

E_c = the channel erosion factor and the other terms are as defined previously.

The channel erosion term, E_c, is similar to a sediment delivery ratio used to predict sediment yield at an outlet. For some of the watersheds used for this study the E_c term was larger than unity because the erosion quantities from channel bed and banks were large and the sediment yield was greater than the gross upland erosion. Therefore, the E_c term was created because a sediment delivery ratio is usually considered to be less than one. Problems encountered in evaluating the other terms of the USLE for conditions of the Southwest United States were also described.

Williams and Berndt (1976) modified the USLE for predicting sediment yield from watersheds:

$$Y = 11800 \, (Qq_p)^{0.56} KCPLS \qquad (2.24)$$

where:

Y = sediment yield from an individual storm, kg,
Q = storm runoff volume, m^3,
q_p = peak runoff rate, m^3/sec, and
KCPLS are as defined in the USLE.

The coefficient was obtained from fitting the equation to data from Texas and Nebraska (Williams, 1975). A sediment delivery ratio was considered not necessary when the rainfall energy term of the USLE was replaced by the runoff term as shown in Equation (2.24). The application of Equation (2.24) required evaluations of the *K*, *C*, *P* and *LS* terms that were different than the methods specified for the USLE (Williams and Berndt, 1972).

A modification of the USLE (Foster *et al.*, 1973) was used by Onstad *et al.* (1976) as the major component in a sediment yield model for small watersheds.

$$A = (0.224) \, WKCPSL \qquad (2.25)$$

where:

W is a hydrologic term and the other terms are as defined in the USLE.

$$W = aR_{st} + (1 - a) \, 0.40 \, Qq_p^{1/3} \qquad (2.26)$$

where:

R_{st} = storm rainfall factor (*EI* units of the USLE),
Q = runoff volume, mm,
q_p = peak runoff rate, mm/hr, and
a = a coefficient ($0 \leqslant a \leqslant 1$) that represents the relative importance of rainfall energy compared with runoff energy for detaching soil.

A value of 0.5 was used for *a* in an earlier study by Onstad and Foster (1975). Equations (2.25) and (2.26) were the results of analyses to describe the source of soil loss with respect to interrill and rill areas.

The first two modifications described here briefly are extensions of the purpose of the USLE in that they were developed to estimate watershed sediment yield. All three modifications are preliminary and limited by region because limited data verification has been accomplished.

2.3 GULLY AND CHANNEL EROSION

The surface channels formed when rills combine and develop to the extent that they cannot be eliminated by normal tillage operations are defined as gullies. Gullies are usually deep, steep-walled upland channels and commonly occur in areas of deep friable subsoils. A gully is often characterized by an over-fall at the gully head that advances upstream. Channel erosion is that soil loss due to the scour induced by the flow or that due to side slope instability. The importance of gully and channel erosion has been observed by several authors. One of the early studies of gully erosion was by Woodburn (1949) in north central Mississippi. He concluded that gully erosion for that area could be estimated as approximately 0.0126 mm of soil loss per m² of active gully surface area per year. Piest *et al.* (1976) noted that erosion of drainageways in the loess region of the central United States has increased during the past few decades. They estimated annual erosion rates of 0.45 kg/m² due to channel degradation. Drainageway and gully erosion accounted for approximately 50 per cent of the total watershed erosion. The accelerated erosion was linked to an intensification of cultivation of the upland watershed where runoff volumes have increased by a factor of 2 or 3 and peak flow rates by more than a factor of 10. The measurements and estimates from 113 watersheds in various parts of the United States were presented by Glymph (1957). Data were presented to show that upland erosion varied from 11 to 100 per cent of the sediment yield and gully erosion was from zero to 89 per cent of the sediment yield. In a sample of twenty watersheds, Striffler (1964) found that eroding stream banks contributed approximately 33 per cent of the total measured sediment load in the streams.

Historically, estimates of gully and channel erosion have been obtained from detailed cross sectional surveys of channel sections taken at periodic intervals throughout the study period. Others have used aerial photographs to estimate changes in channel dimensions over time. There is, however, little information available on techniques to quantitatively predict gully, channel or bank erosion. Lane (1955) presented a proportional relationship for channel erosion in non-cohesive stream channels. Others have used numerous bed load transport equations to obtain a rough estimate of the bed load transport. Using the results of these equations, scour or deposition rates in the channel can be estimated.

Quantitative descriptions of gully growth have been primarily concerned with defining the advancement of the gully head or the increase in areal extent of the gully channel. These effects are of concern because gullies void farmland by dissecting fields and thereby interfere with efficient operations. For the deep loess area of western Iowa, Beer and Johnson (1963) expressed gully growth as:

$$G_A = 81.41 \, R_I^{0.0982} A_t^{-0.0440} L_g^{0.7954} L_w^{-0.2473} e^{-0.0014 \Delta p} \qquad (2.27)$$

where:

G_A = gully surface growth, m²,
R_I = index of surface runoff, mm,
A_t = terraced area of watershed, m²,
L_g = length of gully at the beginning of period, m,
L_w = length from end of gully to watershed divide, m,

$e = 2.7183 =$ base of natural logarithm, and

$\Delta p =$ deviation of precipitation from normal, mm.

Thompson (1964) studied gully head advancement at locations in Minnesota, Iowa, Alabama, Texas, Oklahoma, and Colorado and developed an empirical equation:

$$R = (7.13 \times 10^{-5}) A^{0.49} S^{0.14} P^{0.74} E \qquad (2.28)$$

where:

$R =$ gully head advancement for the time period of interest, m,

$A =$ drainage area above gully head, m^2,

$S =$ slope of approach channel above gully head, as a percentage,

$P =$ summation of rainfall from 24-hour rains equal to or greater than 12.7 mm for the time period of interest, mm, and

$E =$ clay content of the eroding soil profile, as a percentage.

The United States Soil Conservation Service (1966) procedure estimates average annual gully head advance:

$$R = (5.25 \times 10^{-3}) A^{0.46} P^{0.20} \qquad (2.29)$$

where:

$R =$ average annual gully head advance, m,

$A =$ drainage area above gully head, m^2, and

$P =$ the total rainfall from 24-hour rains equal to or greater than 12.7 mm for a time period, converted to average annual basis, mm.

Although no widely accepted mathematical model defining gully or channel erosion is available, Piest *et al.* (1975) have described the mechanisms observed in their study of gully and drainageway erosion in deep loessial soils. Tractive forces of runoff along the gully boundary were not a major cause of erosion in some gullies. Freezing-thawing and wetting-drying on gully banks cause changes in the forces affecting bank stability and lead to mass wasting. Mass wasting is the failure of the bank near the toe with progressive undercutting leading to eventual cave-in of the overhanging portion. This soil debris is periodically cleaned from the gully during periods of high peak runoff events.

Diseker and McGinnis (1967) have developed equations to predict erosion from unprotected road ditch banks. Included in their analysis were rainfall impact energy, average soil moisture, minimum average monthly temperature below 0°C, number of days below freezing, number of freeze-thaw cycles, and bank aspect. The rainfall energy term and the monthly temperature coefficient and slope geometry were significantly correlated to the annual sediment yield from the road banks. Equations were also presented for monthly sediment yields. Meyer *et al.* (1975) used the Universal Soil Loss Equation to estimate road bank erosion for 104 sites on a northern Wisconsin watershed. Soil loss from the 194 sites totalling 12,200 m² was estimated to be 969,000 kg/year.

2.4   SEDIMENT YIELD

Sediment yield is the total sediment outflow from a watershed or drainage basin during any given time. All of the soil loss is not delivered to the stream system because it is deposited at various locations in the watershed. The material that is carried to some point of interest, then, is the sediment yield.

The best method for obtaining sediment yield values is direct measurement of the suspended sediment and bed load for the point of interest in the watershed. This procedure is costly and, in order to obtain good evaluations, a long term effort that may not be justified. Therefore, it is often necessary to estimate sediment yield using predictive methods and data available for the project area.

Three general categories of procedures to estimate sediment yield are:

(1) predictive equations,
(2) gross erosion and sediment delivery ratio computations, and
(3) suspended sediment load or reservoir sediment deposition measurements.

As described previously, few verified theoretical relationships are available to describe soil loss; this dilemma continues as one follows the erosion process to deposition or sediment yield. Therefore, of necessity, sediment yield predictions are empirical.

2.4.1   Predictive equations

Predictive equations are based on watershed parameters and normally have only regional applicability. These equations are usually statistical equations developed from measured watershed parameters. The watershed variables often used are amount or intensity of rainfall, amount or peak rate of runoff, temperature, drainage area size, slope or relief parameter, soil descriptions, and land use descriptions. Additionally, several geologic and time parameters are found in some equations. The number of equations available for estimating sediment yield is quite large; therefore, they will not be presented in detail. However, a summary of several sediment yield prediction equations is presented in Table 2.11. The region or area as well as the size range of the drainage areas for which an equation was developed is presented. In some instances more than one equation was developed and presented by an author. The parameters used as dependent and independent variables in the prediction equations are listed by broad categories. For a particular equation there may be several descriptions of one category. For instance, the equation presented by Williams (1971) has two precipitation variables: weighted intensity and precipitation amount. Other

Table 2.11. Summary of sedime

Reference	Region	Watershed size Range, km^2	No. of equations presented	Dependent variable[a]
Ackerman and Corinth, 1962	Illinois	23–2347	1	SA
Anderson, 1976	Northern California		1	SA
Betson, 1963	Chestuee Creek, Tennessee	295	4	SY
Dendy and Bolton, 1976	Contiguous United States	3–70953	6	SY
Flaxman, 1972	11 Western States	<129	1	SA
Fournier, 1960 and 1962	Africa South of Sahara	>2000	1	SY
Hadley and Schumm, 1961	Upper Cheyenne River Basin	16–1391	3	SA
Herb and Yorke, 1976	Montgomery County, Maryland	0.9–25	32	SY
Hindall, 1976	Wisconsin, 6 provinces	<2590	4	SY
Johnson and Hanson, 1976	Reynolds Creek, Idaho	32–234	1	SS
Striffler, 1963	Tobacco River, Michigan	3–347	14	SS and SY
Williams, 1971	Brushy Creek, Texas	0.5–18	1	SS
Yorke and Herb, 1976	Montgomery County, Maryland	0.9–25	3	SY
Modified Universal Soil Loss Equations:				
Holberger and Truett, 1976	Primarily Eastern USA	1–4	2	SY
Onstad et al., 1976	Treynor Watersheds, Iowa	0.3–0.34	1	SY
Williams, 1975	Brushy Creek, Texas	0.01–18	1	SY

[a] SA—Sediment accumulation in reservoir,
   SS—Suspended sediment,
   SY—Sediment yield.

authors, such as Herb and Yorke (1976), use several variables of a particular category and also present several equations. Not all the variables numbered are necessarily used in each equation.

Although the suspended sediment load and reservoir sediment accumulation equations could be included in a later discussion, they are presented in Table 2.11 because they are equations based on watershed parameters. The sediment accumulation in reservoirs is not strictly sediment yield because the reservoir trap efficiency will effect the percentage of the sediment yield that remains in the reservoir. Reservoir parameters are a part of three of the sediment accumulation equations. Gross erosion or a gross erosion index is present in two of the equations.

Three equations of Table 2.11 are designated as Modified Universal Soil Loss Equations. They are essentially the USLE modified to determine sediment yield. Holberger and Truett (1976) added as parameters the distance from construction site to receptor stream and the percentage of land area in construction to describe sediment yields from construction sites. These two terms are, then, similar to a delivery ratio. Onstad et al. (1976) and Williams (1975) modified the erosivity term of the USLE by using volume and peak rate of runoff.

yield prediction equations

Number of independent variables or parameters by category											
Precipitation	Runoff	Temperature or frost	Drainage area	Slope or relief	Stream length	Soil	Land use	Geologic	Time	Reservoir	Gross erosion
	1			1						2	1
3	1			1	1	2	12	10	1	2	
2									2		
	2		2								
1		1		1		2					
2			1	1							
			1	1						1	1
6	6						1		2		
1	5	1	1	1	1	1	1				
	2										
	2		1	1	2	10	5				
2	2			1		1	2				
				1			2				
2				1	1	1	3				
2	2			1		1	2				
	2			1		1	2				

2.4.2 Gross erosion and sediment delivery ratio

The gross erosion and sediment delivery ratio technique of estimating sediment yield is a two step procedure. First, the gross erosion in a drainage area is computed. The gross erosion includes interill, rill, gully, and stream erosion. Then, a sediment delivery ratio is determined. The sediment delivery ratio is the fraction of the gross erosion that is expected to be delivered to the point of the drainage area under consideration. Sediment yield is obtained by multiplying the gross erosion by the delivery ratio.

Sediment delivery ratio equations have been developed from studies of watersheds in particular regions. As with predictive equations, most sediment delivery ratio equations have limited regional applicability. Several of the equations presented in the literature are summarized in Table 2.12 with an indication of the scope of their applicability and the independent variables used in each equation. The sediment delivery ratio is dependent upon drainage area size and watershed characteristics as described by relief, stream length and the bifurcation ratio (Strahler, 1964). Although not expressed directly in the equations of Table 2.12, sediment delivery ratio is also influenced by the sediment source and its proximity to the stream, the transport system, and the texture of the eroded material (Renfro, 1975).

Table 2.12. Summary o

Reference	Region	Watershed size range, km²	No. of equations presented
Mutchler and Bowie, 1976	Pidgeon Roost Creek, Mississippi	5–8	1
Renfro, 1975	Texas and Oklahoma	1–252	2
Roehl, 1962	Southeast Piedmont	2–432	4

[a] SDR—Sediment Delivery-ratio.
[b] Defined in Strahler, 1964.

The gross erosion and sediment delivery ratio computation method of sediment yield estimation is in some respects similar to the predictive equation method. The similarity is most evident in the case of the modified Universal Soil Loss Equations and for other watershed parameter equations that include gross erosion.

The United States Soil Conservation Service (1971) has developed a general sediment delivery ratio versus drainage area relationship from data of earlier studies. It is intended that this relationship be used only if local or regional relationships are not established and time is not available to develop a sediment yield relationship for the project area. The relationship shows that the sediment delivery ratio varies approximately inversely as the 0.2 power of the drainage area. The wide scatter of data used in the development of this relationship indicates that additional variables affect the relationship. Estimates of the delivery ratio may be obtained from Table 2.13. The use of these estimates should be tempered with a consideration of other factors that may affect the

Table 2.13. General sediment delivery-ratio estimates[a]

Drainage area, square kilometres	Sediment delivery-ratio
0.05	0.58
0.1	0.52
0.5	0.39
1	0.35
5	0.25
10	0.22
50	0.153
100	0.127
500	0.079
1000	0.059

[a] From United States Soil Conservation Service (1971). Reference points taken from logarithmic plot.

Dependent variable[a]	Independent Variables Used				
	Watershed area	Runoff volume	Relief	Length	Bifurcation ratio[b]
SDR	×	×			
SDR	×		×	×	
SDR	×		×	×	×

sediment delivery ratio of a particular site. A higher delivery ratio should be used when the eroding soil is very high in silt or clay and lower if the eroding soil is coarse textured. The conditions of the channels and delivery system should also be evaluated to alter the general relationship of Table 2.13 if desirable.

2.4.3 Suspended load or reservoir deposition extrapolation

Suspended sediment load or reservoir deposition measurements may be available for adjacent watersheds or for watersheds in the same major land resource area. Topography, soils, and land use should be similar for the measured watershed and the watershed for which sediment yield is to be estimated.

Suspended sediment load is determined by measuring the suspended sediment transported by a stream. Stream discharge is determined by stream gauging. The measured sediment concentration is converted to sediment load, in mass per time period, using the average concentration and the volume of flow for a time period. The sediment load versus stream discharge rates for many sample periods are then plotted on logarithmic paper to provide the basis for a sediment rating curve for a stream. A flow duration curve may than be used with the sediment rating curve to determine the average annual suspended sediment yield. This determination does not include the bed load portion of the sediment yield. Estimates of bed load must be made if bed load is thought to be a significant portion of the total sediment load. Bed load may vary from zero to nearly all of the total load depending on the sediment sources and transport capability of the stream.

Reservoirs of known age and sedimentation history are excellent data sources for determining sediment yields. The sediment accumulation over a known time span can be used to obtain the average annual sediment yield. However, reservoir deposition and sediment yield are not synonymous and the reservoir trap efficiency must be used to account for the difference. The trap efficiency of a reservoir is the portion of the total sediment delivered to the reservoir that is retained in the reservoir. Trap efficiency has been related to the reservoir storage capacity—mean annual inflow ratio (Vanoni, 1975). Average annual sediment yield, capacity–inflow ratio, and storage capacity data are available for watersheds in the United States (Agricultural Research Service, 1973).

Annual sediment yield from either of the two measured sources may be transferred to an adjacent watershed of interest in the same physiographic area. To transfer the sediment yield directly, the drainage areas should not be different in size by a factor greater than two. For drainage areas that differ by a factor greater than two, the United States Soil Conservation Service (1971) recommends that the following relationship for humid areas east of the Rocky Mountains be used to transfer sediment yield estimates:

$$S_e = S_m \left(\frac{A_e}{A_m}\right)^{0.8} \tag{2.30}$$

where:

S_e = sediment yield of the unmeasured watershed,
S_m = sediment yield of the measured watershed,
A_e = drainage area of unmeasured watershed, and
A_m = drainage area of the measured watershed.

Detailed discussions of suspended sediment load and reservoir accumulation measurement and computational procedures are available in Vanoni (1975).

2.5 SOIL LOSS ESTIMATION AND PHYSICAL MODELLING

The Universal Soil Loss Equation and similar soil loss estimation techniques were developed primarily to provide a planning tool for conservation technicians; therefore, ease of application was a predominant consideration in development. The general acceptance of the USLE by both technicians and scientists has demonstrated the extent to which this technique has been successfull as a useful field tool and in providing reasonably accurate estimates of soil loss. These methods are empirical relations combined in equations to predict average soil loss from fields with specific combinations of land use and management. They do not satisfy the need for a detailed model that simulates soil erosion as a dynamic process nor do they describe soil movement along a slope.

Analytical work continues to describe the soil erosion processes mathematically using known physical laws. Where specific phenomena cannot be fully developed because adequate physical relationships are not known, empirical coefficients or relationships are used. These empirical substitutes as well as the physical theory must be verified with laboratory and field research. Concepts of these processes have been proposed most recently by Meyer and Wischmeier (1969), Foster (1971), Foster and Meyer (1972, 1975), Onstad and Foster (1975), David and Beer (1975), Meyer *et al.* (1976), Foster *et al.* (1977a and b), and Foster and Huggins (1977). These concepts are developed further by Kirkby (Chapter 6).

REFERENCES

Ackerman, W. C., and Corinth, R. L. (1962). An empirical equation for reservoir sedimentation. *Int. Assoc. Scient. Hydrol. Pub.*, **59**, pp. 359–366.

Agricultural Research Service (1973). *Summary of Reservoir Sediment Deposition Surveys Made in the United States Through 1970*, Miscellaneous Publication No. 1266, United States Department of Agriculture, Washington, D.C.

Agricultural Research Service (1975). Control of Water Pollution from Cropland. *Volume I, A manual for guideline development*, Report ARS-H-5-1, United States Department of Agriculture, Washington, D.C.

Anderson, H. W. (1976). Reservoir sedimentation associated with catchment attributes, landslide potential, geologic faults, and soil characteristics. *Proceedings of the Third Federal Inter-Agency Sedimentation Conference*, PB-245-100, Water Resources Council, Washington, D.C., pp. 1-35-1-46.

Ateshian, J. K. H. (1974). Estimation of rainfall erosion index. *J. Irrigation and Drainage Div.*, *Am. Soc. Civil Engrs*, **100**(IR3), 293–307.

Ayres, Q. C. (1936). *Soil erosion and its Control*, McGraw-Hill, New York, N.Y. 365 pp.

Bailly, M. C., Malvos, M. C., Sarrailh, M. J. M., Rakotomanana, M. J. L., Rampanana, L., and Ramanahadray, M. (1976). Etude de la sensibilité des sols de Madagascar a l'érosion. *Bois et Forêts des Tropiques*, **169**, 15–28.

Beer, C. E., and Johnson, H. P. (1963). Factors in gully growth in the deep loess area of Western Iowa. *Trans. Am. Soc. Agric. Engrs*, **6**, 237–240.

Betson, R. P. (1963). Building a non-linear sediment yield model. In: *Proceedings of the Federal Inter-Agency Sedimentation Conference* (Agricultural Research Service Miscellaneous Publication No. 970), United States Department of Agriculture, Washington, D.C., pp. 65–72.

Browning, G. M., Parish, C. L., and Glass, J. (1947). A method for determining the use and limitations of rotation and conservation practices in the control of soil erosion in Iowa. *J. Am. Soc. Agron.*, **39**, 65–73.

Carter, C. E., Greer, J. D., Braud, H. J., and Floyd, J. M. (1974). Raindrop characteristics in South Central United States. *Trans. Am. Soc. Agric. Engrs*, **17**, 1033–1037.

Cormary, Y., and Masson, J. (1964). Etude de conservation des eux et du sol au Centre de Recherches du Génie Rural de Tunisie, application à un projet-type de la formule de perte de sols de Wischmeier. *Cahiers ORSTOM, Pédologie*, Volume 11, no. 3.

David, W. P., and Beer, C. E. (1975). Simulation of soil erosion. *Trans. Am. Soc. Agric. Engrs*, **18**, 126–133.

Dendy, F. E., and Bolton, G. C. (1976). Sediment yield–runoff–drainage area relationships in the United States. *J. Soil and Water Conserv.*, **31**, 264–266.

Delwaulle, J. C. (1973). Résultats de six ans d'observations sur l'érosion au Niger. *Bois et Forêts des Tropiques*, **150**, 15–37.

Diseker, E. G., and McGinnis, J. T. (1967). Evaluation of climatic, slope, and site factors in erosion from unprotected roadbanks. *Trans. Am. Soc. Agric. Engrs*, **10**, 9–11, 14.

Ellison, W. D. (1947). Soil erosion studies. *Agric. Engng*, **28**, 145–146, 197–201, 245–248, 297–300, 349–351, 402–405, 442–450.

Elwell, H. A. (1977). Soil Loss estimation system for southern Africa. *Department of Conservation and Extension, Research Bulletin No. 22*. Salisbury, Rhodesia.

Elwell, H. A., and Stocking, M. A. (1976). Vegetal cover to estimate soil erosion hazard in Rhodesia, *Geoderma*, **15**, 61–70.

Environmental Protection Agency (1973). *Comparative Costs of Erosion and Sediment Control, Construction Activities.* EPA-430/9-73-016, United States Environmental Protection Agency, Washington, D.C.

Flaxman, E. M. (1972). Predicting sediment yield in the Western United States. *J. Hydraulics Div., Am. Soc. Civil Engrs,* **98**(HY12), 2073–2085.

Foster, G. R. (1971). The overland flow process under natural conditions. In: *Biological Effects in the Hydrologic Cycle, Proceedings of the Third International Seminar for Hydrology Professors, Purdue University, Lafayette, Indiana,* pp. 173–185.

Foster, G. R., and Huggins, L. F. (1977). Deposition of sediment by overland flow on concave slopes. In: *Soil Erosion: Prediction and Control, Proceedings of the National Conference on Soil Erosion,* Soil Conservation Society of America, Ankeney, Iowa, pp. 167–180.

Foster, G. R., and Meyer, L. D. (1972). A closed-form soil erosion equation for upland areas. In: *Sedimentation* (H. W. Shen, ed.). H. W. Shen, Fort Collins, Colorado, pp. 12.1–12.19.

Foster, G. R. and Meyer, L. D. (1975). Mathematical simulation of upland erosion by fundamental erosion mechanics. In: *Present and Prospective Technology for Predicting Sediment Yields and Sources*—Proceedings of Sediment-Yield Workshop, United States Department of Agriculture Sedimentation Laboratory, Oxford, Mississippi, November 1972 (Agricultural Research Service Report ARS-S-40), United States Department of Agriculture, Washington, D.C., pp. 190–207.

Foster, G. R., Meyer, L. D., and Onstad, C. A. (1973). Erosion equations derived from modeling principles. *Unpublished Paper No. 73-2550.* American Society of Agricultural Engineers, St. Joseph, Michigan.

Foster, G. R., Meyer, L. D., and Onstad, C. A. (1977a). An erosion equation derived from basic erosion principles. *Trans. Am. Soc. agric. Engrs,* **20**, 678–682.

Foster, G. R., Meyer, L. D., and Onstad, C. A. (1977b). A runoff erosivity factor and variable slope length exponents for soil loss estimates. *Trans. Am. Soc. Agric. Engrs,* **20**, 683–687.

Foster, G. R., and Wischmeier, W. H. (1974). Evaluating irregular slopes for soil loss prediction. *Trans. Am. Soc. Agric. Engrs,* **17**, 305–309.

Fournier, F. (1960). *Climat et erosion: la relation entre l'érosion du sol par l'eau et les précipitations atmospheriques,* Presses Universitaires de France, Paris. 201 pp.

Fournier, F. (1962). Map of erosion danger in Africa South of the Sahara, explanatory note. Commission for Technical Cooperation in Africa, European Economic Community. April, 1962.

Glymph, L. M., Jr. (1957). Importance of sheet erosion as a source of sediment. *Trans. Am. Geophys. Union,* **38**, 903–907.

Hadley, R. F., and Schumm, S. A. (1961). Sediment sources and drainage-basin characteristics in Upper Cheyenne River Basin. *US Geolog. Surv. Water Supply Paper,* **1531-B**, 137–198.

Herb, W. J. and Yorke, T. H. (1976). Storm-period variables affecting sediment transport from urban construction areas. *Proceedings of the Third Federal Inter-Agency Sedimentation Conference,* PB-245-100, Water Resources Council, Washington, D.C., pp. 1-181–1-192.

Hindall, S. M. (1976). Prediction of sediment yields in Wisconsin streams. *Proceedings of the Third Federal Inter-Agency Sedimentation Conference,* PB-245-100, Water Resources Council, Washington, D.C., pp. 1-205–1-218.

Holberger, R. L., and Truett, J. B. (1976). Sediment yield from construction sites. *Proceedings of the Third Federal Inter-Agency Sedimentation Conference,* PB-245-100, Water Resources Council, Washington, D.C., pp. 1-47–1-58.

Hudson, N. W. (1961). An introduction to the mechanics of soil erosion under conditions of subtropical rainfall. *Rhodesia Science Association Proceedings*, **49**, 14–25.

Hudson, N. W. (1961). *Soil conservation*. Cornell University Press, Ithaca, New York. 320 pp.

Johnson, C. W., and Hanson, C. L. (1976). Sediment sources and yields from sagebrush rangeland watersheds. *Proceedings of the Third Federal Inter-Agency Sedimentation Conference*, PB-245-100, Water Resources Council, Washington, D.C., pp. 1-71–1-80.

Lane, E. W. (1955). The importance of fluvial morphology in hydraulic engineering. *Proc. Am. Soc. Civil Engrs*, **81**, 745-1–745-17.

Laws, J. O. (1940). Recent studies in raindrops and erosion. *Agric. Engng*, **21**, 431–433.

Laws, J. O., and Parsons, D. A. (1943). The relation of rain drop size to intensity. *Trans. Am. Geophys. Union*, **22**, 709–721.

Lloyd, C. H., and Eley, G. W. (1952). Graphical solution of probable soil loss formula for the northeastern region. *J. Soil and Water Conserv.*, **7**, 189–191.

McCool, D. K., Molnau, M., Papendick, R. I., and Brooks, F. L. (1977). Erosion research in the dryland grain region of the Pacific Northwest: recent developments and needs. In: *Soil Erosion: Prediction and Control, Proceedings of the National Conference on Soil Erosion*, Soil Conservation Society of America, Ankeney, Iowa, pp. 50–59.

McCool, D. K., Papendick, R. I., and Brooks, F. L. (1976). The universal soil loss equation as adapted to the Pacific Northwest. *Proceedings of the Third Federal Inter-Agency Sedimentation Conference*, PB-245-100, Water Resources Council, Washington, D.C., pp. 2-135–2-147.

McCool, D. K., Wischmeier, W. H., and Johnson, L. C. (1974). *Adapting the Universal Soil Loss Equation to the Pacific Northwest*, Unpublished paper no. 74-2523, American Society of Agricultural Engineers, St. Joseph, Michigan.

Masson, J. M. (1972). L'érosion des sols par l'eau en climat Mediterranéen, méthods expérimentales pour l'étude des quantités érodés a l'échalle du champ. *La Houille Blanche*, **8**, 673–678.

Masson, J. M. (1976). *Mesure de l'agressivite des pluies en rapport avec l'erosion des sols*, Laboratoire d'Hydrologie Mathématique, Université des Sciences et Techniques du Lanquedoc, Montepellier, France.

Meyer, G. J., Schoeneberger, P. J., and Huddleston, J. H. (1975). Sediment yields from roadsides: an application of the universal soil loss equation. *J. Soil and Water Conserv.*, **30**, 289–291.

Meyer, L. D., De Coursey, D. G., and Romkens, J. M. (1976). Soil erosion concepts and misconceptions. *Proceedings of the Third Federal Inter-Agency Sedimentation Conference*, PB-245-100, Water Resources Council, Washington, D.C., pp. 2-1–2-12.

Meyer, L. D., and Wischmeier, W. H. (1969). Mathematical simulation of the process of soil erosion by water. *Trans. Am. Soc. Agric. Engrs*, **12**, 754–758.

Musgrave, G. W. (1947). The quantitative evaluation of factors in water erosion—a first approximation. *J. Soil and Water Conserv.*, **2**, 133–138.

Mutchler, C. K., and Bowie, A. J. (1976). Effect of land use on sediment delivery ratios. *Proceedings of the Third Federal Inter-Agency Sedimentation Conference*, PB-245-100, Water Resources Council, Washington, D.C., pp. 1-11–1-21.

Olson, T. C., and Wischmeier, W. H. (1963). Soil-erodibility evaluations for soils on the runoff and erosion stations. *Proc. Soil Sci. Soc. Am.*, **27**, 590–592.

Onstad, C. A., and Foster, G. R. (1975). Erosion modeling on a watershed. *Trans. Am. Soc. Agric. Engrs*, **18**, 288–292.

Onstad, C. A., Larson, C. L., Hermsmeier, L. F., and Young, R. A. (1967). A method of computing soil movement throughout a field. *Trans. Am. Soc. Agric. Engrs*, **10**, 742–745.

Onstad, C. A., Piest, R. F., and Saxton, K. E. (1976). Watershed erosion model validation for Southwest Iowa. *Proceedings of the Third Federal Inter-Agency Sedimentation Conference*, PB-245-100, Water Resources Council, Washington, D.C., pp. 1-22–1-34.

Piest, R. F., Beer, C. E., and Spomer, R. G. (1976). Entrenchment of drainage systems in Western Iowa and Northwestern Missouri. *Proceedings of the Third Federal Inter-Agency Sedimentation Conference*, PB-245-100, Water Resources Council, Washington, D.C., pp. 5-48–5-60.

Piest, R. F., Bradford, J. M., and Wyatt, G. M. (1975). Soil erosion and sediment transport from gullies. *J. Hydraulics Div., Am. Soc. Civil Engrs*, **101**(HY1), 65–80.

Renard, K. G., and Simanton, J. R. (1975). Thunderstorm precipitation effects on the rainfall-erosion index of the universal soil loss equation. *Hydrology and Water Resources in Arizona and the Southwest, American Water Resources Association, Arizona Section and Arizona Academy of Science, Hydrology Section, Proceedings of April 11–12 Meeting, Tempe, Arizona*, **5**, 47–55.

Renard, K. G., Simanton, J. R., and Osborn, H. B. (1974). Applicability of the universal soil loss equation to semiarid rangeland conditions in the southwest. *Hydrology and Water Resources in Arizona and the Southwest, American Water Resources Association, Arizona Section and Arizona Academy of Science, Hydrology Section, Proceedings of April 19–20 Meeting, Flagstaff, Arizona*, **4**, 18–31.

Renfro, G. W. (1975). Use of erosion equations and sediment-delivery ratios for predicting sediment yield. In: *Present and Prospective Technology for Predicting Sediment Yields and Sources*—Proceedings of Sediment-Yield Workshop, United States Department of Agriculture Sedimentation Laboratory, Oxford, Mississippi, November 1972 (Agricultural Research Service Report ARS-S-40), United States Department of Agriculture, Washington, D.C., pp. 33–45.

Roehl, J. W. (1962). Sediment source areas, delivery ratios and influencing morphological factors. *Int. Assoc. Scient. Hydrol. Pub.*, **59**, 202–213.

Roose, E. (1977). Use of the universal soil loss equation to predict erosion in West Africa. In: *Soil Erosion: Prediction and Control, Proceedings of the National Conference on Soil Erosion*, Soil Conservation Society of America, Ankeney, Iowa, pp. 60–74.

Singer, M. J., Huntington, G. L., and Sketchley, H. R. (1977). Erosion prediction on California rangeland: research developments and needs. In: *Soil Erosion: Prediction and Control, Proceedings of the National Conference on Soil Erosion*, Soil Conservation Society of America, Ankeney, Iowa, pp. 143–151.

Smith, D. D. (1941). Interpretation of soil conservation data for field use. *Agric. Engng*, **22**, 173–175.

Smith, D. D., and Whitt, D. M. (1947). Estimating soil losses from field area of claypan soil. *Proc. Soil. Sci. Soc. Am.*, **12**, 485–490.

Smith, D. D., and Whitt, D. M. (1948). Evaluating soil losses from field areas. *Agric. Engng*, **29**, 394–396.

Smith, D. D., and Wischmeier, W. H. (1957). Factors affecting sheet and rill erosion. *Trans. Am. Geophys. Union*, **38**, 889–896.

Smith, J. O. and Ports, M. A. (1976). Maryland highway erosion and sediment control: evaluation and future directions. In: *Land Application of Waste Materials*, Soil Conservation Society of America, Ankeney, Iowa, pp. 262–275.

Strahler, A. N. (1964). Geology, Part II. Quantitative geomorphology of drainage basins and channel networks. In: *Handbook of Applied Hydrology* (V. T. Chow, Ed.), McGraw-Hill, New York, N.Y.

Striffler, W. D. (1963). Suspended sediment concentrations in a Michigan trout stream as related to watershed characteristics. In: *Proceedings of the Federal Inter-Agency Sedimentation Conference* (Agricultural Research Service Miscellaneous Publication No. 970), United States Department of Agriculture, Washington, D.C., pp. 144–155.

Striffler, W. D. (1964). *Sediment, Streamflow, and Land use Regulations in Northern Michigan*, United States Forest Service Research Paper LS-16, United States Department of Agriculture, Washington, D.C.

Thompson, J. R. (1964). Quantitative effect of watershed variables on rate of gully-head advancement. *Trans. Am. Soc. Agric. Engrs*, 7, 54–55.

United States Soil Conservation Service (1966). *Procedures for Determining Rates of Land Damage, Land Depreciation, and Volume of Sediment Produced by Gully Erosion*, Technical Release No. 32, United States Department of Agriculture, Washington, D.C.

United States Soil Conservation Service (1971). Sediment sources, yields and delivery ratios. *SCS National Engineering Handbook* (Section 3, Chapter 6), United States Department of Agriculture, Washington, D.C.

United States Soil Conservation Service (1973). *A Method for Estimating Volume and Rate of Runoff in Small Watersheds*, Report SCS-TP-149, United States Department of Agriculture, Washington, D.C.

United States Soil Conservation Service (1975). Universal soil loss equation. *SCS Technical Note Conservation Agronomy No. 32*, United States Department of Agriculture, West Technical Center, Portland, Oregon.

United States Soil Conservation Service (1976). *Erosion and Sediment Control Guide for Urbanizing Areas in Hawaii*, United States Department of Agriculture, Honolulu, Hawaii.

Van Doren, C. A., and Bartelli, L. J. (1956). A method of forecasting soil loss. *Agric. Engng*, 37, 335–341.

Vanoni, V. A. (Ed.) (1975). *Sedimentation Engineering* (American Society of Civil Engineering Manuals and Reports on Engineering Practice No. 54), ASCE, New York, N.Y.

Williams, J. R. (1971). Prediction of sediment yields from small watersheds. *Trans. Am. Soc. Agric. Engrs*, 14, 1157–1162.

Williams, J. R. (1975). Sediment-yield prediction with universal equation using runoff energy factor. In: *Present and Prospective Technology for Predicting Sediment Yields and Sources*—Proceedings of Sediment-Yield Workshop, United States Department of Agriculture Sedimentation Laboratory, Oxford, Mississippi, November 1972 (Agricultural Research Service Report ARS-S-40), United States Department of Agriculture, Washington, D.C., pp. 244–252.

Williams, J. R., and Berndt, H. D. (1972). Sediment yield computed with universal equation. *J. Hydraulics Div., Am. Soc. Civil Engrs*, 98(HY12), 2087–2098.

Williams, J. R., and Berndt, H. D. (1976). *Sediment Yield Prediction Based on Watershed Hydrology*, Unpublished paper No. 76-2535, American Society of Agricultural Engineers, St. Joseph, Michigan.

Wischmeier, W. H. (1959). A rainfall erosion index for a universal soil-loss equation. *Proc. Soil. Sci. Soc. Am.*, 23, 246–249.

Wischmeier, W. H. (1960). Cropping-management factor evaluations for a universal soil-loss equation. *Proc. Soil Sci. Soc. Am.*, 24, 322–326.

Wischmeier, W. H. (1974). New developments in estimating water erosion. *Proceedings of the 29th Annual Meeting of the Soil Conservation Society of America*, SCSA, Ankeney, Iowa, pp. 179–186.

Wischmeier, W. H. (1976). Use and misuse of the universal soil loss equation. *J. Soil and Water Conserv.*, 31, 5–9.

Wischmeier, W. H., Johnson, C. B., and Cross, B. V. (1971). A soil erodibility nomograph for farmland and construction sites. *J. Soil and Water Conserv.*, 26, 189–193.

Wischmeier, W. H., and Mannering, J. V. (1969). Relation of soil properties to its erodibility. *Proc. Soil. Sci. Soc. Am.*, 33, 131–137.

Wischmeier, W. H., and Smith, D. D. (1958). Rainfall energy and its relationship to soil loss. *Trans. Am. Geophys. Union*, **39**, 285–291.

Wischmeier, W. H., and Smith, D. D. (1965). *Predicting Rainfall-erosion Losses from Cropland East of the Rocky Mountains*, Agriculture Handbook No. 282, United States Department of Agriculture, Washington, D.C.

Wischmeier, W. H., and Smith, D. D. (1978). *Predicting Rainfall Erosion Losses*, Agriculture Handbook No. 537, United States Department of Agriculture, Washington, D.C.

Wischmeier, W. H., Smith, D. D. and Uhland, R. E. (1958). Evaluation of factors in the soil-loss equation. *Agric. Engng*, **39**, 458–462.

Woodburn, R. (1949). Science studies a gully. *Soil Conservation*, **15**(1), 11–13.

Woodward, D. E. (1975). Discussion of: 'Estimation of Rainfall Erosion Index', by J. K. H. Ateshian (*ASCE* **100**(IR3), 293–307, 1974). *J. Irrigation and Drainage Div., Am. Soc. Civil Engrs*, **101**(IR3), 245–247.

Yorke, T. H. and Herb, W. J. (1976). Urban-area sediment yield—effects of construction site conditions and sediment-control methods. *Proceedings of the Third Federal Inter-Agency Sedimentation Conference.* PB-245-100, Water Resources Council, Washington, D.C., pp. 2-52–2-64.

Zingg, A. W. (1940). Degree and length of land slope as it affects soil loss in runoff. *Agric. Engng*, **21**, 59–64.

Soil Erosion
Edited by M. J. Kirkby and R. P. C. Morgan
© 1980 John Wiley and Sons Ltd.

CHAPTER 3

Measuring soil loss and experimental studies

J. de Ploey

Instituut voor Aardwetenschappen, Leuven, Belgium

and

D. Gabriels

Department of Soil Physics, State University of Ghent, Belgium

3.1 PROBLEMS AND BASIC PRINCIPLES

3.1.1 Purpose

Until recently, soil erosion research has been conducted in the various disciplines of the earth sciences from different points of view according to the specific formal object of each discipline. Classic geomorphology, insofar as it set out to explain the genesis of relief forms, was more concerned with the topographical effect of soil erosion than in the analysis of the erosion mechanism as such. Soil scientists measured erosion and examined the erosion mechanism from the point of view of erosion control and soil preservation. Civil engineers saw soil erosion as a negative element and studied the geometric forms that occur in the construction industry.

The recent evolution of research has been convergent and the earth sciences are focusing more and more on investigation of the erosion mechanism. Starting with a multitude of observations, qualitatively and quantitatively described, one seeks more now a stage of empirical generalization and, ultimately, the formation of theories and the construction of models. As Hempel says (1965, p. 173), 'Scientific research in its various branches seeks not merely to record particular occurrences in the world of our experience; it tries to discover regularities in the flux of events and thus to establish general laws which may be used for prediction, postdiction (explaining past data in terms of given observations) and explanation.' Hempel's definition, applied to soil erosion, offers a good point of departure to examine the ultimate purpose of measurements and experiments and this is methodologically important.

The flux of events

For soil erosion, a three-phase system is initiated by endogenous and/or exogenous forces. Soil consists mostly of a solid phase combined with liquid and gas phases. This system can be set in motion directly by the force of gravity or be eroded by the intervention of the kinetic energy of water, wind, and ice. Chemical agents can also be factors. The measurement of soil erosion aims at learning the quantity of soil that is moved in a given space over a certain distance during a specific period of time. Measurement can consist of post factum volumetric recording or synchronic–dynamic recording.

Volumetric recording is concerned with the measurement of the volumetric effect of soil erosion after a process or combination of processes has been active. For example, a good estimate can be made from topographical data of the volume shifted in landslides or of the amount of debris eroded into badlands.

Synchronic–dynamic, or simply, dynamic, recording directly measures the amount of debris transported by the erosion agent. These measurements naturally also lead to volumetric considerations. Thus Fournier (1960) calculated for the different continents, the (minimal) amount of annual ablation in $t/km^2/$ year on the basis of the measurement of suspended material transported annually by the major rivers. In the same way, ablation from steep slopes due to run-off can be calculated from the amount of sediment gathered in collecting troughs.

Tracers and markers are used to facilitate or make possible volumetric and dynamic recording. Erosion pins can be used in post factum volumetric recording of local ablation.

Stratigraphic geomorphology evaluates the volumetric changes of the relief in geological time though to date it has been mostly limited to volumetric estimates that involve the magnitude of correlative deposits. Dynamic recordings, however, are necessary if one wants to proceed to morphogenetic explanations in which the causal phenomena are given more attention than the topographic end effect.

The establishment of general laws

This implies that the comparable results of measurements and experiments ought to be able to lead to empirical generalization so that it is possible to proceed to theory formation in which comprehensive laws, expressed in hypothetic terms, are formulated. Then, starting from general laws, predictions and postdictions can be inferred, for example in the context of a hill-slope process, and explanations given.

A necessary condition for all of this is that the empirical data be comparable, that is, that the experiments be as precise as possible, repeatable, and accurately described.

Prediction, postdiction, and explanation

Under the category of prediction fall soil loss formulae such as that of Wischmeier and Smith (1965). Developed in the United States, this comprehensive law allows the farmer and soil conservationist to calculate the ablation of fields in order to be able to take appropriate erosion control measures. The formula (Mitchell and Bubenzer, Chapter 2, equation (2.7)):

$$A = R.K.L.S.C.P$$

includes a rain factor R, a soil erodibility factor K, two topographic factors, slope length L and slope gradient S, and two technical cultivation factors, cropping management C, and erosion control practice P.

The universal soil loss equation was designed to predict soil loss from sheet and rill erosion, which must be distinguished from field sediment yield as the equation does not account for soil deposition in depressions within the field.

Nearly two decades of widespread use in soil conservation district programmes have confirmed its reliability when used as designed. But sometimes the equation is also used for purposes that it was not designed for, simply because it seems to meet the need better than any other tool available (Wischmeier, 1976).

Important work has also been done in Rhodesia on test plots for the prediction of soil erosion (Hudson, 1958; Hudson and Jackson, 1959). In the USA, a soil loss formula for wind erosion comparable to that of Wischmeier and Smith has been devised by Woodruff and Siddoway (1965). Terzaghi and Peck (1967) give examples of predictive empirical formulae for mass transport and the equilibrium of slopes that have been worked out by civil engineers.

The purpose of these erosion formulae is to express the relation between the volumetric effect of erosion and the parameter values of specific representative terrain and climatic factors. The value of these formulae is naturally determined by the accuracy of the resulting prediction, which can be tested by field measurements. It is also important that the numerical values of each of the parameters be readily determinable.

Stratigraphic geomorphology places more emphasis on postdiction since it is concerned with how the relief developed during a geological period. Quantitative approaches are, at the most, only as precise as is the knowledge of the ecology of the palaeomilieu. A hillslope process–response model can be applied when the palaeogeographic conditions are judged to be sufficiently known for meaningful parameter values to be inserted in the erosion formula.

Practical and usable erosion formulae, which indicate the connection between the amount of ablation and the numerical values of the various terrain factors, are therefore still not satisfactory as far as explaining the erosion mechanism is concerned. This requires that the erosion phenomenon be described in elementary physical or physical–chemical terms, for example, those by which water and air flow are explained or with which equilibria of soil masses are discussed.

Experimental research here offers new perspectives for systematic factor analysis. Research into the properties of soils and correlative deposits is also important as has been emphasized by Jungerius, Koster, and Kwaad (1973).

3.1.2 Measurements and experiments

Measurements

Soil erosion is measured, volumetrically and dynamically, at carefully chosen and representative measurement sites or measurement stations. In direct measurements, the measured quantity is directly related to a scale. Thus, the amount of annual ablation can be measured by erosion pins and expressed in metres. This is direct volumetric recording. Direct dynamic recording would be, for example, the determination of runoff erosion of a slope by capturing water and sediment in collectors, the results being expressed in grams and litres. Indirect measurement consists of measuring the effect of the unknown quantity and from this measured effect calculating the quantity itself. In this way, Laws (1941) photographically measured the terminal velocity of rain drops, and Hudson (1963) measured their diameter by the so-called flour-pellet method by which flour pellets are weighed to determine the rain drop diameter.

Measurements involve error. Chance and systematic measurement errors must be indicated, either by noting their absolute values or their relative values in percentages of the measured value. Important, too, is the critical problem of the possible distortion of the erosion medium caused by implantation of the measurement apparatus.

Measurements can apply to the total erosion effect or to the study of an aspect of soil erosion as in the measurement of elementary erosion factors. For example, the length and degree of slope of a test plot must first be measured when slope erosion is to be determined by direct dynamic recording with collectors.

Indirect measurement can also be directed to the determination of significant values presumed to be representative of the erosion phenomenon. Thus the plasticity index of specific soil types is measured in order to determine the degree that this index is representative of erodibility, for example, in determining the susceptibility of certain soil types to sliding.

Experiments

In general, an experiment creates a situation designed to lead to an explanation of a particular phenomenon. Specifically as far as soil erosion is concerned, this means that the erosion phenomenon is completely or partially simulated in the field or in the laboratory in order to investigate systematically the erosion mechanism. It is presumed that the experimenter can control at least one significant erosion factor so that its role in the total phenomenon can be tested.

The experiment can be repeated and the setup changed in order to better evaluate the role of the various (known) erosion factors. The erosion process is specified by a complex of independent and dependent variables that together condition the cause and effect relations and are analysed in the experiment. The analysis usually rests on repetition of the experiment in which each factor is tested in sequence as a variable in an otherwise constant medium. This naturally assumes that the experimenter has complete control over all factors introduced in the experiment, a problem that is essentially of a technical nature.

Important in each experiment is whether it concerns a total or a partial simulation of natural events and the extent to which the structure of the experimental setup is equivalent to the structure of the natural process. There is a fundamental difficulty with equivalence control: the structure of the natural process is only partially known and cannot therefore be perfectly simulated. Still, one can presume that even an imperfect simulation can often produce positive knowledge insofar as an essential understanding of the manner in which specific factors and mechanisms interact is acquired. The study of a particular erosion mechanism does not necessarily imply the simulation of a local natural system observed in the field. In the test channel of a wind tunnel, the deflatability of different sandy texture classes can be tested without actually simulating an observed regional situation. The laboratory experiment can be situated in a general context.

In the setup of an experiment, it is important to pay careful attention to the boundary conditions, and this is particularly important when various factors are to be manipulated as variables. It must be specified how each factor is situated in space and time and which energy level extremes are to be tested, presuming that one has an adequate understanding of the boundary conditions of the natural system. It is often advisable to test immediately the extreme low and high energy levels in the initial runs of the experiment in order to have a quick idea of the range within which the phenomenon varies and the significance of the variations. When the experimental model is limited in size—and this is always the case in the laboratory—then the degree must be determined to which this limitation interferes with the equivalence of the test, for example, the degree to which the marginal effects influence the measurements in flumes. Equivalence control is also appropriate here, and especially so when changes of scale— mostly scale reductions regarding mass, time, and length— are contemplated. Further changes of scale must be applied with the greatest of care, the associated problems being well-known as, for example, in connection with the model laws for fluvial flumes (Graf, 1971, pp. 385–398).

As we have mentioned, laboratory experiments can be situated in regional and non-regional contexts. Rain simulator tests directed to the study of thin-film hydraulics are non-regional. A field experiment is situated, obviously, in the framework of a local ecosystem. Field experiments can lead to the formulation of universal laws to the extent that the ecosystem exhibits universal character-istics and the obtained results are statistically justified and suitable for extrapola-tion.

Field experiments, as a rule, imply measurements, but not all measurements are situated in an experimental context. For an experiment to be an experiment, it is presumed that at least one factor in the dynamic system is controlled by the experimenter. Thus we can speak of a dynamic experiment when tracers (coloured sediment, mobile plates, radioactive tracers and so on) are used in the study of an erosion mechanism. To a degree, the experimenter does control the erosion phenomenon insofar as he chooses the tracer material and the place where it is introduced into the flow. Of course, this choice influences the transference of the tracer material during the erosion phase and thus ultimately also the inter-pretation of the erosion mechanism. In such a tracer experiment conducted in the field, there is minimal interference on the part of the investigator, but it still can be justifiably called an experiment. The border between measurement and experiment is, however, not always very clear. For example, in studying slides, one can 'measure' the shear strength of a specific soil type. But the in-vestigator not only chooses the sample to be measured, but also partially deter-mines the physical characteristics of that sample—thickness, water content, and so on. The investigator is thus involved in the process itself and, in this sense, we can speak of an experiment. It is, however, a partial experiment in that it is not a global study of the erosion system.

The testing setup must be so structured that the test is repeatable by the same experimenter or others. Repetition of the experiment is generally necessary in order to understand the roles of the various factors. If a number of factors, f_i, are to be tested, each with a number of variable conditions, c_i, then the number of tests necessary, N, is the summation of $f_i . c_i$. The amount of time available can be a limiting factor for N, and thus c_i must be carefully chosen in the planning of the experiment. No limits, in principle, can be placed on f_i, as this would simply mean that the impact of specific factors on the total erosion phenomenon would not be investigated. Such should be avoided as much as possible, par-ticularly because there are always certain factors that are not tested either because they are unknown or because their simulation in the experimental model is not technically possible. In this regard, it can also be noted that, when preparing an equivalent test setup, one has to rely on one's own ingenuity since the available standard apparatus is generally only applicable to limited portions of the setup. Techniques, too, may need to be developed as the investigation proceeds.

Regarding measurements and experiments in general, it is strongly advisable that the protocol be kept as simple as possible and the number of variations be limited to what is absolutely necessary. As a rule, it may be presumed that the probability of cumulative measurement errors and non-equivalence occurring increases as the procedures followed become more complicated.

Sampling and extrapolation of results

Measurements are taken and experiments are performed in the field either in order to resolve a regional erosion problem or in order to conduct a pure process analysis in an area that appears to be particularly representative of a specific erosion mechanism. In both cases, sampling problems arise as it must be decided which site or sites are the most representative of the region and appropriate for the study of the erosion phenomenon. A proper choice often requires a thorough knowledge of the morphography and the morphodynamism of the area. In its most detailed form, a preliminary study can rely on interpretation of aerial photographs and on adequate geomorphological mapping. A thorough reconnaissance of the terrain is absolutely indispensable.

The number of sites or stations that can be established is almost always limited for financial reasons. In view of the broad range of potential field data, therefore, one cannot depend on pure random sampling, which Gregory (1968) defines as 'an unbiased and representative cross section of the body of data.' Samples are taken in clusters, or by stratified sampling. For example, when one examines runoff erosion in a given area with varying slopes but with a uniform pedobotanic system, stratified sampling would consist of erecting three measurement stations: one on a steep slope, one on a gentle slope, and one on an average slope. Testing on the extreme slopes is important in order to determine the boundary conditions of the erosion phenomenon, assuming that, during the measurement period, one will also have the opportunity to gauge the effect of extreme weather conditions, which often requires a period of years.

In a regional erosion survey, regional prediction of erosion is the primary consideration and erosion maps are prepared. Such a project presumes a network of observation points and measurement stations chosen in function of regional boundary conditions, cartographic extrapolation possibilities, and available means. The Soil Conservation Service of the USDA has done pioneer work in this area and has now embarked on an erosion inventory of the United States, Puerto Rico, and the US Virgin Islands. The inventory is based on 72,500 primary sample units representing 0.6 per cent of the total area. Information will be gained on the relative importance of various processes and sources of erosion. Sheet, rill, and wind erosion are being determined in 1977, and gully erosion, and erosion of streambanks, roadsides, and construction areas are to be inventoried in 1978. Additional information will relate these to such things as land use, land capability, and conservation practices. A future benefit of the inventory will be the ability to identify changed erosion rates and to relate these to changed conditions on these same areas (Holeman, 1977). But the available means often do not permit complete mapping of the phenomena. One must then resort to a representative study of a test region whereby general qualitative observations are combined with limited measurements. Such a survey is very useful provided that the test region is sufficiently representative of the surrounding area or, possibly, of a subcontinental unit. The study of Rapp *et al.* (1972) of the Uluguru Mountains in Tanzania is a striking example of this method as regards East African isolated mountains with abundant rainfall.

Should the emphasis be on pure process analysis, then the choice of measurement stations and experimental sites can be made by regional reconnaissance focusing on the places where the specific process seems to be active in its clearest and most varied forms. The process is then analysed and the observed and measured physical and physical–chemical interactions are properly expressed in formulae, possibly including field parameters. Strictly speaking, such formulae have only regional application. General application is possible only when, in other test areas, it can be established that exactly the same interactions occur between the various factors and parameters, and provided that appropriate values can be determined for specific regional field parameters.

3.2. FIELD AND LABORATORY MEASUREMENTS AND EXPERIMENTS

3.2.1 Introduction

During the last decade and at a quickened tempo in recent years, new techniques have been developed and existing techniques improved for the adequate measurement of erosion and the realistic simulation of specific processes. Whether it be a question of volumetric or dynamic field measurements, direct or indirect measurements, or field experiments or global or partial laboratory experiments, it may be taken for granted that investigators have tried, with the simplest possible setups, to achieve the most complete and exact results in the context of problems defined as precisely as possible. The latter consideration implies that a critical review of the designed techniques is necessary relative to the intended purposes and the results to be achieved. Evaluation of results given in the literature still occasionally poses problems to the extent that the investigators themselves very often do not state them clearly. This is one result of the state of the art, with the exception of geotechnical methods, still being in the experimental stage. Almost every laboratory has developed its own techniques, which in turn rapidly evolve, and these techniques have yet to be standardized. Therefore, it does not seem desirable to give, perhaps prematurely, a definitive value judgment on many of the techniques that have been proposed. It is more appropriate here, in view of stimulating further research, to consider what problems have been treated up to now and the main characteristics of the measurement devices and experimental setups that have been used. So we shall now turn to a somewhat systematic discussion of the areas of rainwash, mass movement, and wind erosion.

3.2.2 Ablation and sedimentation by rainwash

General survey

To the extent that soil erosion research is practically oriented, i.e. directed to land management and soil conservation, the mapping of erosion phenomena is an important task. General regional erosion surveys are also an important preliminary to the proper selection of the location of measuring stations, such as erosion plots.

In principle, erosion, and sedimentation phenomena are registered on basic maps that give the various components of the physical environment, i.e. topography, hydrography, soil geography, vegetation, and soil use. The purpose of such maps is to show ablation and sedimentation as much as possible in the broader framework of the entire regional ecosystem. In some areas, rainwash phenomena do not develop in isolation but are associated with other processes such as mass movement (slides, flows) and wind erosion. In such situations, it is desirable to pass on to global mapping of slope erosion and possibly to fluvial dynamics. Among ablation and sedimentation phenomena most often mapped are (1) sheetwash, (2) rill wash when channels are washed out to a minimum depth of a few centimetres, (3) combined systems of sheet and rill wash where runoff channels can develop that extend to colluvial fans, (4) gully erosion when the gullies are of one metre or more in depth and can no longer be filled in by ordinary agricultural techniques, (5) badlands, and (6) colluviation by runoff on basal slopes. Naturally, all these phenomena offer significant micromorphological variability and changing dynamics that one tries to record and accommodate on large scale and detailed maps.

Rapp (1975), on the basis of his experience in Africa, states that erosion surveys can begin with a three-level approach for catchment studies: ground truth, aerial photographs, and satellite images. Satellite images do seem to have been very useful in the study of recent wind erosion and dune formation in arid and semi-arid regions (Van Zuidam, 1971; Rapp, 1974; Mainguet, 1975). It is not yet clear, however, how far orbital photography also allows for dynamic mapping of rain erosion phenomena. This certainly applies to classic aerial photo analysis such as that exhibited in the various publications of the International Institute for Aerial Survey and Earth Sciences (ITC) in Holland. It must be noted that a detailed typology of ablation and sedimentation by rainwash is only possible provided that field work is included. It has been shown that maps can be made using only aerial photographs that, under good conditions, reflect the six principal phenomena mentioned above but without much specification of micromorphography or local dynamics. The mapping of colluvial deposits is also often problematic.

Whether the basic data (maps, aerial photos, satellite images) are comprehensive or limited, as is the case, for example, in many developing countries, the most efficient erosion survey is still best done by a small and highly mobile group of specialists who can do the mapping based on extensive field reconnaissance. In 1933, the Soil Erosion Service (SES) was established in the USA. In the following year, a team under the direction of Bennett began the Reconnaissance Erosion Survey. In two months, the 115 specialists of the team had made an erosion survey map of the entire United States. Later detailed studies have altered little of the very reliable basic structure of this map (Held and Clawson, 1965). The legend for this map with its 9-fold division of erosion phenomena is qualitative but universally applicable because it is simply structured and can be applied to specific regions provided that the terms, 'light, moderate, and severe' can be appropriately defined. National erosion surveys urgently need to be made for many countries with limited means. Belaid (1970) quickly and efficiently constructed a usable erosion map of Tunisia, an example of what can be done.

It is clear that in the second phase of the survey more sophisticated mapping can be done that places soil erosion in its entire ecological context (Rapp and others, 1972; Journaux, 1975). In this connection, we should also mention the more detailed morphodynamic studies of catchment basins that are considered to be representative. Of interest, too, is mapping for the prediction of soil erosion. For this purpose, Stocking and Elwell (1973) made a soil erosion hazard map of Rhodesia based on quantitative and qualitative data regarding five erosion factors: erosivity of rainfall, relief, cover, soil erodibility, and human occupation. The map of Rhodesia was divided into grid squares of 184 km^2. The values of each individual factor were divided into five categories ranging from low to high erosion hazard. Each category was given a numerical value from I to V depending on its potential to influence erosion. Each specific area was allocated an erosion hazard rating for each individual factor and finally, the sum total of these values for the area gave a combined erosion hazard rating.

Measurement of soil loss and sedimentation rates

After having performed the general survey, one can make a justifiable choice of a representative basin or slope type where measurements of erosion and sedimentation can be taken in order to analyse the sediment budget, possibly with the intention of examining explicitly the mechanisms at work.

Studies such as those of Leopold *et al.* (1966) on badlands in semi-arid New Mexico and Rapp *et al.* (1972) on tropical catchments in Tanzania offer a complete overview of the various techniques that are applied in volumetric and dynamic recording. Interesting in this regard is also the investigations done by Polish geomorphologists in the Carpathians (Gerlach, 1970).

In the choice of a measurement site, account must first of all be taken of the availability of regional data relative to the ecosystem. Consideration must be given to the possible proximity of a weather station (for data on precipitation input, evaporation losses, etc.) as well as to hydrological, soil-geographical, and biogeographical information on the area. It is also worthwhile to give attention to the general activity of fauna, the impact of which on soil erosion has been recently emphasized by Yair (1974) and Imeson and Jungerius (1975) in studies done, respectively, in the Negev Desert and in a Luxembourg forest. If a permanent station is to be set up, then rain gauges and a rain recorder are necessary for account has to be taken of spottiness in rainfall in small watersheds as an element of the partial area contribution (Sharon, 1970, 1972). If one is concerned with discontinuous observations of rainstorms, especially in areas with high evaporation, then a crest-stage rain gauge is valuable, such as was used by Kirkby and Kirkby (1974) in Arizona. The highest water level is read from a ring of cork powder that is deposited on the interior wall of an inner vessel, which is inverted to avoid splash erosion.

Volumetric recording envisages only the determination of microtopographic effects that result from ablation or sedimentation. Among the simple and efficient methods available, we mention first the use of erosion pins, or the pin-with-washer method. With them, the vertical component of the contribution of erosion or sedimentation is measured on stakes implanted in slopes, rills, or gullies. Schumm (1967) emphasizes that even this simple method must be applied with care. The pins must be long enough to be firmly fixed in the ground and not subject to surface creep. The recommended length can vary between a few decimetres and one metre. The pins must be strong but as thin as possible as, it must be remembered, every obstacle alters the local stream and erosion pattern and the larger the object, the more the change. The pins ought to be smooth so that they are less sensitive to possible frost heaving. It is also desirable that the amount of scour or fill with respect to the average ground surface be measured. The pins are slipped through a large washer and driven vertically into the ground with the bottom of the washer flush with the ground surface. The washer should be a few centimeters in diameter with the central hole somewhat larger than the thickness of the pin so that the washer will descend as erosion washes away the soil underneath. With weak runoff, e.g. a unit discharge of around 1 litre/dm/hour, this will perhaps not occur and then only pins are used such as thin welding rods or even ordinary long iron nails, which rust and therefore are less visible to undesirable visitors. The pins are numbered and, according to the terms of the problem, are arranged (1) in a grid system on a test slope, (2) along transverse testlines, or (3) on isolated test sites. Leopold *et al.* (1966) give a number of examples of the use of the pin-with-washer method, such as in the study of channel evolution of rills and gullies by means of pin sections.

Pins or stakes can also be inserted a known distance from the vertical sides and from the headcut of gullies to measure the rate of channel enlargement and headcut retreat. One can then go on to plane table mapping and, over longer time intervals, aerial photography, as Nir and Klein (1974) demonstrated in Israel. Obviously, volumetric computations are based not only on planimetric data but also on depth measurements. Thus, Leopold *et al.* (1966) could show in New Mexico that sedimentation not only builds up in arroyo beds and large gullies, but also in small headwater rills. This must be taken into account when pin sections and stakes are set out to determine the sediment budget for a given basin. Monumented cross sections can be measured over a period of time if fixed boundaries can be set as benchmarks.

In order to measure the microtopographic evolution of badland slopes, Campbell (1974) designed a movable contour plotting frame on which the height changes are measured at 25 points in a 1 metre square grid with an accuracy of 0.5 mm. On the basis of these data, a computer drawn contour map is made for each plot.

Painted and numbered rocks can be used to visualize the transport of relatively large debris on slopes, in gullies, and in rivers. Leopold *et al.* (1967) recommend cement-based paints as being more durable than oil-based paints, and they describe a method of following the transport of coarse debris in rills and gullies by placing numbered and coloured rocks at fixed intervals in the channel. On pediments in Arizona, Kirkby and Kirkby (1974) measured the total mass transport of coarse debris with the painted erosion line method. Lines of 15–60 metres by 18 mm were sprayed with enamel paint on the gravel and pebbles. After rain, measured with a crest-stage rain gauge, the movement of the sediment per size class was measured from the erosion line with a steel tape. The total mass transport equals the product of the distance covered and the cross sectional area, summed up for all particles, and is expressed in $cm^3/cm/yr$. The computed transport is a minimal value because of the non-recoverability of particles that increases as the movement is greater and the period of time longer.

Dynamic recordings are based on the registration of both flow and sediment and can thus lead to volumetric computations of ablation and sedimentation. Important pioneer work on this subject was in the field of rain erosion by soil scientists in cooperation with the US Department of Agriculture. Wischmeier and Smith (1958, 1965) in the United States and Hudson (1957) in Rhodesia have explained the goals and methodology of this kind of research. Polish geomorphologists were among the first to employ field measurements (Gerlach, 1967; Słupik, 1970). Many researchers have made use of collectors of the Gerlach trough type by which flow and sediment are collected either on open slopes or from bounded plots. On open runoff plots, measurements can be taken directly if the flow lines run mostly parallel to the slope lines and one has an idea of the size of the collected impluvium on the basis of the length of the slope segment, which is measured from the drainage divide. This presumes a very careful selection of the test slope.

Gregory and Walling (1973) give a rather extensive survey of the instrumentation that can be used for the setting up of runoff plots. Up till now, there is little standardized commercial equipment and each laboratory is obliged to search for new methods or to improve existing methods in function of the particular problem and available means. Simple setups are preferred provided they are efficient, inexpensive, and, primarily, give the fewest measurement errors. As a rule, more and more sophisticated equipment will be necessary if one wishes to delve further into the erosion mechanism and not remain limited to considerations of the end result of erosion and sedimentation. For precipitation input, for example, one can be content with a non-recording rain gauge, but it is obvious that a more precise analysis of the processes requires an autographic rain gauge to supply information on rainfall intensity.

There is also the problem of measuring splash erosion in the field and in the laboratory. The basic element of a splash sampler is the splashboard on which sediment is collected to be transferred, possibly with the aid of a water sprayer, to a container. The specific splash erodibility is expressed in transferred mass per surface unit per time unit, for example, $kg/m^2/sec$. One must also know, therefore, the surface area from which the splash is captured, taking into account that a portion of the splash erosion is not recorded because the transferred particles from the surface do not all reach the splashboard. In the field, as in the laboratory, a setup should be chosen whereby the splash from a narrow strip of sediment is intercepted by two splashboards. These can be placed parallel with the contour lines so that both upslope and downslope components can be intercepted. The limit values of splash erodibility can be learned by successive measuring of continually narrower strips. Finally, by extrapolation on a splash diagram, one can graphically determine the value of the splash erosion for a unit strip, for example, with a width of one centimetre. In field measurements, one must take care to see that the received splash erosion values are not influenced by rain falling at a slant. In the laboratory, the total splash erosion—the lateral splash as well as that at the lower and upper limits—is measured in long, narrow flumes. A representative splashboard model for field research has been designed by Ellison (1944). Quantitative data on splash erosion mechanisms and transport have been compiled by Moeyersons and De Ploey (1976).

Overland flow also causes erosion, which is called runoff erosion. Such can be tested by placing tracer sand under a small shield so that splash erosion is eliminated and only runoff erosion occurs. De Ploey (1967, 1969) in Zaire used radioactive tracer sand labelled with Scandium[46] with which one can also make autoradiographic recordings. Fluorescent sand can be used with a black light to observe dispersal at night. In this manner, Leopold *et al.* (1966) observed the dispersal of sediment from erosion lines. Radioactively marked coarse sediment can be used for transport studies in gullies and rivers (Courtois and Mussot, 1969). The main advantage of radioactive marking is that buried sediment can be traced.

The Gerlach trough is the proto-model for the measurement of overland
flow and sedimentation (Gerlach, 1967). The trough is fixed to the A-horizon
of the slope by means of a metal flange and protected from splash erosion by
a movable lid (Figure 3.1). Undesired distortions can occur as the horizontal

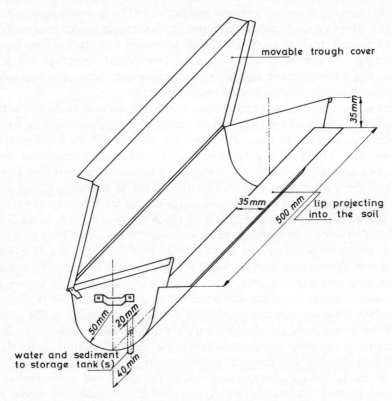

Figure 3.1. A Gerlach trough (After Gerlach)

metal flange is driven in, especially if the ground is stony or full of roots. Leopold
and Emmett (1967) have therefore suggested that in some cases the trough
should be built into a poured concrete collecting platform placed in such a way
that it corresponds to the slope surface. Flow and sediment are drained *via*
plastic hoses to a container dug into the downward slope. The eroded concentrations are determined by filtering, drying, and weighing in the laboratory.
Chronometry can be determined by building in a water level recorder (float
recorder, tipping bucket, etc.). The runoff hydrograph can then be compared
with pluviometric diagrams and the infiltration rate can be computed from

the difference between the rain intensity and the runoff rate. Generally, one or more screens are set in the trough to collect organic material and coarse debris. Filtration of fine suspended sediment can be accelerated by flocculation (Hudson, 1957).

Volumetric conversions can be carried out provided the bulk density of the eroded material is known. 1.4 and 1.5 are often taken as average values (Fournier, 1960; Rapp *et al.*, 1972). Bulk density can also be directly determined in the field with the aid of a densitometer. One can also perform the desired sediment-petrographic analyses of the eroded substratum and of the collected sediment, provided one knows the soil composition.

Gerlach suggests that the troughs be lined up in sets and distributed in echelon over the slope in order to evaluate the influence of the length factor. As already mentioned, this method is applicable when the carefully chosen slope is isotropic and the flow lines run parallel so that the impluvium can be computed within acceptable limits of error. With splash erosion, litter flow, and discontinuous runoff, the length factor normally has no influence on the average captured quantities of flow and sediment. Still, account must be taken of the often non-uniform production of runoff over a specific area. In this case, bounded runoff plots that go as far as the drainage divide seem to be necessary. Small and completely bounded plots allow for the study of specific phenomena such as the runoff threshold and the specific erodibility of the substratum. The construction of a backboard naturally prevents the passage of flow and sediment from the slope segment above it. Results acquired from small runoff plots (290 cm × 140 cm) have been discussed by authors such as Soons and Rainer (1968) in New Zealand. These authors point out that in the construction of a plot the vegetation cover can be disturbed so that measurements can only be taken after the vegetation is reestablished. It is possible to estimate the influence of the length factor with small plots by extrapolating values obtained from a series of plots with lengths increasing geometrically, for example by comparison of the lengths $1L$, $2L$, $4L$ or $1L$, $3L$, $9L$. This method is suitable for very long slopes where it is impossible to lay out a runoff plot to the drainage divide.

Fractional-acre runoff plots, with the objective of evaluating erosion, have been used in the United States since 1917 to measure soil and water losses due to rainfall. Although location of the plot area depends primarily on the research objectives, some site requirements are essential to satisfactory erosion measurements.

The objectives of erosion study projects in the past have generally included a crop or management comparison and an evaluation of soil and water losses as influenced by the climate and soil. However, as the use of plotsize rainfall simulators, such as the rainulator, becomes more widespread, most runoff plots under natural rainfall will be used primarily to evaluate erosion as influenced by the climatic and soil conditions of a particular geographic area.

Therefore, the new plots used for climatic evaluation are located on a major soil type and on a prevailing slope (Mutchler, 1963).

In the course of several decades now of specialized research in the field on soil erosion and conservation problems in the United States it was found that there is no satisfactory substitute for runoff plots as they supply basic data which may be secured only by actual measurement of the quantities of soil and water lost by runoff and erosion.

In the study of erosion on farmland, American agricultural engineers have compared the erosion of determined runoff plots with the erosion on standard plots.

A standard plot (unit plot) is selected in such a way that a measure of soil loss can provide a quantitative, experimentally determined, value of the soil erodibility factor K (Mitchell and Bubenzer, Chapter 2). According to Wisch-meier and Smith (1965) a unit plot has a uniform lengthwise slope of 9 per cent, in continuous fallow, tilled up, and down the slope, and is 72.6 feet long. With this length, widths in multiples of 6 feet give multiples of 1/100 acre in area. Continuous fallow, for this purpose, is land that has been tilled and kept free the vegetation for a period of at least 2 years or until prior crop residues have

Figure 3.2. Runoff plots installation at Morris, Minnesota, USA (After Mutchler and Young)

decomposed. During the period of soil-loss measurements, the plot is ploughed and placed in conventional corn seedbed condition each spring and is tilled as needed to prevent vegetal growth or serious surface crusting. When all of these conditions are met, each of the factors L (slope-length), S (slope-gradient), C (cropping-management) and P (erosion-control practice) has a value of 1.0 and K equals A/R (A: computed soil–loss and R: rainfall factor).

The conditions listed above were selected as unit values in the soil loss equation because they represent the predominant slope length and the median gradient on which past erosion measurements in the United States have been made, and the designated management provides the surface condition least influenced by differences in climate and local cropping systems.

Direct measurements of K on well replicated unit plots as described should reflect the combined effects of all the variables that significantly influence the ease with which a soil is eroded by rainfall and runoff.

A step-by-step procedure for establishing runoff plots has been published by Mutchler (1963). Figure 3.2 illustrates a typical plot installation at Morris,

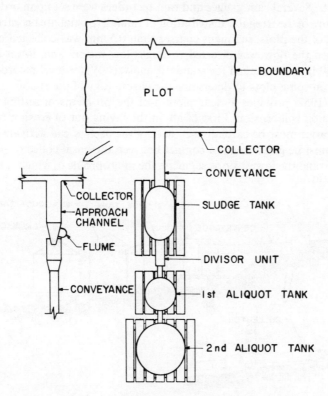

Figure 3.3 Typical plan of runoff plot equipment (After Mutchler and Young)

Minnesota, USA. Similar plots were constructed on tropical soils in Nigeria (Lal, 1976).

The equipment needs for runoff plots are (see Figures 3.3, 3.4, and 3.5):

1. boundaries around the plot to define the measured area;
2. collecting equipment to catch and concentrate runoff from the plot:
3. conveyance equipment to carry runoff to a sampling unit;
4. sampling unit to aliquot the runoff and soil loss into manageable quantities; and
5. storage tanks to hold aliquot portions of runoff and soil loss for analysis.

From erosion rates measured in this way, Wischmeier and Smith (1965) derived the universal soil loss equation to predict global rain erosion on farmland. Yair (1974) at Sde Boker in the Negev Desert laid out an 11,335 m^2 experimental site on the side of the valley of a first order basin to investigate the mechanism and rhythm of slope erosion as a function of the dynamics of the entire local ecosystem. The wall of the valley was divided into 10 bounded erosion plots. Several rain gauges and rain recorders were set up in order to get a good picture of the irregular distribution of rainfall (partial area contribution). At the base of the plots, sediment coarser than 0.5 mm was collected on metal screens. Then the flow was directed through flow meters and, lower down, a sample of 30 litres was taken for standard analysis. Water level recorders were also set up on some plots to determine the chronology of the runoff.

Hudson (1957) provides a clear picture of the problems of setting up field experiments on soil erosion. First of all, in the laying out of erosion plots, the plot boundaries must be carefully set as they themselves can activate erosion. Attention must be given to good contact between the basal collector (Gerlach-type trough) and the lowest slope segment, the microprofile of which can change

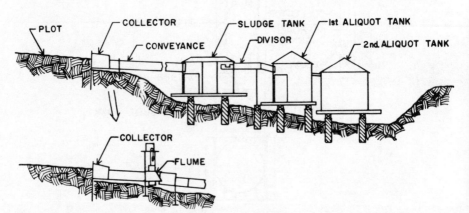

Figure 3.4. Typical profile through the centre line of a set of runoff equipment (After Mutchler and Young)

Figure 3.5. Close-up of the multi-slot divisor (After Mutchler and Young)

by erosion and sedimentation. Floating material is collected separately on screens. With large runoff plots, one or more dividers are set before and between the tanks so that only a well-defined fraction of the possibly excessive amount of runoff and sediment is collected and analysed. The difficulty consists in so constructing the divisors that they only separate a specific aliquot at all stages of flow. Geib (1933) was the first to discuss the construction of a divisor box with 5, 9, and 11 rectangular slots. Smith and Wischmeier (1962 consider the multislot divisor as the best and also mention the Coshocton-type rotating vane sampler, shown in Figure 3.6 (Parsons, 1954).

Research into the dynamics of rills and gullies can make use of the methods of grid systems and erosion lines with erosion pins and pins-with-washers, as already mentioned. Water level recorders coupled with weirs can be set up in order to calculate the flow regime in a main channel. Schick (1967) proposes a model for a bedload trap and a suspended-load sampler of the 'Hayim type' with sampling bottles at different depths. Graf (1971) provides an extensive description of all sorts of sediment measuring devices and the application of various tracing methods with radioactive, paint, and fluorescent tracers.

Observations of relatively short duration at measurement or experimental stations can be instructive for the analysis of erosion and sedimentation mechanisms. Insofar as one intends, however, to determine the denudation rate in a

Figure 3.6. Coshocton rotating vane sample splitter. The disc
is 30–40 cm in diameter

regional context and to go on to prediction, it is important that the observation period be sufficiently long to acquire information about the effects of less frequent, though still very effective, extreme events. For the purpose of acquiring a preliminary understanding of them, it seems necessary to extend the observation period over at least a few years.

In a drainage basin, erosion as well as sedimentation occurs. Rivers evacuate quantities of bedload and suspended-load, the primary origin of which is generally not very clear. A river can just as well be carrying its own alluvium as slope debris. Local erosion and sedimentation will proceed at various points in a drainage basin simultaneously. Thus, sedimentation can be dominant on well-defined colluvial slopes, on alluvial plains, in closed basins and man-made reservoirs. For all these sites, the question is how the morphogenetic balance develops and to what extent debris is transported by the main river. This means that computations of the denudation rate, which are based on the fluvial transport of suspended load, must be very carefully interpreted. In most cases, the global denudation is underestimated. In principle, however, there is also the possibility of overestimating to the extent that slope erosion may be insignificant in a basin and the river is eroding essentially its own alluvium. The formula developed by Fournier (1960) is well-known. On the basis of the rainfall regime and the relief of subcontinental areas, it predicts the transport of suspended-load (*DS*):

$$\log DS = 2.65 \log p^2/P + 0.46 \log H^2/S - 1.56$$

where:

DS = suspended sediment yield (t/km^2/yr),
H = mean relief of the basin or the difference between the main altitude and the mean altitude (m),
S = catchment area (km^2), and
p^2/P = climatic index with p = rainfall in the wettest month in mm, and P = mean annual rainfall in mm.

Fournier used this formula to compute continental denudation rates on the basis of an average bulk density of subsoils equal to 1.4. This formula is interesting as it provides a picture of the spatial differentiation of the denudation. Misleading conclusions, however, can arise in areas with large relief energy where intensive rainwash is accompanied by the building up of extended colluvial slopes.

Alluvial and colluvial deposits can be evaluated in a regional morphogenetic balance provided their extent and chronology are known. For recent deposits radiochronological and archaeological data are important. Man-accelerated hill slope erosion is discussed in general by Butzer (1974) and specifically for the Mediterranean basin by Vita-Finzi (1969).

Important for the estimation of sediment yields in drainage basins and for the investigation of the relation between erosion and the ecosystem are the measurements of sedimentation rates in reservoirs, including stock ponds and large reservoirs (Langbein and Schumm, 1958; Leopold *et al.*, 1966; Rapp *et al.*, 1972). Rapp and his colleagues have recently completed a survey in the Dodoma catchments in Tanzania on the basis of reservoir studies combined with catchment erosion surveys and sediment sampling in streams (Figure 3.7). Sedimentation rates in reservoirs were measured by periodic re-levelling of known cross-profiles on the floors. Transects were made by manual soundings from a rubber boat and by levelling on the dry part of the bottom. The global sedimentation upstream and downstream of the reservoir was precisely measured by aerial photo analysis and levelling. A contour map was made of the bottom of the reservoir. The evolution in time of the reservoir volume was calculated by adding up the partial volumes, V:

$$V = h(A + B)/2$$

where:

V = volume in m^3,
h = contour interval in m,
A = area of the upper contour in m^2, and
B = area of the lower contour in m^2.

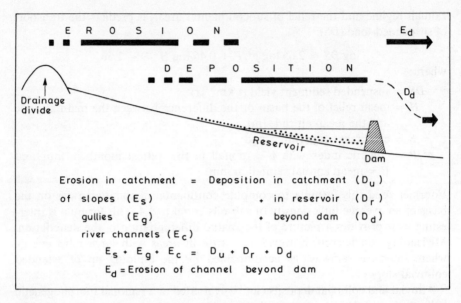

Figure 3.7. Diagram of erosion and deposition zones in the Dodoma catchments, Tanzania (After Rapp)

To the total of the *V*-volumes was also added the volume below the lowest contour, which equals the product of the area of the lowest contour and the mean depth from that contour to the bottom. Each reservoir volume was then compared with the initial volume of the full supply level. The annual sediment yield in m^3/km^2 can then be calculated from the reduction of the volume of the reservoir and correlated with the surface area of the basin. As Figure 3.7 makes clear, significant sedimentation can occur both upstream and downstream from the reservoir because sediment is carried away over the spillway in times of flood. In Tanzania, Rapp also determined that, in many catchments with pediments, less than half of the amount of the eroded soil is deposited in a reservoir because large amounts are deposited as thin sandy sheets on lower pediments, sand fans, and alluvial plains. Langbein and Schumm (1958) state that, as a rule, sediment yields decrease with increased catchment areas, reflecting the flatter gradients of the drainage basin. For the computation of the sediment yield in m^3/km^2, Rapp assumed the average dry bulk density of the eroded material to be 1.5 g/cm^2. Fournier (1960) assumed 1.4 g/cm^3 for the intercontinental scale.

Analysis of processes

Volumetric and dynamic recording of denudation and sedimentation rates raise the questions of the specific erosion and sedimentation mechanisms and of the interactions between determining factors. With the latter, laboratory research can be as important as field experiments. For rainwash, the use of rain simulators is central. Simulated rainfall offers advantages that we can summarize as follows (cf. Meyer, 1965): (1) it yields more rapid results, especially where the testing of extreme and thus less frequent conditions, as with rainfall in arid regions, is concerned; (2) it is more controlled inasmuch as one can take appropriate measurements with selected intensities and durations; (3) it is more adaptable and, in the laboratory, specific parameters such as the influence of the degree of slope can be tested in a system in which all factors but one are held constant in successive runs.

Hudson (1964) has clearly summarized how, particularly in meteorological studies, an understanding is attained of the structure of rainfall and of the nature of the parameters that specify the erosivity of rainfall. Intensity, drop size distribution, and drop fall velocity are the most important factors here. For what concerns intensity–duration–frequency information, one has to rely on data from the Weather Bureau. Drop size distribution is generally measured by the filter paper method or with the flour pellet method. For all rainfall in a turbulent atmosphere, the drop diameter has been determined to fluctuate between 0 and 6 mm (Blanchard, 1948). A typical bell-shaped frequency curve is obtained when the volume percentages are plotted against drop size. For drizzling rains, the median diameter is around 1 mm while with an intensity of 50 mm/h diameters of the order of 2.5 mm occur. Hudson (1963) has demonstrated that the median diameter can again decrease when the intensity exceeds 100 mm/h.

According to the measurements of Laws (1941) and Gunn and Kinzer (1949), the relationships between the drop diameter D and the terminal velocity V in m/s are as follows:

D	0.5	1.0	2.0	3.0	4.0	5.0	6.0 mm
V	2.1	4.0	6.5	8.1	8.8	9.1	9.2 m/s

The terminal velocity can be determined by various methods such as high speed photography, which is generally useful in the study of runoff phenomena in the laboratory.

The erosivity of rainfall is proportional to $D^a . V^b$ and thus knowledge of both these parameters is important in the construction of a rain simulator. It is desirable, if not necessary, that the $D^a . V^b$ values approximate those of natural rainfall. Whether erosivity correlates better with V^2 (kinetic energy) than with V (momentum) is still an open question (Ellison, 1944; Rose, 1960).

There are two main types of rain simulators: drip screens and sprays (Hudson, 1964; Bryan, 1968). With drip screens, thread droppers (Ellison and Pomerene, 1944), hypodermic needles mounted on the bottom of an aluminium container (Ekern and Muckenhirn, 1947), and also small capillary tubes (Adams *et al.*, 1957) have been used. All these types produce rainfall with rather large drops—

between 3.5 and 5 mm. Because of this, they are not particularly suitable for erodibility studies, though acceptable for partial experiments such as those relating to erosion mechanisms and thin film hydraulics.

With the spray simulator, the drop former is a nozzle, and a mechanism must be provided to apply the spray in the desired manner. Since 1936, a large number of nozzle simulators have been tested, notably by researchers of the US Soil Conservation Service. The difficulty with this type of simulator is preventing too high an intensity giving rise to too much dispersal so that the average drop diameter is too small. Bryan (1968, 1970) has discussed this problem, among others, with the performance of the 'Edmonton pattern' rainfall simulator. In recent prototypes, the nozzle is directed downwards and the pressure is able to be regulated so that the impact velocity of the drops is comparable with the terminal velocity V of raindrops. In order to get appropriate drop sizes, high rates of discharge are applied and intermittent spray is used (Meyer and McCune, 1958; Bubenzer and Meyer, 1965). One of the best recent models is a simulator with a rotating disk made by Morin *et al.* (1967). Nozzles of the full cone spray type are selected which give a spray above a rotating metal disk with a section aperture. Morin's simulator comes very close to producing the drop size distribution and the impact energy of natural rains for intensities of up to 50 mm/h on an area of around 1 m^2.

Both drip screen and spray types of simulators have the advantage of being rather simple to construct, and are therefore inexpensive. Field experiments with a simulator can however turn out to be rather expensive if a truck is needed to transport the water supply and the apparatus. For large field plots, sprinklers have been used. Schachori and Seginer (1962) have discussed this technique quite extensively and associated field studies in Israel have been discussed by Seginer *et al.* (1962). It should be pointed out that, in field tests, it is desirable to wet down a buffer zone around the test area in order to prevent lateral percolation from the plot to the drier surrounding areas, thus giving a false picture of infiltration.

As yet, no light, portable, and reasonably inexpensive simulator has been devised for field research. Costin and Gilmour (1970) describe a simulator that consists of a hand operated boom and nozzle from which rainfall is supplied to a 2 × 3 metre plot from drums of water by means of a mobile engine-driven centrifugal pump. The mean drop diameter produced is 1.3 mm, which is somewhat on the small side.

In field and laboratory measurements and experiments it is always presumed that the most important mechanical and chemical characteristics of the materials to be tested are known, such as textural composition, chemical-mineralogical composition, bulk density, porosity, and water content. The structural dynamics are significant and can be analysed by micromorphological studies. The evolution of the water content is also very important, and thus pre-wetting techniques are necessary. 24 hours before the test, for example, the ground can be saturated in order to obtain conditions comparable to field capacity (Hudson, 1964).

For the study of thin film hydraulics and runoff erosion mechanisms, the laboratory offers definite advantages that are partially the same as the principal advantages of simulator research as summarized by Meyer (1965). Only in the laboratory is it possible to test adequately the role of each parameter in a setup in which all factors but one are held constant in a sequence of test runs. Along with this, meaningful laboratory research always implies critical evaluation of the equivalence of the setup with respect to the natural system. Laboratory tests depend essentially on the use of simulators and flumes with variable dimensions. When the length factor of the slope plays a secondary role (discontinuous processes) relatively small flumes are recommended. The length effect can actually be obtained by supplying terminal runoff to the upper end of the flume. Particularly important is the manner in which the material to be studied is dumped into the flume in order to get a filling equivalent to the field conditions. This poses particular problems for soil material as it is impossible to bring an undisturbed sample from the field to the laboratory, and certainly when it concerns a relatively large volume that must be deposited in the flume in one thick slab. Therefore, one must look to conventional, though in a sense arbitrary, preparation of disturbed samples that permit the study of the erodibility of soils under comparable conditions. Bryan (1968) suggests in this regard, as did Moldenhauer and Long (1964), that splash erosion and washoff erosion be measured on rather large undisturbed aggregates acquired by passing air-dried samples gently through a 6.19 mm aperture square-hole sieve.

The base of the flume must be able to be drained and this can be achieved by placing a layer of glass beads on top of a single layer of tissue in the base of the pan. The water table levels can be measured by inserting small piezometric tubes. Pre-wetting has to be applied according to the assumed soil moisture conditions at the beginning of rainfall, for example, the field capacity. Terminal runoff can be pumped to the upper end of the flume to increase the runoff discharge and to simulate afterflow and melting water flow. With a dirty water pump, one can also introduce an unknown sediment load. A screw mechanism allows the appropriate slope to be given to the flume. With all tests, the load is measured by determining the concentrations, in grams per litre, by pouring off and/or by evaporating the captured runoff. Tracer material can be used to clarify the movement of water and sediment. Savat (1975) used silver discs less than one micrometre thick to measure indirectly the surface velocities of the water by determining the length of strips formed on film during time exposures. Approximate values of average flow velocity can be obtained by using dyes with low diffusion rates such as Rhodamine, which is quite conspicuous. Measurements of discharge volume can be made simply with a beaker and a chronometer. Should continuous recording be desired, a floating recorder coupled to a potensiometric recorder can be used. Given the unit discharge Q

and the average velocity v, then the average thickness d of the waterfilm is Q/v. Direct measurement of d can be done with a point gauge micrometer. Emmett (1970) gives a margin of error of 150 μm for this method. More sophisticated apparatus for the determination of thickness, for microphotographic velocity measurements, and the visualization of streaming are described by Wenzel and Mathews (1970), Shen and Li (1973), and Phelps (1975). Hot film probes have been used to measure the various velocity components of turbulent flow (Raichlen, 1973; Resch, 1970). It is important to know the temperature dependent viscosity (v) of the water, an elementary piece of data that determines, among other things, the Reynolds number ($Re = v \cdot d/v$).

A general technical problem in measurement is that every measuring apparatus introduced in the flow causes autonomous disturbances. It is hoped that in the future more possibilities will be offered by optically based techniques (photography, laser techniques, and so on).

3.2.3 Mass movements

As Carson and Kirkby (1972, p. 101) clearly state, the actual processes of mass transport can seldom be labelled as pure slide, flow, or heave movements. Actual mass transport is mostly a mixture of all these transport modes, which are found, alternatively or synchronically, in areas of elastic, plastic, and viscous media. Rheological approximation of transport phenomena implies that the transport model provides for the combination of these three basic states. Thus, particularly since Sharpe (1938), a classification system of mass transport processes has been elaborated that recognizes that, in the field, the processes often overlap in space and time. Washburn (1967) in Greenland, and Higashi and Corte (1971) with experiments, have clearly demonstrated that the solifluction process is inseparably linked to the phenomenon of frost creep. A landslide can quickly turn into a debris flow or a mudflow when liquefication occurs because of a rapid decrease of the coefficient of dynamic friction. A rock avalanche can become a pure rockfall, and rivers can be fed by mudflows during a period of heavy erosion. All of these states must be taken into account when one analyses phenomena in the field on a regional level.

Every regional study that ultimately envisages a quantitative investigation of mass transport and of mass transport mechanisms must be based on an overall and thorough knowledge of the geography and geology of the area, or, if one prefers, a thorough survey of the regional ecosystem as it occurs in its morphological–geological context. In handbooks, mass transport phenomena are necessarily described and analysed in the framework of rather stereotyped and geometric models. In the field, these serve only as background knowledge that must give way to flexible and appropriate interpretations of the observed complex phenomena. In discussing slope stability, competent engineers such as Terzaghi and Peck (1967) repeatedly emphasize that experience coupled with a thorough field study takes priority in the prediction and postdiction of mass transport phenomena.

First, let us consider how, in a field survey, purely static data can be gathered of the most obvious phenomena such as slides and flows. Recording and mapping of the topographical effect of mass transport is primary here. The following techniques can be used.

(a) Aerial photography

Aerial photos give a three-dimensional view of an extensive area. The borders of landslides and flows, more and less recent, are apparent from all kinds of spatial and linear features such as discontinuities in the topography and vegetation, scarps, irregularly undulating surfaces with ponded depressions, shifting of tracks and roads, and constrasts between old and new vegetation. Aerial photo analysis combined with the study of the available orohydrographic maps permit mass transport to be examined in the physiographic units that constitute the region. From aerial photos of the scale of 1 : 5000, detailed contour maps of slide and flow topography with contour intervals of 0.4 to 0.5 metres can be made. For this scale, flight altitudes of 600 to 750 metres are required for classic aerial cameras (of focal length 100 to 150 mm). With a photo scale of around 1 : 25,000, one can map with contour intervals of 2 m at best and with an accuracy of around 1 m, which is generally insufficient for mass movement topography. Topographic mapping in localities with prominent relief and where repeated surveys may have to be made can also be based on stereo-photogrammetric analysis of terrestial photographs. Thompson (1966, pp. 919–998) describes these techniques. Belcher (1960, pp. 403–456) gives an interesting exposition on actual photo-interpretation.

(b) Field observations

In a general reconnaissance survey, attention is first given to microtopographic recording of slides and flows (Figure 3.8). Long and cross profiles can be made simply with a measuring rod, inclinometer (Meridian), tape, and compass (Temple and Rapp, 1972; Lundgren and Rapp, 1974). This is combined with good photographic documentation that also portrays the vegetation. The soil profiles present in scarps and scars are described and possibly sampled in order to determine the physical characteristics of the substrata (bulk density, grain size composition, water content, and shear strength characteristics) and to compare them with the deposits of slides and flows. Trenches can be dug to trace typical structures such as slide surfaces and flow structures. Precipitation records are important as well as information that can sometimes be gathered from interviews with local people.

Knowledge of the lithological structure of the substrata is particularly important for, among other things, the distribution of the regolith with respect to the

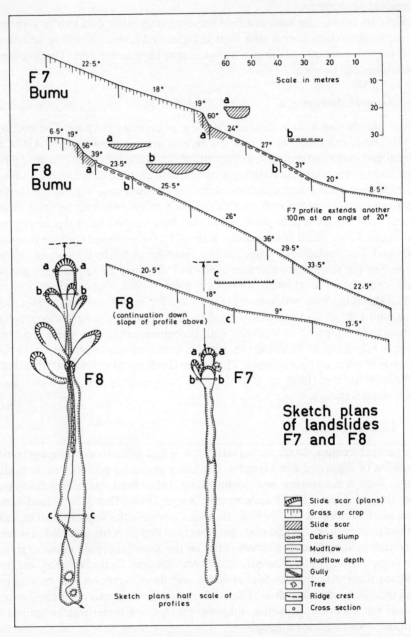

Figure 3.8. Profile analysis of complex landslide-mudflow types in the Bumu area, Tanzania (After Rapp)

substrata, the lithological variations and stratification, and the spatial distribution of the strata, faults, and joints. These latter elements play a dominant role in the water economy of the slope, which in turn has a great impact on the stability of the substrata. This is because the hydrostatic and hydrodynamic pressures (artesian pressures) are mostly irregularly distributed in the substrata and significantly determine the irregular distribution of the resistance to mass transport. Particular attention must be given in the field investigation, with possible augmentation with borings, to the location of relative plastic and/or aquiferous layers. Zaruba and Mencl (1969) give a number of detailed examples of the morphological and hydrogeological context within which landslides and flows occur.

(c) *Mapping*

The observations discussed above give data for topographical and topographic-geological plans, cross sections, and longitudinal profiles. Very clear thematic maps can be constructed with legends correlated to those of the classic geomorphological maps. In the survey made by Rapp *et al.* (1972) there are numerous simple but relevant examples of maps indicating topography, hydrography, geology, vegetation, land use, and various phenomena such as slides, mudflows, channel sedimentation, and debris fans.

Obviously, the above-mentioned techniques can be applied repetitively in active areas so that a dynamic survey is generated by which causal relations are gradually clarified and by which a basis is provided for the possible selection of the sites for one or more measurement stations.

Naturally, it is not easy to go on to the computation of transport and denudation rates on slopes determined by mass transport because, in contrast with soil wash, it is a question not only of surface transport but also of subsurface transport, which is vertically differentiated. Moreover, the intensity and frequency of the processes vary significantly from place to place so that extensive measurements are necessary if, with a generally limited team of observers, one wants to calculate the denudation rate for one large type slope or a small drainage basin. Rapp (1960) did this for the 18 km^2 basin of Kärkevagge, Sweden. He calculated the average per year displaced volumes and transport distances for rockfalls, avalanches, earth-slides, creep, and running water. The transport distance can be expressed as vertical, horizontal, or inclined components. Thus Rapp calculated the transport in ton-metres/year and the relative mass transfer in tons/km^2/year for each of the main processes mentioned above. The latter expression is suitable for the mutual comparison of different areas.

Global volumetric data, which estimate the topographical effect of mass transport, can only be acquired provided the primary topography existing before the process became active is known or if an artificial reference surface is established. Thus Temple and Rapp (1972) in Tanzania calculated slide volumes,

$V = V_1 + V_2$, where V_1 equals the combination of cylindrical and spherical volume representations of the slide head, and V_2 is equal to the conic middle and lower section of the scar (Figure 3.9).

$$V = \pi ab \sin \theta (0.021a + 0.146 L)$$

where:

a = the maximum width,
b = the maximum depth,
θ = the angle between the slope and the vertical, and
L = the total measured length of the slide scar.

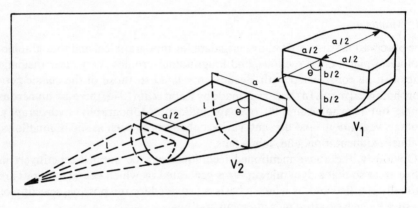

Figure 3.9. A calculation of slide volumes (After Rapp)

In Northern Scandinavia, Rapp (1960, 1974) measured fresh volumes of debris flows and rockfall deposits by calculations based on cross sections of talus cones and slopes. He used a measuring tape, a band inclinometer, and a 2 m measuring rod. The reference surface can be the snow cover in winter or wire nets installed over talus slopes.

Raw regional estimates of landslide volumes can be made by working with planimetric data based on aerial photos supplemented with ground truth regarding the average depths of slide heads and scars. Aerial photo observations are best in areas where there are clear contrasts in the density and age of the vegetation.

Local eyewitness accounts sometimes offer important information on the location and frequency of rockfalls and slides. Rates of escarpment retreat can be calculated over a geological period provided the geologic framework (absolute dating) is sufficiently known so that the position of the original escarpment and its age can be specified. With such information, Yair and Gerson (1974) arrived at a minimal rate of 0.1 to 0.2 m/1000 years for escarpment retreat in the extremely arid environment of Sharm el Sheikh on the Southern Sinai Peninsula.

Dynamic recordings relative to debris and debris mantle shifting and to rates of transport can lead to volumetric calculations only if one has data on the subsurface conditions of the transport, namely, the depth of the transport and the variations of movement with the depth. To the study of surface transport of stony material (block creep, movements in periglacial patterned ground) simple and efficient measurement methods can be applied. The position of painted or not painted blocks and stones is taken with reference to immobile benchmarks (solid bedrock, concrete benchmarks). The evolution of painted rock lines can be followed over a period of years by the use of a string and plumb line (Rudberg, 1967). Jahn and Cielinska (1974) studied block movements in the Sudety Mountains in this way, and Pissart (1964, 1973) obtained some interesting results regarding the movement of patterned grounds in the Chambeyron Massif in the Alps by a combination of rock painting and photographic recording. Movements from a fixed reference line can also be registered from a theodolite platform provided that the position of the theodolite is carefully determined with a plumb bob relative to a benchmark. Dynamic recording of marked blocks and stones provides relative results provided that the benchmarks are well-fixed and the measurement operation causes no, or only minimal, disturbance in the direct surroundings of the moving rocks.

More complicated, and occasionally more problematic, are the methods used for monitoring subsurface movements such as creep, solifluction, and mudflows. To do this, marker elements (stakes, rods, flexible inserts, coloured sand, etc.) are generally implanted in the moving medium. Setups have thus been developed to measure vertical soil movements in and or periglacial environments whereby the movements are followed by a horizontally placed flange that rests on the soil or is slightly buried. The flange is fitted with a vertical iron rod from which the movements are read off episodically (the motometer, Jahn, 1961), or transferred through a system of mechanical levers to a recording pen and registered on a rotating drum (Yaalon and Kalmar, 1972). In Greenland, Washburn (1967) apparently successfully measured frost creep and solifluction on slopes by setting out target lines with, among other things, cone targets mounted on wooden pegs, the movements of which were measured with a theodolite. Also successful in the measurment of creep movements seems to be the method of drilling small (1 to 5 cm) auger holes and filling them with tracer materials, generally test pillars that consist of a series of small cylinders made of wood, plastic, or modelling clay (Rudberg, 1967; Jahn and Cielinska, 1974). After a specified period, the profile of movement is measured by digging up the test pillar and calculating its displacement with reference to fixed points or to a plumb line. It seems preferable to implant stakes and test pillars vertically in the slope in order to avoid false rotations as far as possible. This is suggested by Carson and Kirkby (1972, pp. 272–300) who discuss in detail soil creep moves as a global mass rather than by diffusion. There is also the problem of knowing if the insert moves with the soil as a solid, or if the latter only receives a portion of the total average shear, causing the measured downhill deformations to be underestimated minimum values. When adequate verification techniques for this are lacking, one may only assume that the measurements provide relative values that are intraregionally and interregionally comparable insofar as the measurement methods are identically the same. It will be necessary, in the near future, to standardize measurement methods. A continuous record of creep movements seems to be able to be made using flexible tube inserts in combination with an inclinometer and measurements of the curvature based on strain gauge devices (Williams, 1957; Hutchinson, 1967). Wooden pegs can also be used to measure surface mudflow movements. With them, Prior and Stephens (1971) used a water-level recorder, installed beside a mudflow, to get continuous recording of the movements on a vertically mounted chart drum. The movements of the peg were transmitted *via* a nylon line to a measuring wheel and then, by means of gears, to a pen.

For the prediction of slope stability, the determination of threshold hillslopes, and the postdiction of geomorphological events, one investigates the relation between the slope angle and the soil mechanical properties of the granular substrata and, in particular, the value of the residual angle of shearing resistance (Carson and Petley, 1970; Rouse, 1975). For this the following equation applies:

$$\alpha_L = \arctan\left(1 - \frac{m \cdot \gamma_w}{\gamma_t}\right) \tan \phi'_r$$

where:

α_L = the limiting angle,
γ_w = the bulk density of water,
γ_t = the saturated bulk density of the soil,
ϕ'_r = the residual angle of shearing resistance, and
m = the position of the water table with respect to the slope surface and the slip surface (m varies between 0 for a completely drained subsoil and 1 for a completely saturated subsoil).

So if $m = 0$, then there is, in principle, an equilibrium condition, $\alpha_L = \phi'_r$. If $m > 0$, then α_L decreased to the degree that a portion of the granular regolith, 'immersed in water', shows a lesser shear resistance. Knowledge of γ_t (determined in place with a densitometer), of ϕ'_r (shear box or triaxial test), and of m (installation of piezometer tubes) thus allows the study of the stability of slopes for landslides as the above mentioned authors have done. If the necessary apparatus to determine these parameters is not available, significant analysis can still be done on regoliths that consist of sand–silt–clay mixtures, namely, (1) determination on slopes of the evolution of the water content; (2) grain size analysis of the regolith; (3) determination of the plasticity index of well-chosen samples of the regolith. From many studies of the relation between the clay content and the plasticity index value it seems to be that in many areas the value of ϕ'_r decreases when the clay content in granular soils increases and/or when the value of the plasticity index increases (Skempton, 1964; Terzaghi and Peck, 1967, p. 122; Rouse, 1975). The simple determination of grain size composition and of the plasticity index can thus have a directive significance for the knowledge of relative ϕ'_r values of slopes in a particular area.

Shear strength measurements in the field can be made with the common vane shear test. Bore hole shear test devices are also used, such as those with which Lhones and Handy (1968) investigated the stability of slope angles in friable loess. The standard penetration test is used for the determination of the angle of shearing resistance ϕ' in sandy sediment, though for clayey sediment the cone penetrometer is better.

To determine c' and ϕ'_r values of well to moderately drained sandy to sandy–clay sediments, use a slow, drained, direct shear test. The role of the water table in the calculation of α_L is represented by the factor, $0 < m < 1$. Shear strength measurements for clays must often be made under undrained conditions with measurements of pore water pressures. For this, a triaxial compression test apparatus is used. Extensive information regarding these texts may be found in Terzaghi and Peck (1967) and Scott (1969). It must be realized that laboratory tests of shear strength, often done on remoulded samples, must be seen as experiments and also that in the evaluation of field conditions, much attention must be paid to structural elements such as changes of the lithofacies, fissures, and joint systems.

It seems that field experiments will soon be able to contribute to process analysis. In this context, Statham (1976) describes dropped stone experiments made to check on a scree slope rockfall model.

Soil creep and solifluction have already been studied in the laboratory, mostly on small samples (Kirkby, 1967; Higashi and Corte, 1971; De Ploey, 1973). Journaux and Coutard (1972, 1976) performed tests on large samples ($25 \, m^3$) and discuss, together with Pissart (1972) and Corte and Higashi (1972), the various techniques that are used to register the movements resulting from frost-thaw cycles such as pressure transducers, displacement transducers and strain gauges.

Regarding soil creep, the question arises whether the mechanism corresponds with a random diffusion of soil particles (Culling, 1963) or with shear movements. One may expect that laboratory experiments will throw more light on this and also upon the question of how far creep movements correspond with a rheological event, described in terms of viscosity, plasticity, and elasticity (Fedder, 1972). Laboratory tests have to become more important as there is doubt about the validity of some field indicators of creep (Carson and Kirkby, 1972). For example, one of the difficulties is how far the introduction of measurement mechanisms and measurement techniques. It seems that inserts allow for good, though not repeatable, recording of soil movements provided that the soil apparatus disturbs the ground movements. Measurements of surface creep (block creep, splash creep, runoff creep) present less problem in this sense because the material can be easily marked and ready use can be made of optical methods such as photography and laser beam reflection (Moeyersons, 1975; De Ploey and Moeyersons, 1975).

3.2.4 Wind erosion

Measurement of deflation, transport, and accumulation rates

Aerial photography and satellite imagery offer, relatively, the greatest possibilities for the mapping of aeolian landscapes. Satellite images have led recently to new hypotheses regarding sand transport and dune building in the Sahara, both modern and that of the Quarternary Period (Mainguet and Callot, 1974; Mainguet and Canon, 1976). For this continental scale mapping, consideration was taken of the direction of the axes of barchans, yardangs, elongated dunes, and ridge and corridor patterns. According to McKee (1973), sand dune areas are visible on the LANDSAT MSS bulk imagery in all spectra from band 4 (green) to and including band 7 (infrared). On black and white images, it is quite possible to confuse sand bodies with cloud formations, drifting snow, or salt deposits, all of which come out equally bright. On colour composite prints, however, sand appears yellow whereas other features are clearly white.

Aerial photos can be used for more than reconnaissance surveys of desert basins as Smith (1969) did in Northern Africa. Finkel (1959) has shown how aerial photos of scales of 1:20,000 to 1:60,000 can be used for detailed studies of dune movement, provided that one has accurate ground points for horizontal and vertical control. On aerial photos of Southern Peru for the period of 1955–1958, the shifting of barchans was measured with an accuracy of 1.0 metres on the ground. In conjunction with this, Finkel, using a theodolite and a stadia rod, measured the topographic parameters such as H, the height of the slip face; L, the length of the horns measured from the centre of the crest in the direction of the wind; and W, the width across the horns. Thus, the empirical relations between these parameters could be compared. Plane-table and alidade surveys were also made of barchans in California by Norris (1966). The topographical data were referred to a benchmark consisting of a 0.6 m (2 ft) length of 20 mm ($\frac{3}{4}$ in) pipe driven flush with the ground surface. Volume calculations of the barchan dunes were made by dividing them into elementary tetrahedron volumes.

Tsoar (1974) measured the progress of barchans with reference to poles stuck into the ground in front of the dune at the base of the slipface. The distance was measured between the poles and the sharp line of contact of the slip face and the ground surface.

In Mexico, Bull (1974) measured the aeolian aggradation of small playa mounds by planting welding rods in two lines across the mounds normal to each other. Welding rods can also be used as erosion pins to measure deflation. Along with them, fluorescent sand tracers can be used to detect the places where the deflation is most intensive.

Tsoar found that the advance rates of the barchans of El Arish were 10 to 30 per cent smaller than the values calculated according to Bagnold's (1954) formula. In Peru, however, Finkel found very good agreement. Thus, Bagnold's formula appears to be usable. The rate of advance of a barchan is given by

$$C = \frac{q}{H.\gamma}$$

where:

C = the rate of advance,
γ = the bulk density of loosely packed sand,
q = the rate of sand flow, and
H = the height of the barchan brink.

Also

$$q = B(V_z - V_t)^3$$

and

where:

$$B = \left(\frac{0.174}{\log (z/k')}\right)^3 c \sqrt{\frac{d.\rho}{D.g}}$$

z = the height at which the wind velocity is measured,
k' = the height to which the wind velocity rays converge during sand-driving,
d = the average diameter of grains transported by the wind,
D = the standard grain diameter of 0.025 cm,
ρ = the air density,
c = the empirical coefficient, 1.8 for naturally graded dune sand,
V_z = the wind velocity at height z, and
V_t = the threshold speed for transporting sand grains measured at height k'.

In order to calculate the rate of sand movement, account must thus be taken of the time that the wind surpasses the threshold velocity. For El Arish, Tsoar took the threshold velocity to be 625 cm/s. For the hours that the wind had a greater velocity, the transport was calculated on the basis of equivalent storms with a velocity of 712 cm/s (the median value of the lowest group of wind velocities which was taken to be 625–800 cm/s).

In Libya, Bagnold (1954, pp. 77–84) measured wind velocities in the open air using a set of pitot tubes connected to a multiple manometer for the pressure readings. A ground trap for surface creep was set up and saltating grains were captured in a small saltation collector. Figures 3.10 and 3.11 show respectively

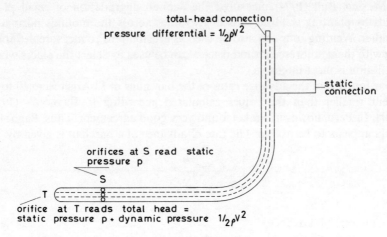

Figure 3.10. The pitot-static tube for measuring wind velocity

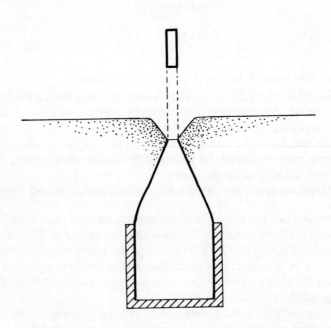

Figure 3.11. Ground trap for surface collection of wind-
blown data

the construction of a pitot-static tube and of a ground trap for surface transport. Bagnold's saltation collector has to be oriented to the wind during a storm. In order to be able to capture dune sand from all directions during an unexpected storm when the observer was absent, De Ploey (1977), at the Kalmthout (Kempenland, Belgium) Dune Station, set up a collector consisting of a 60 cm high pile of circular cakepans with a diameter of 19 cm. Grains were trapped in the pans from all directions in 2.5 cm deep grooves (Figure 3.12). A psammograph was also set up at this dune station to record the chronology of the sand storms. Grains in saltation passed through a 3 mm vertical slit of an external wind-oriented cylinder and were captured on adhesive tape fixed to a drum moved by an internal clock (Figure 3.13).

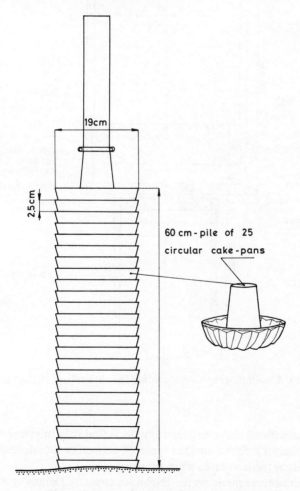

Figure 3.12. A saltation collector for dune sands

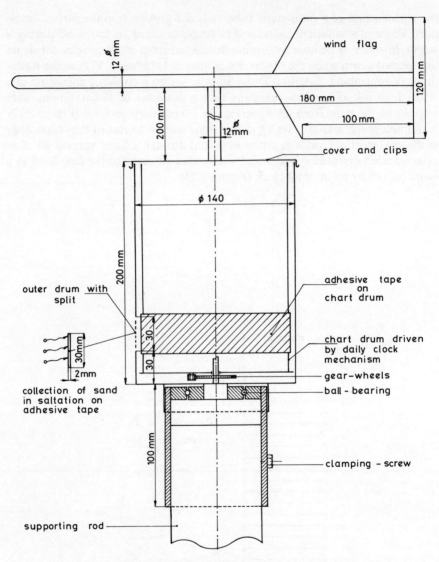

Figure 3.13. A psammograph for recording the chronology of sand storms

One problem with all collecting instruments is that they interfere with the air stream. The efficiency of the various types of collectors therefore needs to be tested by deflation trials in large wind tunnels.

Dust collections were made on the Barbados coast by Delany *et al.* (1967) on a 14 m wooden tower. Used for dust traps were 1 m² meshes made of 1/2 mm diameter mono-filament nylon woven to give an open area of about 50 per cent. The meshes were kept damp so that the impinging dust would stick. The collecting efficiency was around 50 per cent for particles greater than about 1 μm. On the tower was placed a cup-type anemometer which was calibrated against a standard airflow meter placed immediately behind the mesh. This permitted the calculation of the volume of wind passing through the mesh. In California, Marchand (1970) used dust traps consisting of 700 cm² rectangular pans filled with marbles and mounted on 10 cm × 10 cm posts about 1 m above ground level.

Analysis of processes

Field surveys of deflation and sedimentation by wind can lead to the formulation of interesting hypotheses concerning regional erosivity of wind and erodibility of soils and sediments, though the most fundamental contributions to the knowledge of the latter element must come from wind tunnel studies. As with rainulator tests for rain erosion, the wind tunnel offers the possibility of testing, under controlled conditions, the role of a large number of variable factors that determine aeolian erosivity and erodibility. Such studies have already been Bagnold (1954) and Chepil and Woodruff (1963).

Although the closed circuit tunnel offers more control possibilities for the physical characteristics of the circulating air stream and is also less noisy, it is also much more expensive than the 'straight-through' open circuit tunnel, which has the following essential components: the entrance or contraction cone, the test section, the diffuser, and the fan with straightener vanes. Basic information on wind tunnel design can be found in Pope and Harper (1966). Figure 3.14 shows a simple open circuit tunnel that has been set up in the Laboratory for Experimental Geomorphology in Louvain. Empirically determined favourable relations determined the dimensions of this tunnel: the test section is 1.50 m long with a slightly diverging cross-section of 800 cm². The limited length of the test section only allows the study of the fluid threshold of deflation. A satisfactory sand flow can only be attained if this section is at least 15 m long; it then becomes possible to study the circumstances of the impact threshold (Knottnerus, 1974). Techniques such as the weighing of the test section or a portion thereof permit the determination of the amount of deflation on the tray. The start of the first deflation (fluid threshold) can be visualized in a relatively weak laser ray wherein the saltating grains begin jumping. Smoke streams or helium bubble generation can be used for flow visualization.

Obviously with all wind tunnel tests, an extensive soil mechanical description of the sediment or soil material that fills the trays must be available. One of the main goals of future research needs to be the determination of the range of erodibility that characterizes sediment as a function of variable characteristics such as structure and surface roughness.

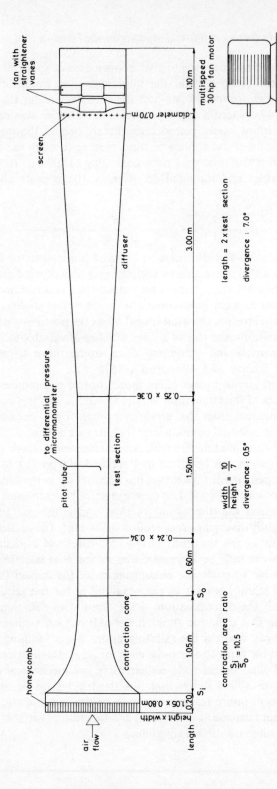

Figure 3.14. An open-circuit wind tunnel with a concentration cone, a test section and a diffuser with a terminal drive

Partial experiments will have to be devised to specify specific mechanisms. To this purpose, Allen (1970) constructed a rotating drum to test the avalanching of sands on the slip faces of dunes.

REFERENCES

Adams, J. E., Kirkham, D., and Nielsen, D. R. (1957), A portable rainfall simulator infiltrometer and physical measurements of soil in place. *Proc. Soil. Sci. Soc. Am.*, **21**, 473–477.

Allen, J. R. (1970), The avalanching of granular solids on dune and similar slopes. *J. Geology*, **78**(3), 326–351.

Bagnold, R. A. (1954), *The Physics of Blown sand and Desert Dunes*, Methuen, London.

Belaid, R. (1970). Notice de la carte de l'érosion en Tunisie. *Min. Agric. Division Sols*, **2**, 65–76.

Belcher, D. J. (1960). Photo interpretation in engineering. In: *Manual of Photographic Interpretation*, American Society of Photogrammetry, Falls Church, pp. 403–456.

Blanchard, D. C. (1948), Observations on the behaviour of water drops. *Gen. El. Res. Lab., Project Cirrus*, 7.

Bryan, R. B. (1968), Development of laboratory instrumentation for the study of soil erodibility. *J. Earth Sci.*, **2**, 38–50.

Bryan, R. B. (1970). An improved rainfall simulator for use in erosion research. *Can. J. Earth Sci.*, **7**(6), 1552–1561.

Bubenzer, G. D., and Meyer, L. D. (1965). Simulation of rainfall and soils for laboratory research, *Trans. ASAE*, **8**(1), 72–73.

Bull, W. B. (1974). Playa processes in the volcanic craters of the Sierra Pinacate, Sonora, Mexico, *Z. für Geomorph.*, Suppl. Bd. **20**, 117–129.

Butzer, K. W. (1974). Accelerated soil erosion: a problem of man-land relationships. *Perspectives on Environment* (I. R. Manners and M. W. Mikesell, Eds), Washington, pp 57–78.

Campbell, J. A. (1974). Measurements of erosion on badlands surfaces, *Z. für Geomorph.*, Suppl. Bd. **21**, 138–150.

Carson, M. A., and Kirkby, M. J. (1974). *Hillslope Form and Process*, Cambridge University Press.

Carson, M. A., and Petley, D. J. (1970). The existence of threshold hillslopes in the denudation of the landscape, *Trans. Inst. Brit. Geog.*, 71–95.

Chepil, W. S., and Woodruff, N. P. (1963). The physics of wind erosion and its control, *Adv. in Agron.*, **15**, 211–302.

Costin, A. B., and Gilmour, D. A. (1970). Portable rainfall simulator and plot unit for use in field studies of infiltration, runoff, and erosion. *J. Applied Ecology*, **7**, 193–200.

Courtois, G., and Mussot, R. (1969). Etude des débits de charriage sur le Cady en utilisant les traceurs radioactifs. *Bull. Ass. Géogr. francais*, **373**, 445–458.

Culling, W. E. H. (1963). Soil creep and the development of hillside slopes. *J. Geology*, **71**, 127–162.

Delany, A. C., Parkin, D. W., Griffin, J. J., Goldberg, E. D., and Reimann, B. E. (1967). Airborne dust collected at Barbados. *Geochimica and Cosmochimica Acta*, **31**, 885–909.

De Ploey, J. (1967). Erosion pluviale au Congo occidental. In: *Isotopes in Hydrology*, International Atomic Energy Authority, Vienna, pp. 291–301.

De Ploey, J. L. (1969). L'érosion pluviale: expériences à l'aide de sables traceurs et bilans morphogéniques. *Acta Geogr. Lovaniensia*, **7**, 1–28.

De Ploey, J. (1973). A soil-mechanical approach of the erodibility of loess by solifluction. *Revue Géom. dyn.*, **22**(2), 61–70.

De Ploey, J., and Moeyersons, J. (1975). Runoff creep of coarse debris: experimental data
and some field observations. *Catena*, **2**, 275–288.

De Ploey, J. (1977). Some experimental data on slopewash and wind action with reference
to Quaternary morphogenesis in Belgium. *Earth Surface Processes*, **2**(2–3), 101–
115.

Ekern, P. C., and Muckenhirn, R. J. (1947). Water drop impact as a force in transporting
sand. *Proc. Soil Sci. Soc. Am.*, **12**, 441.

Ellison, W. D. (1944). Studies of raindrop erosion. *Agric. Engng*, **25**, 131–136 and 181–182.

Ellison, W. D., and Pomerene, W. H. (1944). A rainfall applicator. *Agric. Engng*, **25**,
220.

Emmett, W. W. (1970). The hydraulics of overland flow on hillslopes. *US Geol. Surv.
Prof. Paper*, **662-A**.

Fedder, D. (1972). Rheological equation of a silt, determined with the aid of a newly
developed test apparatus. *Engng Geology*, **6**, 43–68.

Finkel, H. J. (1959). The barchans in Southern Peru. *J. Geology*, **67**, 614–647.

Fournier, F. (1960). *Climat et érosion: la relation entre l'érosion des sol par l'eau et les
précipitations atmospheriques*, Presses Universitaires de France, Paris.

Geib, H. V. (1933). A new type of installation for measuring soil and water losses from
control plates. *J. Am. Soc. Agron.*, **25**, 429–440.

Gerlach, T. (1967). Hillslope troughs for measuring sediment movement. *Revue Géom.
dyn.*, **4**, 173.

Gerlach, T. (1970). Etat actuel et méthodes de recherches sur les processus morpho-
génétiques actuels sur le fond des étages climatiques et végétaux dans les Carpathes
Polonaises. *Studia Geom. Carpatho-Balcanica, Polska Ak. Nauk*, **4**, 47–63.

Graf, W. H. (1971). *Hydraulics of Sediment Transport*, McGraw-Hill, New York.

Gregory, S. (1968). *Statistical Methods and the Geographer*, Longmans, London.

Gregory, K. J., and Walling, D. E. (1973). *Drainage Basin Form and Process*, Arnold,
London.

Gunn, R., and Kinzer, O. D. (1949). The terminal velocity of fall for water droplets in
stagnant air. *J. Meteorol.*, **6**, 243–248.

Held, R. B., and Clawson, M. (1965). *Soil Conservation in Perspective*, John Hopkins
Press, Baltimore.

Hempel, C. G. (1965). *Aspects of Scientific Explanation*, Free Press, New York.

Higashi, A., and Corte, A. E. (1971). Solifluction: a model experiment, *Science*, **171**,
480–482.

Holeman, J. N. (1977). The National Erosion Inventory in Progress by the Soil Conserva-
tion Service, U.S. Department of Agriculture. Paper presented at the IFIP Working
Conference on Modelling and Simulation of Land, Air and Water Resources Systems,
Ghent, Belgium, August 30–September 2, 1977.

Hudson, N. W. (1957). The design of field experiments on soil erosion. *J. Agric. Engng
Res.*, **2**(1), 56–65.

Hudson, N. W. (1958). Runoff and soil loss from arable land in Southern Rhodesia.
7th Techn. Sess. Int. Un. Conserv. Nature, Athens.

Hudson, N. W., and Jackson, D. C. (1959). Results achieved in the measurement of
erosion and runoff in Southern Rhodesia. *3rd Inter-African Soils Conf.*, Dalba, pp.
575–583.

Hudson, N. W. (1963). Raindrop size distribution in high intensity storms. *Rhod. J.
Agric. Res.*, **1**(1), 6–11.

Hudson, N. W. (1964). A review of rainfall simulators. *Res. Bull. Dept. Conserv.*, Salisbury, 7.

Hutchinson, J. N. (1967). The free degradation of London Clay Cliffs. *Proc. Geotech. Conf. Oslo*, 1, 113–118.

Imeson, A. C., and Jungerius, P. D. (1974). Landscape stability in the Luxembourg Ardennes as exemplified by hydrological and (micro)pedological investigations of a catena in an experimental watershed. *Catena*, 1, 273–296.

Jahn, A. (1961). Quantitative analysis of some periglacial processes in Spitsbergen. *Zeszyty Naukowe, Uniw. Wroclawski, Ser.* B, 5.

Jahn, A., and Cielinska, M. (1974). The rate of soil movement in the Sudety Mountains, *Abh. Ak. Wissenschaften Göttingen, Math. Phys. Kl.*, *III*, 29, 86–101.

Journeaux, A., and Coutard, J. P. (1972). Etude en simulation de l'action du gel sur des grands modeles de sol. *Bull. Centre de Géom. Caen*, 13–14–15, 35–62.

Journeaux, A., and Coutard, J. P. (1976). Les expériences sur l'action du gel dans les sols. *Bull. Centre Géom. Caen*, 21, 13–20.

Journeaux, A., (1975). Légende pour une carte de l'environement et de sa dynamique, *Publ. Fac. Lettres Sci. Hum. Caen.*

Jungerius, P. D., Koster, E. A., and Kwaad, F. J. P. (1973). *Fysische Geografie*, Oosthoeks Uitgeversmaatschappij, Utrecht.

Kirkby, M. J. (1967). Measurement and theory of soil creep. *J. Geology*, 75, 359–378.

Kirkby, A. V. T., and Kirkby, M. J. (1974). Surface wash at the semi-arid break in slope. *Z. für Geomorph.*, Suppl. Bd. 21, 151–176.

Knottnerus, D. J. C. (1974). Soil blowing research by means of a wind tunnel. *Inst. Soil Fertility (Project 263)*, Haren, The Netherlands.

Lal, R. (1976). Soil erosion problems on an alfisol in Western Nigeria and their control. *IITA Monograph*, No. 1.

Langbein, W. B., and Schumm, S. A. (1958). Yield of sediment in relation to mean annual precipitation. *Trans. Am. Geophys. Union*, 39(6), 1076–1084.

Laws, J. O. (1941). Measurements of the fall velocity of waterdrops and raindrops. *Trans. Am. Geophys. Union*, 22, 709–721.

Leopold, L. B., Emmett, W. W., and Myrick, R. W. (1966). Channel and hillslope processes in a semi-arid area, New Mexico. *US Geol. Surv. Prof. Paper*, 352-G.

Leopold, L. B., and Emmett, W. W. (1967). On the design of a Gerlach trough. *Revue de Géom. dyn.*, 4, 172.

Leopold, L. B., Emmett, W. W., and Myrick, R. M. (1967). On some variations in painting rocks. *Revue de Géom. dyn.*, 4, 160–161.

Lohnes, R. A., and Handy, R. L. (1968). Slope angles in friable loess. *J. Geology*, 76(3), 247–258.

Lundgren, L., and Rapp, A. (1974). A complex landslide with destructive effects on the water supply of Morogoro town, Tanzania. *Geogr. Ann.*, 56-A(3–4), 251–260.

Mainguet, M., and Callot, Y. (1974). Air photo study of typology and interrelations between the texture and structure of dune patterns in the Fachi-Bilma Erg, Sahara, *Z. für Geomorph.*, Suppl. Bd. 20, 62–68.

Mainguet, M. (1975). Etude comparée des ergs, à l'échelle continentale (Sahara et déserts d'Australie). *Bull. Assoc. Géogr. Francais*, 424–425, 135–140.

Mainguet, M., and Canon, L. (1976). Vents et paléovents du Sahara. Tentative d'approche paléoclimatique. *Revue de Géom. dyn.*, 18, 241–250.

Marchand, D. E. (1970). Soil contamination in the White Mountains, Eastern California. *Bull. Geol. Soc. Am.*, 81, 2497–2505.

McKee, E. D. (1973). A study of morphology, provenance, and movement of desert sand seas in Africa, Asia, and Australia. *Symposium-NASA on Significant Results obtained from the ERTS-1*, NASA SP-327, pp. 291–304.

Meyer, L. D. (1965). Symposium on simulation of rainfall for soil erosion research. *Trans. Am. Soc. Agric. Engrs*, 8, 63–65.

Meyer, L. D., and McCune, D. L. (1958). Rainfall simulator for runoff plots. *Agric. Engng*, **39**(10), 644–648.

Moeyersons, J. (1975). An experimental study of pluvial processes on granite gruss. *Catena*, **2**, 289–308.

Moeyersons, J., and De Ploey, J. (1976). Quantitative data on splash erosion, simulated on unvegetated slopes. *Z. für Geomorph.*, Suppl. Bd. **25**, 120–131.

Moldenhauer, W. C., and Long, D. C. (1964). Influence of rainfall energy on soil loss and infiltration rates I. Effect over a range of texture. *Proc. Soil Sci. Soc. Am.*, **28**, 813–917.

Morin, J., Goldberg, D., and Seginer, I. (1967). A rainfall simulator with a rotating disk. *Trans. Am. Soc. Agric. Engrs*, **10**, 74–79.

Mutchler, C. K. (1963). *Runoff Plot Design and Installation for Soil Erosion Studies*, Agricultural Research Service Report ARS-41-79, United States Department of Agriculture, Washington, D.C.

Nir, D., and Klein, M. (1974). Gully erosion induced by changes in land use in a semi-arid terrain (Nahal Shiqma, Israel). *Z. für Geomorph.*, Suppl. Bd. **21**, 191–201.

Norris, R. M. (1966). Barchan dunes of Imperial Valley, California, *J. Geology*, **74**, 292–306.

Parsons, D. A. (1954). Coshocton-type runoff samplers. *Laboratory Investigations*, Soil Conservation Service, Report SCS-TP-124, United States Department of Agriculture, Washington, D.C.

Phelps, H. O. (1975). Shallow laminar flows over rough granular surfaces. *J. Hydraulics Div.*, *Proc. Am. Soc. Civil Engrs.*, **101**(HY3), 367–384.

Pissart, A. (1964). Vitesses des mouvements du sol au Chambeyron (Basses Alpes). *Biul. Peryglacjalny*, **14**, 303–309.

Pissart, A. (1972). Vitesse des mouvements de pierres dans les sols et sur des versants périglaciaires au Chambeyron (Basses Alpes). *Congrés et Coll. Univ. Liège*, **67**, 251–268.

Pissart, A. (1973). Résultats d'expériences sur l'action du gel dans le sol, *Biul. Peryglacjalny*, **23**, 101–113.

Pope, A., and Harper, J. J. (1966). *Low-speed Wind Tunnel Testing*, Wiley, New York.

Prior, D. B., and Stephens, N. (1971). A method of monitoring mudflow movements, *Engng Geology*, **5**, 239–246.

Raichlen, F. (1967). Some turbulence measurements in water. *J. Engng Mechanics Div.*, *Proc. Am. Soc. Civil Engrs*, EM-2, 73–97.

Rapp, A. (1960). Recent development of mountain slopes in Kärkevagge and surroundings, northern Scandinavia. *Geogr. Ann.*, **42**, 73–200.

Rapp, A., Berry, L., and Temple, P. (1972). Studies of soil erosion and sedimentation in Tanzania, *Geogr. Ann.*, **54A**(3–4), 105–379.

Rapp, A. (1974). Slope erosion due to extreme rainfall, with examples from tropical and arctic mountains, *Abh. Ak. Wiss. Göttingen, Math.-Phys. Kl. III*, **29**, 118–136.

Rapp, A. (1975). Soil erosion and sedimentation in Tansania and Lesotho. *Ambio*, **4**, 154–163.

Resch, F. J. (1970). Hot-film turbulence measurements in water flow. *J. Hydraulics Div.*, *Proc. Am. Soc. Civil Engrs*, (HY3), 787–797.

Rose, C. W. (1960). Soil detachment caused by rainfall. *Soil Sci.*, **89**, 28–35.

Rouse, W. C. (1975). Engineering properties and slope form in granular soils. *Engng Geology*, **9**, 221–236.

Rudberg, S. (1967). On the use of painted rocks aligned along a contour. *Revue Géom. dyn.* **4**, 161.

Savat, J. (1975). Discharge velocities and total erosion of a calcareous loess: a comparison between pluvial and terminal runoff. *Revue Géom. dyn.*, **4**, 113–122.

Schachori, A., and Seginer, I. (1962). Sprinkling assembly for simulation of design storms as a means for erosion and runoff studies. *Bull. Int. Ass. Sc. Hydr.*, **7**(4), 57–72.

Schick, A. (1967). Bedload trap, *Revue Géom. dyn.*, **4**, 182–183.

Schumm, S. A. (1967). Erosion measured by stakes. *Revue Géom. dyn.*, **4**, 161–162.

Scott, C. R. (1969). *Soil mechanics and foundations*, Elsevier, Amsterdam.

Seginer, I., Morin, J., and Schachori, A. (1962). Runoff and erosion studies in a mountainous terra-rossa region in Israel. *Bull. Int. Ass. Sc. Hydr.*, **7**(4), 79–92.

Sharon, D. (1970). Areal patterns of rainfall in a small watershed as affected by wind and meteorological conditions. *IASH-Unesco Symp. on the results of research on repres. exper. basins*, Wellington.

Sharon, D. (1972). The spottiness of rainfall in a desert area. *J. Hydrology*, **17**, 161–175.

Sharpe, C. F. S. (1938). *Landslides and Related Phenomena*, Columbia University Press, New York.

Shen, H. W., and Li, R. M. (1973). Rainfall effect on sheet flow over smooth surface, *J. Hydraulic Div., Proc. Am. Soc. Civil Engrs*, (HY5), 773–791.

Skempton, A. W. (1964). Long-term stability of clay slopes. *Geotechnique*, **14**, 77.

Słupik, J. (1970). Methods of investigating the water cycle within a slope, *Studia Geom. Carpatho-Balcanica, Polska Ak. Nauk*, **4**, 127–137.

Smith, H. T. U. (1969). *Photo-interpretation Studies of Desert Basins in Northern Africa*, Air Force Cambridge Research Lab., AFCRL-68-0590, Bedford, Mass.

Smith, D. D., and Wischmeier, W. H. (1962). Rainfall erosion, *Adv. in Agron.*, **14**, 109–148.

Soons, J. M., and Rainer, J. N. (1968). Micro-climate and erosion processes in the Southern Alps, New Zealand. *Geogr. Ann.*, **50A**(1), 1–15.

Statham, I. (1976). A scree slope rockfall model, *Earth Surface Proc.*, **1**, 43–62.

Stocking, M. A., and Elwell, H. A. (1973). Soil erosion hazard in Rhodesia, *Rhod. agric. J.*, **70**(4), 95–101.

Temple, P. H., and Rapp, A., (1972). Landslides in the Mgeta area, Western Uluguru Mountains, Tanzania, *Geogr. Ann.*, **50A**(3–4), 157–194.

Terzaghi, K., and Peck, R. B. (1967). *Soil mechanics in Engineering Practice*, Wiley, New York.

Thompson, M. M. (1966). Manual of Photogrammetry, American Society of Photogrammetry, Falls Church.

Tsoar, H. (1974). Desert dunes morphology and dynamics, El Arish (Northern Sinai), *Z. für Geomorph.*, Suppl. Bd. **20**, 41–61.

Van Zuidam, R. A. (1971). Orbital photography as applied to natural resources survey, *Publ. ITC-Delft, Ser. B-61.*

Vita-Finzi, C. (1969). *The Mediterranean Valleys*, Cambridge University Press.

Washburn, A. L. (1967). Instrumental observations of mass-wasting in the Mesters Vig District, Northeast Greenland. *Med. om Grønland, Komm. Vid. Unders. Grønland*, **B166**, 4.

Wenzel, H. G., and Mathews, M. J. (1970). Photomicroscopic sublayer velocity measurement, *J. Hydraulics Div., Proc. Am. Soc. Civil Engrs*, (HY12), 2467–2480.

Williams, P. J. (1957). The direct recording of solifluction movements. *Am. J. Sci.*, **255**, 705–714.

Wischmeier, W. H. (1976). Use and misuse of the universal soil loss equation. *J. Soil and Water Conserv.*, **31**(1), 5–9.

Wischmeier, W. H., and Smith, D. D. (1958). Rainfall energy and its relationship to soil loss. *Trans. Am. Geophys. Union*, **39**, 285–291.

Wischmeier, W. H., and Smith, D. D. (1965). Predicting Rainfall Erosion Losses from Cropland East of the Rocky Mountains, Agricultural Handbook No. 282, Agricultural Research Service, United States Department of Agriculture, Purdue Agricultural Experiment Station.

Woodruff, N. P., and Siddoway, F. H. (1965). A wind erosion equation. *Proc. Soil Sci. Soc. Am.*, **29**, 602–608.

Yaalon, D. H., and Kalmar, D. (1972). Vertical movement in an undisturbed soil: continuous measurement of swelling and shrinkage with a sensitive apparatus. *Geoderma,* **8,** 231–240.

Yair, A. (1974). Sources of runoff and sediment supplies by the slopes of a first order drainage basin in an arid environment (Northern Negev-Israel). *Abh. Ak. Wissensch. Göttingen, Math-Phys. Kl. III,* **29,** 403–417.

Yair, A., and Gerson, R. (1974). Mode and rates of escarpment retreat in an extremely arid environment (Sharm el Sheikh, Southern Sinai Peninsula). *Z. für Geomorph.,* Suppl. Bd. **21,** 202–215.

Zaruba, Q., and Mencl, V. (1969). *Landslides and Their Control,* Elsevier, Amsterdam, and Academia, Prague.

Soil Erosion
Edited by M. J. Kirkby and R. P. C. Morgan
© 1980 John Wiley and Sons Ltd.

CHAPTER 4

Mechanics of water erosion and their spatial and temporal controls: an empirical viewpoint

R. Evans

Soil Survey of England and Wales, Cambridge, UK

Soil erosion is considered here as the removal of soil at rates in excess of soil formation and is primarily attributable to man and his activities. Rates of geological erosion (Lowdermilk, 1934) can be high in semi-arid regions with rainfall between 250 to 356 mm per annum where vegetation is sparse but high intensity rainfall is sufficiently frequent and widespread to cause erosion (Langbein and Schumm, 1958). Elsewhere, geological rates of erosion are lower (Kinoy, Chapter 1). In arid regions rainfall is too infrequent and localized for erosion to be widespread and amounts of debris moved are very small (e.g. Yair and Klein, 1973); in wetter regions under their natural vegetation cover, the vegetation prevents rain-splash or flowing water from removing soil particles. However, man, indirectly by his animals over-grazing vegetation, or directly by his cultivations causes the soil to be exposed to rainfall.

Raindrops compact the soil surface and splash and flowing water move soil particles; these processes can seal the surface and this lowers the rate at which water passes into the soil. If more rain falls than can infiltrate into the soil, runoff will take place. Sheet wash (or inter-rill flow) occurs where the flow of water is generally unconfined, except between soil-clods, covers much of the surface and continuous channels do not form. As the velocity of flow increases the water incises into the soil and rills form. Rills are continuous channels of narrow width and shallow depth and can be obliterated by ploughing (Soil Survey Staff, 1951) or weathering of the surface (Morgan, 1977). Gullies are wider, deeper, and generally permanent features.

4.1. VEGETATION FACTORS

4.1.1 Fully vegetated slopes

Runoff and erosion from good pasture or woodland are small, often less than 5 and 1 per cent respectively of runoff and erosion from bare soil (e.g. Bennett, 1939). Runoff is low because rates of infiltration of water through vegetated surfaces are high compared to those on bare soil (Woodward, 1943), as vegetated soils often have a better structure and more stable aggregates. When raindrops strike vegetation the energy of the drops is dissipated and there is no direct impact on the soil surface. Some of the water is intercepted by the vegetation and evaporates to the atmosphere; the remainder drips or flows down grass stems or tree trunks to the ground below. The drops falling from the leaves are often larger than the original raindrops because they have coalesced (Stocking and Elwell, 1976). However, there is often a surface of litter beneath grass or trees and this intercepts the drop falling through the plant canopy. The water reaching the soil remains clear, and the soil pores at the surface are not infilled with splashed soil particles.

If runoff does take place the leaves and roots of plants inhibit movement of soil particles. Leaves form a rough surface, impede flow, and reduce the velocity of flowing water and roots bind the soil.

4.1.2 Partly vegetated slopes

Runoff and erosion increase rapidly on soils with less than 70 per cent vegetation cover (Copeland, 1965). In semi-arid regions where vegetation cover is often below 20–30 per cent, runoff and erosion are related to the amount of bare ground, increasing as the proportion of bare ground increases (Branson and Owen, 1970). The amounts of soil eroded do not correlate as well as runoff with proportions of bare ground; this is because the soil moved often does not reach the stream, being trapped downslope by vegetation or deposited in valley floors or on concave breaks of slope.

Over-grazing of pastoral lands leads to the breakdown of the turf cover and exposure of bare soil. Continued removal of the leaves by grazing weakens the root system as carbohydrates produced by photosynthesis are not transported to the root system in sufficient quantity to ensure adequate root growth (Dasmann et al., 1973). Little is known of the stocking rates which lead to instability of the turf cover though Evans (1977) shows that bare ground can be initiated in upland grazings in England at stocking rates higher than one sheep per 0.8 hectare.

In arable fields erosion decreases as the crop cover becomes denser, this decrease being most marked when plants cover more than 30 per cent of the surface (Elwell and Stocking, 1976). Cereals will rarely cover more than 90 per cent of the ground surface. Rates of erosion of clean-tilled row crops are higher, for instance, than under cereals (e.g. Smith *et al.*, 1945). Stocking and Elwell (1976) also note that even with a good ground cover, in tall crops such as maize, more erosion may take place than expected because the height from which large coalesced raindrops fall is sufficient for high terminal velocities to be attained; splash and sheet erosion take place, therefore.

4.1.3   Absence of vegetation

Except in arid regions slopes are rarely completely bare of vegetation unless the land is under arable; bare arable land is most susceptible to water erosion, and large amounts of soil can be removed in one storm or during a wet season. Even in the United Kingdom where soil erosion is not considered a problem (Morgan, 1977) $0.3\,kg/m^2$ of soil have been eroded in one storm and $0.8\,kg/m^2$ lost in five storms in fields in Cambridgeshire and Bedfordshire respectively (Evans and Morgan, 1974). In the United States, erosion rates much greater than this have been recorded, for example $>20\,kg/m^2$ per annum (e.g. Bennett, 1939); similar to those measured in a field in north Norfolk, England (Evans and Nortcliff, 1978). The factors determining soil loss are rainfall intensity and duration, soil particle size distribution and properties related to this such as surface roughness and aggregation, and slope form.

4.2   RAINFALL FACTORS

4.2.1   Momentum and kinetic energy

The interactions of raindrop size, velocity and shape, the duration of the storm and the windspeed, control the erosive power of rainfall. As raindrops increase in size their terminal velocity increases (Figure 4.1, after Gunn and Kinzer 1949; De Ploey and Gabriels, Chapter 3). The momentum of a raindrop, which is its mass × velocity, also increases rapidly with drop size up to about 4 mm diameter, therefore.

The kinetic energy of rainfall, which is the energy of the total number of raindrops at a given intensity, is estimated from the distribution of raindrop size for a given intensity. Carter *et al.* (1974) and Hudson (1963) show that the raindrop size distribution includes a greater proportion of larger drops above 4 mm diameter at intensities between 50–100 mm per hour and more than 200 mm per hour (Figure 4.2); at other intensities there are many smaller (< 2.5 mm) drops. Drops larger than 5.5–6.0 mm are unstable because of turbulence of the air (Blanchard, 1950) and break up; however, at intensities above 200 mm per hour coalescence of smaller drops again takes place. The kinetic energy of rainfall, therefore, is at a maximum at rainfall intensities between 50 and 100 mm per hour and greater than 250 mm per hour (Figure 4.2). Wischmeier and Smith

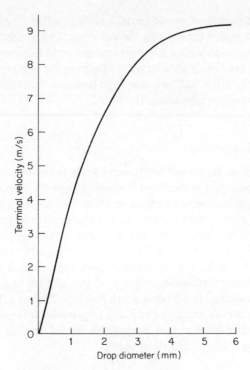

Figure 4.1. Terminal velocities of water droplets in stagnant air (after Gunn and Kinzer, 1949)

(1958) did not allow for this change in distribution of kinetic energy, and over-estimated kinetic energy for rainfalls greater than 100 mm per hour.

Wischmeier (1959) and Wischmeier and Smith (1958) found that of the rainfall factors studied, kinetic energy explained most of the measured soil loss from field-plots. The best measure of energy was the maximum rainfall intensity occurring over 30 minutes (EI_{30}). However, Hudson (1971) notes that this measure was not as good as one which assumed that there is an erosive threshold, of 25 mm rain per hour, and that only rainfalls of more than this amount are of importance in causing erosion. Elwell and Stocking (1975) used various indexes of momentum, energy, and rainfall depth, as have other workers (e.g. Lal, 1976b; Morgan, 1977) but individual indexes do not appear to apply widely. In some instances one index is better for predicting one type of erosion than another, for instance, Morgan (1974) found that Fournier's precipitation index, a measure of rainfall aggressiveness, was better for predicting gully erosion whereas mean annual precipitation was best for predicting amount of rainsplash erosion.

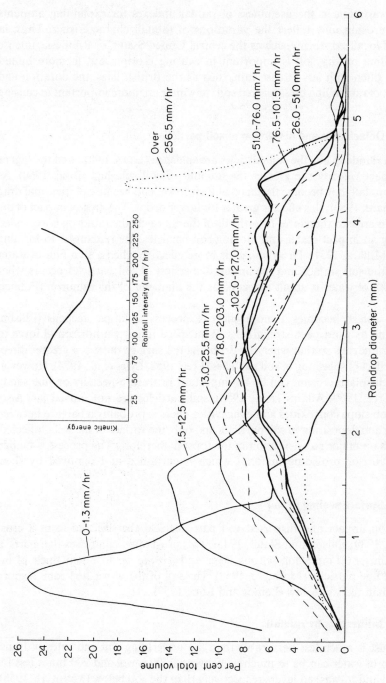

Figure 4.2. Raindrop size distribution and rainfall intensity (after Carter, *et al.*, 1974)

The variation in the usefulness of rainfall indexes for explaining amounts of soil erosion may reflect the variations of rainfall characteristics. Thus, in areas of localized storms such as the central United States, or Rhodesia, intense rain falling on dry soil is important in causing erosion, but in more humid regions often with less intense rain, such as the British Isles, the duration and amount of rain falling on saturated soils in winter are more important in causing erosion.

4.2.2   Detachment and dispersion of soil particles

When a raindrop hits the soil particles are splashed (Laws, 1940); and the higher the impact velocity the greater the amount of soil splashed (Bisal, 1960). As raindrops fall they become flattened at their lower end because of frictional drag (Blanchard, 1950); this effect is greater for larger drops. The area of impact of the drop is increased therefore. Two-thirds of the energy of the raindrop is expended forming an impact crater and moving soil particles, the remainder to forming spray (Mihara, 1951). Drop impact is more effective when a thin film of water covers the soil surface and maximum dispersion of soil particles occurs when the depth of water is about the same as the diameter of the raindrop (Palmer, 1963).

Rain splash becomes a more effective process as slope steepens (Ellison, 1944) and as windspeed strengthens. Windspeed imparts a horizontal force to falling raindrops so that when the drop hits the surface there is a greater directional effect of splashing of soil particles (Free, 1952, Lyles *et al.*, 1974). However, although splash erosion can be an important process, especially on fine sandy soils (Free, 1960; Adams *et al.*, 1958), soil particles are not moved far. Also, except on slope convexities and concavities, there is probably a balance between erosion and deposition of soil particles and only the surface of the soil is affected. There is no major redistribution of the topsoil, therefore. The process is mainly important for providing material which is entrained and removed by sheet wash.

4.2.3   Surface sealing of soil

Raindrop impact and dispersed soil particles seal the soil and form a crust (Duley, 1939; Ellison and Slater, 1945). This crust often comprises two parts, a very thin (*c.* 0.1 mm) non-porous layer and a zone up to 5 mm thick of in-washed fine particles (McIntyre, 1958). The soil of the in-washed zone is more dense than the soil below (Lemos and Lutz, 1957).

4.2.4   Infiltration by rainfall

The crust is much less permeable than the underlying soil and rates of transmission of water can be as much as between 2000 times and 200 times less for the seal and in-washed layers respectively than the soil below (McIntyre, 1958);

though Tackett and Pearson (1965) noted much smaller differences. Hence, the infiltration of rainfall into crusted soils is very slow. Pools of standing water will form and coalesce and sheet flow will begin.

4.3 SHEET WASH, RILLS, AND GULLIES

The interaction of rain splash and sheet wash is important, each process acting separately is less efficient at moving soil particles than when the processes are acting together (Young and Wiersma, 1973). This is because the soil particles are brought into suspension by raindrop splash and then transported by sheet flow. Also, raindrop splash imparts turbulence to laminar flow. Reported velocities of sheet flow from laboratory and field plot studies range from 1.5–4.0 cm per second (Young and Wiersma, 1973) to *c*. 30 cm per second (Ellison, 1947a). Velocities of 16 cm per second are required to erode soil particles of 0.3 mm diameter (Hjulstrøm, 1935) and velocities as low as 2 cm per second will carry the particle in suspension.

Morgan (1977) observed that in the field sheet flow covered up to 50 per cent of the slope surface during a heavy rainfall, but that erosion did not take place uniformly across this slope. Where water was confined between soil clods there was evidence of erosion with soil deposited in small fans below the confinement. In his study of sheet flow Emmett (1970) also noted linear concentrations of flow within sheet wash. He estimated that the depth, up to 3 mm, and velocity of flow were such that the Reynolds number was generally between 1500 and 6000, and that both laminar (< 1500) and turbulent flow (> 6000) took place. Turbulent erosive flow only takes place where flow is confined, therefore, elsewhere flow is probably laminar and non-erosive.

Sheet wash is an effective erosive process, on sandy soils, because it covers large areas of slope, in contrast to the small areas covered by rills (Morgan, 1977). On coherent soils resistant to splash and sheet erosion, erosion will mainly be by rills or gullies. Mosley (1974) has shown in the laboratory that erosion of a silt/sand mix is primarily by rilling.

Rills and gullies form when the velocity of water flow increases, probably at speeds in excess of 30 cm per second (Ellison, 1947a) and flow becomes turbulent. This increase in hydraulic gradient may come about for a number of reasons, for example, an increase in slope gradient; an increase in rainfall intensity; or because surface storage is exceeded and incision takes place at the point of flow. Whether rills or gullies form probably depends on soil factors as well as the velocity and depth of water flow. A flow of 0.6–1.0 m/s or more has been noted in an eroding gully in the Yorkshire Wolds (Foster, 1978); figures very similar to those of flow measured in a major rill (Kar and Bandyopadhyay, 1974).

Rills and gullies remove much larger volumes of soil per unit area than does sheet wash. For instance, in the loessial watersheds of Iowa for a 31 day period

gully erosion in two watersheds removed 343.8 and 358.9 kg/m² per annum
(Spomer *et al.*, 1971) whereas rill and sheet wash erosion removed 24.9 and
68.9 kg/m².

4.4 SOIL FACTORS

4.4.1 Soil particle size distribution

In the United States erosion of arable land occurs often on sandy and silty soils
(e.g. Wischmeier and Mannering, 1969). Less frequently it has been noted that
some clays and clayey soils are also erodible (Bennett, 1926; Ripley *et al.*,
1961).

To assess the importance of particle size as a contributory factor to erosion,
the particle size distributions of 56 soils of lowland England which were known
to be erodible, were plotted on a triangular diagram (Figure 4.3). Particle sizes
were taken from published and unpublished soil profile descriptions of the
Soil Survey. On the diagram are also plotted particle size distributions of erodible
soils from Canada (Ripley *et al.*, 1961), India (Mehta *et al.*, 1963), and the United
States (Lutz, 1934; Megahan, 1975; Middleton, 1930; Peele, 1937; Peele *et al.*,
1945).

There is a wide scatter of points across the diagram. However, generally,
erodible soils have a restricted clay content. Thus, 87.5 per cent of the soils
contain between 9 and 35 per cent clay; and 75.0 per cent between 9 and 30
per cent clay. There are no erodible soils in the sand class, and the sand in the
erodible loamy sand soils is fine sand (53–2000 µm) (compare Mitchell and
Bubenzer, Section 2.2.2).

Soils with more than 30–35 per cent clay are generally coherent and form
stable soil aggregates which are resistant to raindrop impact and splash erosion.
Clays are often cloddy soils and their rough surfaces store much water and are
resistant to sheet and rill erosion. Sands and coarse loamy sands have high
infiltration rates, and even if this is exceeded sand particles of more than 0.3 mm
(300 µm) diameter are not easily eroded by flowing water (Morgan, 1977) or
by raindrop impact.

Bouyoucos (1935) found a direct relationship between his clay ratio, per cent
sand and silt over per cent clay, and amounts of erosion. Other workers consider
the clay ratio a less satisfactory index of soil erodibility than some other indexes
(Bryan, 1968) although Ballal (1954) found a good correlation between the clay
ratio and Middleton's erosivity ratio. Replotting Woodburn and Kozachyn's
(1956) data shows there is a better negative linear relationship of per cent clay
and erosion index than there is for the positive curve of sand content and
erosion index. Also, there is a reasonable linear correlation of clay content and
erosion index for Barnett *et al.*'s (1965) data and for per cent clay and amounts
of erosion for the data of Peele (1937) and Peele *et al.* (1945). However, Rose

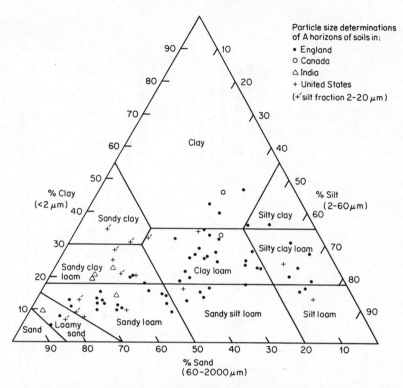

Figure 4.3. Particle size distribution of erodible soils

(1960) and Epstein and Grant (1967) found greater detachment of soil particles by raindrops as clay content of soils increased. Bennett (1926) notes that clays with silica:sesquioxide ratios greater than two are erodible, and Bryan (1971) considers clay mineral type is an important factor controlling the stability of soil aggregates and hence erodibility.

4.4.2 Soil aggregation

The proportion of a soil in water stable aggregates less than 0.5 mm is a good index of erodibility (Bryan, 1974; Rai *et al.*, 1954); the greater the proportion of aggregates <0.5 mm the greater the erodibility of the soil. Generally sandy soils are more erodible to simulated rainfall than clayey soils as the aggregates of these sandy soils slake more readily and seal the soil surface. Soils with higher clay and organic matter contents have more stable aggregates because of the strong bonds between their colloids (Greenland, 1965a, b). Other, less important, factors contributing to aggregate stability are: that divalent (calcium) cations are held more firmly by colloids than are monovalent (sodium) cations; the

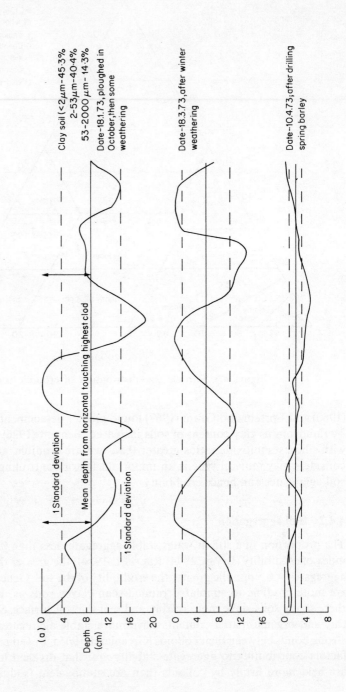

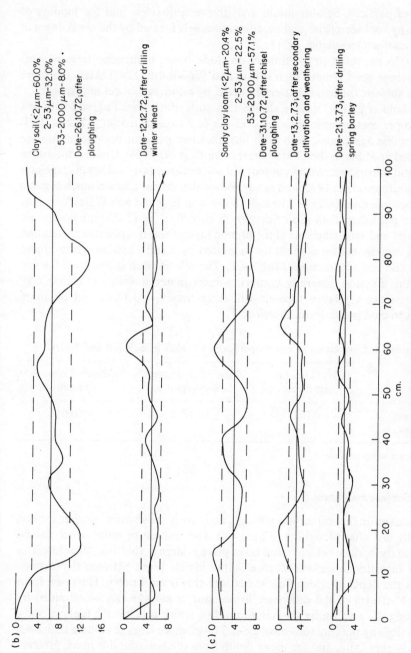

Clay soil (<2 μm – 60.0%
        2–53 μm – 32.0%
        53–2000 μm – 8.0% .

Date – 26.10.72; after
ploughing

Date – 12.12.72, after drilling
winter wheat

Sandy clay loam (<2 μm – 20.4%
          2–53 μm – 22.5%
          53–2000 μm – 57.1%

Date – 31.10.72, after chisel
ploughing

Date – 13.2.73; after secondary
cultivation and weathering

Date – 21.3.73, after drilling
spring barley

Figure 4.4. Surface roughness of three soils

binding of particles by aluminium and iron sesquioxides; and the binding of particles by root secretions and mucillaginous gels formed by the breakdown of organic matter (Greenland, 1971).

Crusts do not form, or are thinner on clods, hence infiltration rates of rainfall into these soils remain high (Borst and Woodburn, 1942). Also, the more cloddy a surface, the less is material moved by raindrop impact and splash.

Greenland et al. (1975) have shown that soils of lowland England with less than two per cent organic carbon (= 3.5 per cent organic matter, Allison, 1973) have unstable aggregates. Soils with less than two per cent organic matter are considered erodible in the United States (Smith et al., 1954). Under continuous cultivation the organic matter content of a soil declines as it oxidises on exposure to air (Skidmore et al., 1975) and sandy soils which do not contain much organic matter become erodible. For 110 soils in lowland England and Wales for which there are published data (Soil Survey) on clay, fine silt (2–20 μm), coarse silt (20–60 μm) and sand fractions of the plough layer, there is a positive correlation between organic matter content (as measured by a soil's loss-on-ignition) and the clay and fine silt fractions (Table 4.1). The relationship is less good for fine silt, as this fraction generally comprises inert material with few coarse clay mineral particles. Clayey topsoils will have better structures and be more resistant to clod breakdown, therefore.

Table 4.1. Correlation of soil properties at 110 sites in England and Wales

	Clay: Loss-on-ignition	Fine silt (2–20 μm): Loss-on-ignition	> 20 μm: Loss-on-ignition
Correlation Coefficient	0.753	0.479	− 0.601

All significant at 0.1 per cent.

4.4.3 Surface roughness

After cultivation the soil surface is rough, and clayey soils have a rougher surface, especially just after ploughing (Figure 4.4). The amount of water which can be stored on the surface before runoff takes place is large at this time, therefore. It is difficult to estimate surface storage as the lowest point between depressions controls the depth at which flow starts, and this is not known. However, from Figure 4 an idea of the difference in amount of storage can be obtained by calculating the area under the horizontal line which touches the highest clods. After ploughing the clay soils have 1.6–2.3 times more storage volume than does the sandy clay loam, and the mean depth from the horizontal is much greater.

The surface is generally not worked for some time after ploughing to allow clods to weather and break down and the field is then harrowed to produce a fine tilth and the seed is drilled. Surface roughness is least after drilling and rolling of the seed bed (Figure 4.4), and differences between soil types are then smallest, the clay having only 1.1 times more storage. A winter sown crop, for example, winter wheat, has a rougher seedbed (Figure 4.4) as there has been less time for the clods to weather. Thus the winter prepared surface has 2.3 times more storage volume than the clay surface prepared in spring. Runoff will occur more quickly where tractor wheelings provide an outlet downslope.

4.4.4 Surface stoniness

Stony soils are less liable to erosion (Lamb *et al.*, 1950). The soil is not only protected by the stones but infiltration is increased as water flows into the soil around the edges of the stones (McIntyre, 1958). If there has been rilling or gullying adjacent to the stony soils, soil pillars or pedestals will form below the stones as soil is removed by splash into the rill.

4.4.5 The soil profile

The soil profile often determines the depth of erosion. Soil horizons below the A layer or plough layer (Ap) are often more compact and less erodible. If resistant bedrock is near the surface only rills will develop; whereas if the parent material is unconsolidated, for example sands and gravels or loess, deep gullies can be cut.

Soils with a well-structured, prismatic B horizon will drain quickly, but soils with an horizon of dense poor structure or a well-developed platy structure will have impeded drainage. Where permeability is impeded at depth this will reduce infiltration of water into the soil. Renner (1936) and Rauzi *et al.* (1968) found that the degree of development of soil structure was an important factor governing the erodibility of soils on rangelands.

4.4.6 Antecedent soil moisture

Soil which is already wet when rain falls upon it will attain its final infiltration rate more quickly. Often, the correlation between erosion and amounts of rain falling in a season or a year is related to this factor. In England erosion often occurs in winter and spring when evaporation of water from the soil is minimal and soils remain saturated for long periods. Attempts have been made to estimate the effects of antecedent soil moisture on erosion (Wischmeier and Smith, 1958). But unless rates of evaporation or transpiration from crops can be calculated with reasonable accuracy, which can be done for England (MAFF, 1967), and these related to the amount of water stored in the profile (Hodgson, 1974) estimates of antecedent moisture will remain crude and its effect on erosion difficult to perceive.

4.5 SLOPE FACTORS

4.5.1 Slope angle

The angle of slope is an important factor governing the efficacy of splash erosion; as slope angle steepens more soil is splashed downslope (Ellison, 1944). The relationships of slope angle to rill and sheet erosion are equivocal, however, although it is assumed in equations for predicting soil loss that slope angle is an important factor controlling severity of erosion (Musgrave, 1954; Wischmeier and Smith, 1965). The average values of the exponents used in these equations are derived from a small number of studies (Wischmeier *et al.*, 1958) carried out in the field and laboratory on plots of small size. Not all studies show an increase in erosion as slope angle increases (Lillard *et al.*, 1941; Neal, 1938) though there is often a marked increase in erosion on slopes of 5–10 per cent compared to erosion on gentler slopes. But, on slopes steeper than this erosion is often less.

4.5.2 Slope length

As slope length increases erosion is assumed to increase in severity (Musgrave, 1954; Wischmeier and Smith, 1965). Though there is some evidence for this (Zingg, 1940), often the evidence is equivocal (Lal, 1976a; Wischmeier *et al.*, 1958; Smith *et al.*, 1954). Most of the data is derived from field plot studies of restricted length; most plot lengths are less than 30 m, and rarely as long as 200 m. Also, these slopes are not related to the form of the land on which they are situated.

4.5.3 Slope form

In England eroded soils are often in fields which enclose land from the watershed to the valley floor and the relief is 'rolling' or gently undulating. The relatively long gentle crest slopes give way to short convex–concave valley side slopes (e.g. Morgan, 1977). Photographs of rill and sheet eroded fields in the United States often show similar landforms (e.g. Ellison, 1947b; Lillard *et al.*, 1941).

For 103 fields in lowland England in which erosion has occurred, slope length from lowest to highest points in the field was measured, the range of height was estimated and, where possible, maximum valley side slope angle was estimated. The measurements were taken from 1:10,000, 1:10,560, and 1:25,000 Ordnance Survey maps with contour intervals of 7.6 m (25 ft) or 5 m. The sites are widely scattered in two 100 × 100 km blocks of land in central and eastern England. (Figure 4.5d).

Most of the fields (83 per cent) were less than 400 m long (Figure 4.5(a)),

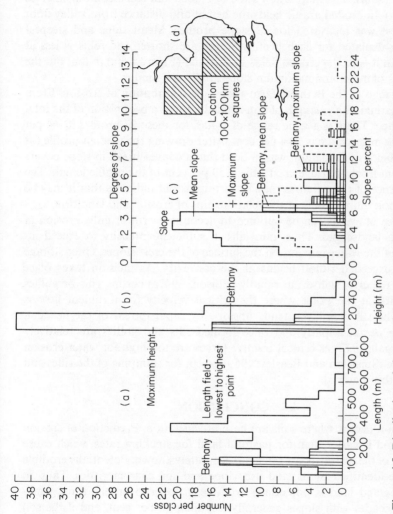

Figure 4.5. Distribution of slope variables for 103 hillsides in Eastern England. Shaded histograms show comparative data for sites near Bethany, Missouri
(a) Total slope length from divide to lower point in field
(b) Total slope relief
(c) Maximum and mean slope gradient
(d) Target population for sites sampled

had a relief of less than 25 m (94 per cent) (b), and mean slopes of less than 6 per cent (83 per cent) (c). Erosion was not recorded in fields shorter than 50 m.

In Smith *et al.* (1945) is a map of the Conservation Experiment Station at Bethany, Missouri. The map is at a scale of 1:5000 and has a contour interval of 1.5 m (5 ft). In eroded arable fields the height and distance from valley floor to ridge crest was measured for 22 slope profiles. Mean slope and steepest slope were calculated for each profile. Slopes are shorter and relief is less at Bethany than it is in the eroded fields in England (Figure 4.5(a) to (c)); but the distributions of mean and maximum slope angles are similar.

A second map of the Bethany farm shows the distribution of erosion. From this can be estimated the length of non-eroded crest as a proportion of the total length of slope. There is a wide range of values for these 22 profiles (0–64 per cent) and the mean value is 20.4 per cent. After drawing the median profile for 103 eroded fields in England it was noticed that a convex break in slope occurs at about 70 m from the ridge crest; this is 20 per cent of the profile length. The correspondence of these figures may be fortuitous but indicates that it may be fruitful to look more closely at the relationships of erosion and landform.

A sequence of events can be outlined to account for rill or gully erosion of these rolling landscapes. Thus, rain falls of sufficient intensity to exceed infiltration and the water remains on the surface of the crest slopes. Once storage capacity is exceeded runoff begins at the convexity. As incision takes place water from the crest will drain rapidly (Ellison, 1947a) cutting rills or gullies into the slope at the point where the critical velocity for turbulent flow is exceeded. The rill or gully extends upslope by rapid retreat of the headcut. The erosion occurs rapidly and once formed rills and gullies are relatively stable (Morgan, 1977) as critical tractive forces are too high for water erosion to take place (Smerdon and Beasley, 1959) except for slumping of the sides and removal of this material.

4.6 CONCLUSION

Water erosion occurs where soils are bare of vegetation. Prediction of erosion of arable land is easier than for pastoral land for stocking rates which cause break down of the turf cover are often not accurately known. Potentially erodible land can be identified by its land use, slope and soil characteristics. Thus, in lowland England land is potentially erodible if down to arable, if on gently rolling landscapes with slopes generally less than 10 per cent, and if the soil contains between 9 and 30 per cent clay. The critical time for erosion is when the surface is smoothest after preparation of the seed bed and when rainfall intensity or duration is such that infiltration is exceeded.

REFERENCES

Adams, J. E., Kirkham, D., and Scholtes, W. H. (1958). Soil erodibility and other physical properties of some Iowa soils. *Iowa State College, Journal of Science*, **32**, 485–540.

Allison, F. E. (1973). *Soil Organic Matter and Its Role in Crop Production.* Elsevier, New York and London. 673 pp.

Ballal, D. K. (1954). A preliminary investigation into some of the physical properties affecting soil erosion of Madhya Pradesh soils. *J. Indian Soc. Soil. Sci.,* 1, 37–41.

Barnett, A. P., Rogers, J. S., Holliday, J. H., and Dooley, A. E. (1965). Soil erodibility factors for selected soils in Georgia and South Carolina. *Trans. Am. Soc. Agric. Engrs,* 48, 393–395.

Bennett, H. H. (1926). Some comparisons of the properties of humid-tropical and humid-temperate American soils; with special reference to the indicated relations between chemical composition and physical properties. *Soil Sci.,* 21, 249–275.

Bennett, H. H. (1939). *Soil Conservation.* McGraw-Hill, New York and London. 993 pp.

Blanchard, D. C. (1950). Behaviour of water drops at terminal velocity. *Trans. Am. Geophys. Union,* 31, 836–842.

Bisal, F. (1960). The effect of raindrop size and impact velocity on sand splash. *Can. J. Soil Sci.,* 40, 242–245.

Borst, H. L., and Woodburn, R. (1942). The effect of mulching and methods of cultivation on runoff and erosion from Muskagua silt loam. *Agric. Engng,* 23, 19–22.

Bouyoucos, G. J. (1935). The clay ratio as a criterion of susceptibility of soils to erosion. *J. Am. Soc. Agron.,* 27, 738–741.

Branson, F. A., and Owen, J. B., (1970). Plant cover, runoff, and sediment yield relationships on Mancos Shale in western Colorado. *Water Res. Res.,* 6, 783–790.

Bryan, R. B. (1968). The development, use and efficiency of indices of soil erodibility. *Geoderma,* 2, 5–26.

Bryan, R. B. (1971). The efficiency of aggregation indices in the comparison of some English and Canadian soils. *J. Soil Sci.,* 22, 166–178.

Bryan, R. B. (1974). Water erosion by splash and wash and the erodibility of Alberta soils. *Geogr. Ann.,* 56A, 159–182.

Carter, C. E., Greer, J. D., Braud, H. J., and Floyd, J. M. (1974). Raindrop characteristics in south central United States. *Trans. Am. Soc. Agric. Engrs,* 17, 1033–1037.

Copeland, O. L. (1965). Land use and ecological factors in relation to sediment yields. In: *Proceedings of the Federal Inter-Agency Sedimentation Conference* (Agricultural Research Service Miscellaneous Publication No. 970), United States Department of Agriculture, Washington, D.C., pp. 72–84.

Dasmann, R. F., Milton, J. P., and Freeman, P. H. (1973). *Ecological Principles for Economic Development* (International Union for Conservation of Nature and Natural Resources, Morges, Switzerland; and the Conservation Foundation, Washington, U.S.A.), Wiley, London. 252 pp.

Duley, F. L. (1939). Surface factors affecting the rate of intake of water by soils. *Proc. Soil Sci. Soc. Am.,* 4, 60–64.

Ellison, W. D. (1944). Studies of raindrop erosion. *Agric. Engng,* 25, 131–136.

Ellison, W. D. (1947a). Soil erosion studies. III Some effects of soil erosion on infiltration and surface runoff. *Agric. Engng,* 28, 245–248.

Ellison, W. D. (1947b). Soil erosion studies. VI Soil detachment by surface flow. *Agric. Engng,* 28, 442–444, 450.

Ellison, W. D., and Slater, C. S. (1945). Factors that affect surface sealing and infiltration of exposed soil surfaces. *Agric. Engng,* 26, 156–157, 162.

Elwell, H. A., and Stocking, M. A. (1975). Parameters for estimating annual runoff and soil loss from agricultural lands in Rhodesia. *Water Res. Res.,* 11, 601–605.

Elwell, H. A., and Stocking, M. A. (1976). Vegetal cover to estimate soil erosion hazard in Rhodesia. *Geoderma,* 15, 61–70.

Emmett, W. W. (1970). The hydraulics of overland flow on hillslopes. *US Geol. Surv., Prof. Paper*, **662-A**, 68 pp.

Epstein, E., and Grant, W. J. (1967). Soil losses and crust formation as related to some physical properties. *Proc. Soil. Sci. Soc. Am.*, **71**, 547–550.

Evans, R. (1977). Overgrazing and soil erosion on hill pastures, with particular reference to the Peak District. *Journal British Grassland Society*, **32**, 65–76.

Evans, R., and Morgan, R. P. C. (1974). Water erosion of arable land. *Area*, 6:221–225.

Evans, R., and Nortcliff, S. (1978). Soil erosion in north Norfolk, *Journal Agricultural Science*, Cambridge, **90**, 185–192

Foster, S. (1978). An example of gullying on arable land in the Yorkshire Wolds. *Naturalist*, **103**, 157–161.

Free, G. R. (1952). Soil movement by raindrops. *Agric. Engng*, **33**, 491–494, 496.

Free, G. R. (1960). Erosion characteristics of rainfall. *Agric. Engng*, **41**, 447–449, 455.

Greenland, D. J. (1965a). Interaction between clays and organic compounds in soils. Part I. Mechanisms of interaction between clays and defined organic compounds. *Soils and Fertilizers*, **28**, 415–425.

Greenland, D. J. (1965b). Interaction between clays and organic compounds in soils. Part II. Adsorption of soil organic compounds and its effect on soil properties. *Soils and Fertilizers*, **28**, 521–532.

Greenland, D. J. (1971). Changes in the nitrogen status and physical conditions of soils under pastures; with special reference to the maintenance of Australian soils used for growing wheat. *Soils and Fertilizers*, **34**, 237–251.

Greenland, D. J., Rimmer, D., and Payne, D. (1975). Determination of the structural stability class of English and Welsh soils, using a water coherence test. *J. Soil Sci.*, **26**, 294–303.

Gunn, R. D., and Kinzer, G. D. (1949). Terminal velocity of water droplets in stagnant air. *J. Meteorology*, **6**, 243–248.

Hjulstrøm, F. (1935). Studies of the morphological activity of rivers as illustrated by the River Fyries. *Uppsala Geological Institute, Bulletin* **25**, 221–527.

Hodgson, J. M. (1974). *Soil Survey Field Handbook* (Technical Monograph No. 5). Soil Survey, Harpenden. 99 pp.

Hudson, N. W. (1963). Raindrop size distribution in high intensity storms. *Rhod. J. Agric. Res.*, **1**, 6–11.

Hudson, N. W. (1971). *Soil Conservation*, Batsford, London. 320 pp.

Kar, A., and Bandyopadhyay, M. K. (1974). Mechanism of rills in an investigation in micro geomorphology. *Geographical Review of India*, **36**, 204–215.

Lal, R. (1976a). Soil erosion on Altisols in Western Nigeria. I. Effects of slope, crop rotation and residue management. *Geoderma*, **16**, 363–376.

Lal, R. (1976b). Soil erosion on Altisols in Western Nigeria. III. Effects of rainfall characteristics. *Geoderma*, **16**, 389–401.

Lamb, J., Carleton, E. A., and Free, G. R. (1950). Effect of past management and erosion of soil on fertilizer efficiency. *Soil Sci.*, **70**, 385–392.

Langbein, W. B., and Schumm, S. A. (1958). Yield of sediment in relation to mean annual precipitation. *Trans. Am. Geophys. Union*, **39**, 1076–1084.

Laws, J. O. (1940). Recent studies in raindrops and erosion. *Agric. Engng*, **21**, 431–433.

Lemos, P., and Lutz, J. (1957). Soil crusting and some factors affecting it. *Proc. Soil Sci. Am.*, **21**, 485–491.

Lillard, J. H., Rogers, H. T., and Elson, J. (1941). *Effects of Slope, Character of Soil, Rainfall and Cropping Treatments on Erosion Losses from Dunmore Silt Loam*, Technical Bulletin No. 72, Virginia Agricultural Experimental Station. 31 pp.

Lowdermilk, W. C. (1934). Acceleration of erosion above geologic norms. *Trans. Am. Geophys. Union*, **15**, 505–509.

Lutz, J. F. (1934). *The Physicochemical Properties of Soils Affecting Erosion*, Research Bulletin No. 212, Missouri Agricultural Experimental Station. 45 pp.

Lyles, L., Dickerson, J. D., and Schmeidler, M. F. (1974). Soil detachment from clods by rainfall: effects of wind, mulch cover, and initial soil moisture. *Trans. Am. Soc. Agric. Engrs*, **17**, 697–700.

Megahan, W. F. (1975). Sedimentation in relation to logging activities in the mountains of central Idaho. In: *Present and Prospective Technology for Predicting Sediment Yields and Sources*. Proceedings of Sediment-Yield Workshop, United States Department of Agriculture Sedimentation Laboratory, Oxford, Mississippi, November 1972 (Agricultural Research Service Report ARS-S-40), United States Department of Agriculture, Washington, D.C., pp. 74–82.

Mehta, K. M., Sharma, V. C., and Deo, P. G. (1963). Erodibility investigations of soil of Eastern Rajasthan. *J. Indian Soc. Soil. Sci.*, **11**, 23–31.

Middleton, H. E. (1930). *Properties of Soils which Influence Soil Erosion*, Technical Bulletin No. 178, United States Department of Agriculture, 16 pp.

Mihara, Y. (1951). *Raindrops and Soil Erosion* (in Japanese with English summary), Bulletin No. 1, National Institution Agricultural Science, Tokyo, Japan. 59 pp.

MAFF (1967). *Potential Transpiration* (Ministry of Agriculture, Fisheries, and Food, Technical Bulletin No. 16), HMSO, London. 77 pp.

Mosley, M. P. (1974). Experimental study of rill erosion. *Trans. Am. Soc. Agric. Engrs*, **17**, 909–913, 916.

Morgan, R. P. C. (1974). Estimating regional variations in soil erosion hazard in Peninsular Malaysia. *Malay Nature Journal*, **28**, 96–106.

Morgan, R. P. C. (1977). *Soil Erosion in the United Kingdom: Field Studies in the Silsoe Area 1973–75*. National College Agricultural Engineering. Occasional Paper No. 4, 41 pp.

Musgrave, G. W. (1954). Estimating land erosion—sheet erosion. *Int. Assoc. Scient. Hydrol. Pub.*, **1**, 207–215.

McIntyre, D. S. (1958). Permeability measurements of soil crusts formed by raindrop impact. *Soil Sci.*, **85**, 185–189.

Neal, J. H. (1938). *The Effect of the Degree of Slope and Rainfall Characteristics on Runoff and Soil Erosion*, Research Bulletin No. 280, Missouri Agricultural Experimental Station, 47 pp.

Palmer, R. S. (1963). The influence of a thin water layer on water drop impact forces. *Int. Assoc. Scient. Hydrol.*, *Pub.* **65**, 141–148.

Peele, T. C. (1937). The relation of certain physical characteristics to the erodibility of soils. *Proc. Soil Sci. Soc. Am.*, **2**, 97–100.

Peele, T. C., Latham, E. E., and Beale, O. W. (1945). *Relation of the Physical Properties of Different Soil Types to Erodibility*, Bulletin No. 357, South Carolina Agricultural Experimental Station, 31 pp.

Rai, K. D., Raney, W. A., and Vanderford, H. B. (1954). Some physical factors that influence soil erosion and the influence of aggregate size and stability on growth of tomatoes. *Proc. Soil Sci. Soc. Am.*, **18**, 486–489.

Rauzi, F., Fly, C. L., and Dyksterhuis, E. J. (1968). *Water Intake on Midcontinental Rangelands as Influenced by Soil and Plant Cover*, Technical Bulletin No. 1390, United States Department of Agriculture, Washington, D.C. 58 pp.

Renner, F. G. (1936). *Conditions influencing erosion on the Boise River Watershed*, Technical Bulletin No. 528, United States Department of Agriculture, Washington, D.C. 32 pp.

Ripley, P. O., Kalbfleisch, W., Bourget, S. J., and Coper, D. J. (1961). *Soil Erosion by Water, Damage, Prevention, Control,* Publication No. 1083, Research Branch, Department Agriculture, Canada. 34 pp.

Rose, C. W. (1960). Soil detachment caused by rainfall. *Soil Sci.,* **98,** 28–35.

Skidmore, E. L., Carstenson, W. A., and Banbury, E. E. (1975). Soil changes resulting from cropping. *Proc. Soil Sci. Soc. Am.,* **39,** 964–967.

Smerdon, E. T., and Beasley, R. P. (1959). *The Tractive Force Theory Applied to Stability of Open Channels in Cohesive Soils,* Research Bulletin No. 715, Missouri University Agricultural Experimental Station. 36 pp.

Smith, D. D., Whitt, D. M., Zingg, A. W., McCall, A. G., and Bell, F. G. (1945). *Investigations in Erosion Control and Reclamation of Eroded Shelby and Related Soils,* Technical Bulletin No. 883, United States Department of Agriculture,Washington, D.C. 175 pp.

Smith, R. M., Henderson, R. C., and Tippit, O. J. (1954). *Summary of Soil and Water conservation research from the Blackland Experiment Station, Temple, Texas,* Bulletin No. 781, Texas Agricultural Experimental Station. 54 pp.

Soil Survey Staff(1951). *Soil Survey Manual.* Handbook No. 18, United States Department of Agriculture, Washington, D.C. 503 pp.

Spomer, R. G., Heinemann, H. G., and Piest, R. F. (1971). Consequences of historic rainfall on western Iowa farmland. *Water Res. Res.,* 7, 524–535.

Stocking, M., and Elwell, H. (1976). Vegetation and erosion: a review. *Scottish Geographical Magazine,* **92,** 4–16.

Tackett, J. L., and Pearson, R. W. (1965). Some characteristics of soil crusts formed by simulated rainfall. *Soil Sci.,* **99,** 407–413.

Wischmeier, W. H. (1959). A rainfall erosion index for a universal soil loss equation. *Proc. Soil Sci. Soc. Am.,* **23,** 246–249.

Wischmeier, W. H., and Mannering, J. W. (1969). Relation of soil properties to its erodibility. *Proc. Soil Sci. Soc. Am.,* **33,** 131–137.

Wischmeier, W. H., and Smith, D. D. (1958). Rainfall energy and its relation to soil loss. *Trans. Am. Geophys. Union,* **39,** 285–291.

Wischmeier, W. H., and Smith, D. D. (1965). *Predicting Rainfall-erosion Losses from Cropland East of the Rocky Mountains,* Agricultural Handbook No. 282, United States Department of Agriculture, Washington, D.C. 47 pp.

Wischmeier, W. H., Smith, D. D., and Uhland, R. E. (1958). Evaluation of factors in the soil loss equation. *Agric. Engng,* **39,** 458–462, 474.

Woodburn, R., and Kozachyn, J. (1956). A study of relative erodibility of a group of Mississippi gully soils. *Trans. Am. Geophys. Union,* **37,** 749–753.

Woodward, L. (1943). Infiltration—capacities of some plant-soil complexes on Utah range watershed lands. *Trans. Am. Geophys. Union,* **24,** 468–473.

Yair, A., and Klein, M. (1973). The influence of surface properties on flow and erosion processes on debris covered slopes in an arid area. *Catena,* **1,** 1–18.

Young, R. A., and Wiersma, J. L. (1973). The role of rainfall impact on soil detachment and transport. *Water Res. Res.,* **9,** 1629–1636.

Zingg, A. W. (1940). Degree and length of land slope as it affects soil loss in runoff. *Agric. Engng.* **21,** 59–64.

Soil Erosion
Edited by M. J. Kirkby and R. P. C. Morgan
© 1980 John Wiley and Sons Ltd.

CHAPTER 5

Erosional processes of running water and their spatial and temporal controls: a theoretical viewpoint

J. B. Thornes

Reader in Geography, London School of Economics, UK

5.1. INTRODUCTION

This chapter is concerned with the mechanics of soil erosional processes and the ways in which they vary in space and through time. Some attention is also paid to the complex response of erosional systems to change as a result of lags operating within the system. Erosion occurs when the forces tending to entrain and transport materials exceed those tending to resist removal. The displacing forces operate through the agencies of wind, rain, and running water. This chapter relates to the last two.

The thickness of soil may, to a first approximation, be regarded as controlled by the relative rates of production and removal of material. Production occurs principally through the weathering processes whereas removal occurs through the entrainment and transport processes. Where the effects of these two groups of processes are equal, soil thickness remains constant (Figure 5.1). Practical soil erosion problems are mainly concerned with the conditions to the right of the equilibrium line. When the soil thickness is zero, the removal rate is limited to the production rate. This is the weathering limited case. Otherwise the rate of removal is limited by the entrainment or transport capacities. Most of the discussion will consider those cases where there is always material available for entrainment and transport.

Nevertheless an appreciation of soil erosion as an agricultural problem depends not only on the loss rate, but on how much soil is still available and this is partly a function of the original soil thickness and hence the 'normal' or 'geological' balance between production and removal. This in turn depends on

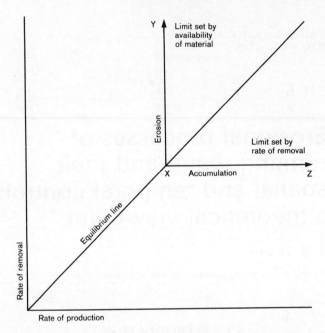

Figure 5.1. The relationship between soil thickness, rate of production of material by weathering and rate of removal by transport processes

chemical weathering processes and their relation to geological type. Moreover, the contemporary properties of soils, which have a bearing on the controls of erosion, such as infiltration rates, shearing strength and susceptibility to dispersion, are related to chemical processes particularly through the breakdown of organic materials. Therefore the chemical processes occurring in the soil are relevant to the soil erosion problem. Like the mechanical processes, the chemical process involved in soil erosion are intimately bound up with the hydrology of the soil, so that processes governing that hydrology need to be considered first.

5.2 BASIC PROCESSES

5.2.1 Infiltration, storage, and the production of runoff

When water is artificially ponded at the surface, the entry of water into the soil is initially very high and falls quite quickly to a steady rate called the final infiltrability, which is closely related to the saturated conductivity of the soil. If rainfall intensity exceeds this infiltration capacity then ponding at the surface occurs leading to the production of excess and eventually overland flow. During this process, except for a few millimetres at the surface, the soil remains unsaturated and so at a negative hydraulic potential and a sharply defined wetting front occurs. Migration of this front defines the final infiltrability. Dunin (1976) calls this type of ponded infiltration 'profile controlled'. It is represented by the characteristic profile curve defined theoretically by Phillip (1969).

Rainfall intensities are sometimes high enough to exceed final infiltrability but usually intensities of this magnitude are relatively rare, except perhaps in semi-arid environments (Kirkby, 1969). Excess produced by ponded type infiltration is called 'infiltration' or 'Hortonian' overland flow after R. E. Horton who envisaged it as the main agent involved in soil erosion (Horton, 1933).

More often intensities are lower than final infiltrability. Rubin (1966) defined two other cases, non-ponded and pre-ponded infiltration. When the ratio of intensity to capacity is less than unity the soil approaches a limiting soil moisture content throughout the wetted zone. The value of the limiting moisture content is that at which the associated capillary conductivity equals the intensity; increasing the intensity produces a higher soil moisture content.

Once the relative intensity exceeds unity a pre-ponding mode occurs in which, although the surface soil moisture is virtually at saturation, the soil pressure

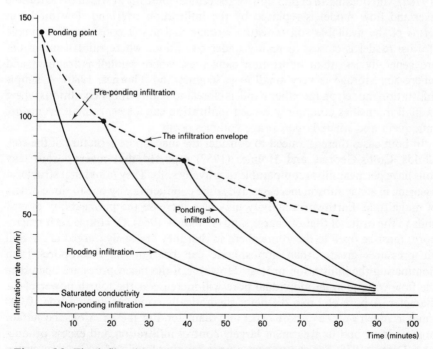

Figure 5.2. The infiltration envelope defining time to ponding for rainfalls of different intensities. Along the horizontal lines infiltration is occurring at a rate equal to the rainfall intensity (after Smith, 1972)

remains negative and so water cannot 'escape' from the soil onto the surface. Only when the soil pressure is zero at the surface may ponding actually occur. Thereafter the rainfall-ponded curve follows the artificially ponded curve (Figure 5.2). The curve defining the time to ponding for a given storage potential with various intensities is called the infiltration envelope (Smith, 1972).

This approach emphasizes the achievement of steady infiltration relative to rainfall intensities. Surprisingly, therefore, we find that some soils, although having very high infiltration rates and capacities, produce excess very quickly (minutes or even seconds). Such situations occur where there is a dramatic reduction in the infiltration rate after a given storage has been filled. This may result from the relative thinness of a highly permeable soil, for example in certain coarse semi-arid soils (Thornes, 1976) or where there is a sharp change in the pore size distribution. In this case, time to ponding is extremely short as a *result* of high infiltration rates rather than despite them. The Philip model has no provision for the limited storage often found. A model which can deal with this situation assumes that all rainfall is infiltrated until the storage is filled and any subsequent rainfall produces overland flow.

A comparison of these extreme cases has been made theoretically by Kirkby (1978a) who obtains an expression for the critical intensity at which the saturated overland flow model is replaced by the infiltration overland flow model in terms of the available soil moisture storage and a soil constant. The simple storage model is closest to reality under conditions where rainfall intensities are generally less than infiltration capacities, where rainfall is frequent and where soil storage is very small (e.g. Scoging and Thornes, 1980). A simple infiltration model, on the other hand, is closest to reality under conditions where rainfall intensities commonly exceed infiltration capacities, rainfall events are infrequent and where large storage volumes occur.

In both cases there is a need to consider the macropore structure of the soil. Childs, Collis-George, and Holmes (1957) reported that macroporous clay soils have permeabilities comparable to gravelly soils. They found that structural fissuring in a clay subsoil increases hydraulic conductivity by one to three orders of magnitude. Earthworm activity may also increase the permeability of clay soils to the order of that of coarse sands (Youngs, 1964). Of course such macropores must be open to the atmosphere so that they are easily purged of air. Soil air pressures greater than atmospheric can prevent the macropores from dominating the infiltration process. However, if the macropores are open then the flow volume in an individual pore will increase as the fourth power of the diameter. A pore of 1 mm diameter theoretically conducts downwards 10,000 times as much water as a pore of 0.1 mm diameter. For these reasons the surface organic layer and its treatment largely control infiltration and excess production. Dixon (1975) has shown that a wide ranging family of infiltration curves can be generated for a given soil profile merely by altering surface roughness and openness. Infiltration rates into tropical soils, where both high storage in the root mat zone and abundant macropores resulting from animal activity combine, are very high and conventional overland flow may consequently never occur (Nortcliff, Thornes, and Waylen, 1979).

When large structural cracks occur, such as inter-pedal joints, rapid downward movement of water through initially air-filled large vertical pores occurs thereby passing by the moist soil inside the peds, and this has been demonstrated using chloride breakthrough curves. Often in clay soils the smaller cracks may be closed off by swelling which leads first to concentration of the flow in large pores and eventually to the flow of water through rather than around the soil prisms. Schumm (1956) found that the shales of the Chadron formation, which are clay rich, actually behaved with a very high permeability, due to dessication cracks, so that little runoff occurred even with extremely high rates of water application. The role of macrostructures is to some extent controlled by the antecedant moisture content not only through storage but also because significant air pressures may develop which may reduce infiltration during the early stages. Linden, Dixon, and Guitjens (1977) found in practical experiments using border irrigation that infiltration rates could be reduced by as much as one third during the first ten minutes. This phenomenon is probably also an important control of infiltration into ephemeral channel beds during flooding.

5.2.2 Overland flow

In uniform open-channel flow on a shallow sloping bed expressions for velocity are normally derived from the Brahms assumption that the downstream component of weight of flowing water is just balanced by the resisting forces resulting from bed friction. For turbulent flow the resistance is assumed proportional to the square of the velocity leading to the Chezy or Darcy–Weisbach resistance formulae. In the former

$$v = c(R \cdot S)^{0.5} \tag{5.1}$$

where c is the Chezy roughness coefficient, R the hydraulic radius, and S the slope. In the Darcy–Weisbach formula

$$v = \left(\frac{2g}{f} \cdot R \cdot S\right)^{1/2} \tag{5.2}$$

where f is the roughness coefficient, which is sometimes related to the particle roughness. Typically (Wolman, 1955), f is given by:

$$\left(\frac{1}{f}\right)^{1/2} = 4.07 \log_{10}\left(\frac{r}{d_{84}}\right) + 2.0 \tag{5.3}$$

in which d_{84} is the particle size at which 84 per cent of the bed particles is finer. A plot of f against the Reynolds number (Re) of the flow shows that in the laminar range the Darcy–Weisbach f decreases with increasing R_e more or less linearly for channel flow. With fully turbulent flow the slope of the relationship decreases and under some conditions becomes negligible.

For laminar flow over smooth surfaces, the relationship is

$$f = 24/R_e \tag{5.4}$$

whereas for laminar flow over rough surfaces

$$f = k/R_e \tag{5.5}$$

where k is a parameter related to the characteristics of the surface and can become very large (40,000) for dense turf (Woolhiser, 1975). It is not feasible to measure the depth of flow over very rough surfaces or in field studies so that the friction factor must be obtained by inference from the steady state detention storage, from analysis of the rising hydrograph or by model optimization techniques. Some typical values are given in Table 5.1.

Table 5.1. Resistance parameters for overland flow (after Woolhiser, 1975)

Surface	Laminar flow	Turbulent flow	
	k	Manning's n	Chezy c
Concrete or asphalt	24–108	0.01–0.013	73–38
Bare sand	30–120	0.01–0.016	65–33
Gravelled surface	90–400	0.012–0.03	38–18
Eroded bare clay	100–5000	0.012–0.033	36–16
Loam soil			
Sparse vegetation	1000–4000	0.053–0.13	11–5
Short grass prairie	7000–10,000	0.10–0.20	6.5–3.6
Bluegrass sod	7000–40,000	0.17–0.48	4.2–1.8

The more general form given by Horton (1945) is

$$v \propto R^m S^n \tag{5.6}$$

in which values of m and n respectively are 2.0 and 1.0 for laminar flow, and 0.67 and 0.5 for turbulent flow. In shallow flows, such as on hillsides, the conditions are dominated by roughness elements. Horton described this type of flow as 'subdivided' and found values of m and n to be >0 and <0.5 respectively.

In earlier work by Horton *et al.* (1934) the basic behaviour of sheet flow was studied through the rating equation

$$q = KD^m \tag{5.7}$$

where q is discharge per unit width, D is flow depth and K and m are constants for a given set of conditions. For fully turbulent flow, using Manning's equation, they argued that m should be 1.66 and for laminar flow it should be 3.0.

The most extensive results are due to Keulegan (1944) who derived a complete solution for spatially varied discharge over a sloping surface and Izzard (1944) who applied it to experimental data. Recently Muzik (1974) has compared experimental hydrographs with those produced from kinematic wave modelling. The equation of continuity for the unsteady spatially varied flow is written as

$$\frac{dq}{dx} + \frac{dy}{dt} = i \qquad (5.8)$$

in which q is the discharge per unit width, y the depth of flow and i the rainfall excess. This is linked with the flow equation

$$q = \alpha y^m \qquad (5.9)$$

where α is a constant which includes slope. For a slope of length L (Figure 5.3) the time (t_s) to equilibrium overland flow is

$$t_s = \left(\frac{L}{\alpha i^{m-1}}\right)^{1/m} \qquad (5.10)$$

The depth of flow for this equilibrium at the bottom of the profile is given by:

$$y = i \cdot t_s \qquad (5.11)$$

and the discharge q at the bottom of the slope is given by:

$$q = (i \cdot t)^m \qquad (5.12)$$

where t is less than the equilibrium, and for longer durations

$$q = L \cdot i \qquad (5.13)$$

which is, of course, the usual form.

These solutions are for rainfalls of infinite duration. For rainfalls of finite duration, the time required for recession t_a after the cessation of the rainfall is given by solution of the equation

$$L = \alpha y^{m-1}\left(\frac{y}{i} + mt_a\right) \qquad (5.14)$$

where L is the length required for equilibrium.

Figure 5.3 shows the result of experimental and theoretical simulation obtained by Muzik for a rainfall intensity of 38.4 mm/hr applied over 50 seconds. The hydrographs show good agreement and similar results have been obtained from field experiments on semi-arid soils in Spain (Scoging, 1978). The value of m in Muzik's experiment which best fitted the observed times to equilibrium was 1.666, exactly the theoretical value anticipated by Horton *et al.* (1934).

Emmett (1970) carried out both laboratory and field experiments on the overland flow process. In the laboratory flows were found to be in the Reynolds

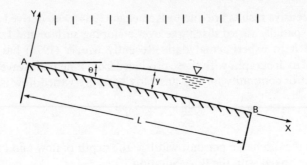

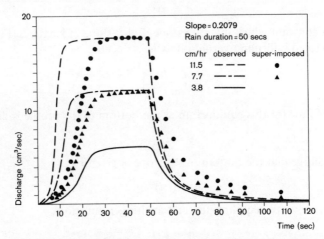

Figure 5.3. Terminology and simulated and observed hydrographs according to the model and experiments by Muzik (1974)

number range 1500–6000. Flows over smooth surfaces, both turbulent and laminar, were found to have exponents very similar to those predicted by Horton *et al*. With rough surfaces the values of *m* were less than those predicted theoretically and this was attributed to a more complex and rapid increase in depth of flow due to roughness. Variations of flow depth under different slope conditions were also found to be quite strong. In the field greater depths of runoff were also experienced. The higher roughness coefficients correlated poorly with ground slope and showed a ten-fold increase when compared with laboratory surfaces.

Flow is usually disturbed by falling rain. Impacting the surface, the raindrops impart momentum to the flowing water and its turbulent intensity increases. These effects tend, for a constant Reynolds number, to increase with rainfall intensity. For a constant rainfall intensity the effects diminish at higher Reynolds numbers. Most laboratory work shows that the friction factor is appreciably larger when runoff is affected by falling rain and Savat (1977) reported that this effect does diminish as slope increases. Increase in the friction factor leads to an increase in the flow depth. This effect in Emmett's experiment was appreciably greater than the 17 per cent measured by Parsons (1949); 50 per cent for Reynolds numbers of 200 and about 65 per cent for Reynolds numbers of 1000. Here the isolated effect of the artificial rain was to double the friction factor over that for flows without rainfall. Generally speaking, with rainfall, as the flow passes from laminar to turbulent, there should be a drop in the flow depth since the effects of the friction factor decreases. Woolhiser *et al.* (1971) found this control was less on natural (gravel) than artificial (Butyl rubber) surfaces under converging flow. Under falling rain it appears that the relationship between Reynolds number and the Darcy–Weisbach friction factor, modelled as $f = k/R_e$, may be obtained from

$$k = k_0 + A(i/2.54)^b \tag{5.15}$$

where i is the rainfall intensity in cm/hr and A and b are empirical parameters, where the coefficient A is of the order of 10 and the exponent b is approximately 1. This result becomes insignificant for vegetated surfaces according to Woolhiser (1975).

The approach to vegetal effects has mainly been empirical (Langbein and Schumm, 1958) or inferential (Burkham, 1976). These approaches, relating total sediment yield or channel scour to vegetation, do not differentiate between the mechanical (roughness) effect and the strictly hydrological effect referred to earlier. Melton (1965) attempted to introduce vegetation effects through a correction factor in the Manning equation, though Ree (1958) has shown that this retardation coefficient is not valid where the vegetation penetrates through the flow, the usual case in overland flow.

The flow presents a drag force on each plant which is proportional (through a drag coefficient) to the average approach velocity of the water to the plant and the projected area of the plant in the streamwise direction. By incorporating this in the total momentum equation for flow under steady uniform conditions the effects of vegetation can be approximately defined. Vegetation density is defined by

$$D = \Sigma Ai/(AL) \tag{5.16}$$

in which ΣAi is the projected area of the vegetation in the streamwise direction, in a channel of length L, with a cross-sectional area of flow A. If this density is held constant as depth of flow changes then the roughness (expressed by the Manning coefficient) increases in proportion to the 2/3 power of the hydraulic radius.

In fact with vegetation which is fully submerged and can bend, the roughness is higher at low flows because the vegetation is unable to bend. As velocities

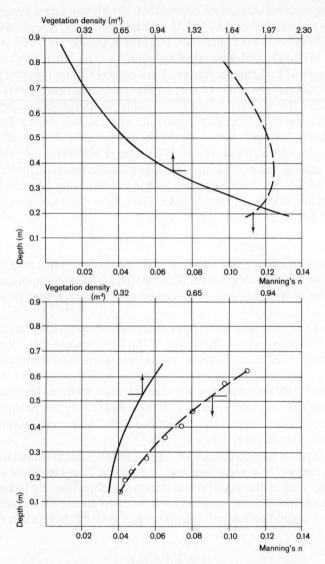

Figure 5.4. The variations in Manning's roughness parameter for two row crops according to depth of flow. The
upper diagram is for a thinly seeded wheat stand in which
the vegetation density decreases with depth (solid line)
whilst the roughness coefficient remains fairly constant.
The lower diagram shows the same two curves for a variety
of sorghum (after Petryk and Bosmajian, 1974). Solid curves
show vegetation density and broken curves Manning's *n*

increase there is a reduction in roughness, as observed by Palmer (1946). The assumption of constant vegetation with depth may be reasonable where, as in flood plains when the depth increases there is a decrease in tree trunk area, but an increase in effective area due to branches and foliage. On hillslopes, under crops such as wheat and sorghum, the effect is more complicated as indicated by Figure 5.4. If the drag coefficient is assumed to be unity, then the velocity is expressed by

$$V = C^{0.5}s^{0.5} \qquad (5.17)$$

where C is a vegetation cover term involving the inverse of vegetation density (Petryk and Bosmajian, 1975).

Tussock vegetation, coarse particles (stoniness) and transverse rills all present roughness elements which normally project through the flow, and in this respect their effect is more akin to islands projecting above stream flow in wide channels. This produces unsteady flow with surging, an effect which is also produced with large volumes of leaf litter. The actual boundary resistance is comparable to that for open channel flow but the form resistance due to these effects is much higher.

5.2.3 Subsurface flow

Subsurface flow is of crucial importance in determining the character of soil erosion through its effects on relative availability of surface water as conventional or saturated flow. It is also important as an agent of weathering which controls the erodibility of materials as well as their hydraulic characteristics through the transport of minerals in solution and through exchange mechanisms. In the gross form, that is the drainage of hill-slopes by pipes, it also determines the location of active gullying (see Section 5.2.4) and finally, fines may be translated mechanically by subsurface flows. Subsurface flow is well documented in *Hillslope Hydrology*, edited by M. J. Kirkby (1978), and a relatively brief treatment is included here, principally as recognition of its importance in all aspects of soil erosion.

Water movement in the soil occurs as a result of potential differences by migration of liquid water and to a lesser degree gaseous diffusion, across the pore spaces which make up the soil. The potential is provided by head differences between locations in the saturated zone and resistance to movement is provided by the pore structure. In a saturated zone, therefore, potential lines tend to be normal to the force of gravity, and flow takes place vertically or parallel to the surface under Darcy's Law, which is

$$u = -K \cdot \frac{dh}{dx} \qquad (5.18)$$

where u is the mean discharge per unit cross-section, h is height above a given level and x the distance over which height is measured. K is the hydraulic conductivity, which comprises a fluid factor, γ/p, with γ the specific weight of the fluid and p the dynamic viscosity, and a medium factor k which is called the intrinsic permeability. Natural values of k for saturated media are of the order of 10^{-12} to 10^{-10} m². The normal range of apparent groundwater velocities corresponding to these values of permeability is from $2\,\text{m}\,\text{yr}^{-1}$ to $2\,\text{m}\,\text{day}^{-1}$.

In unsaturated conditions the collection of pores making up the flow path is only partially saturated and the hydraulic conductivity is a function of the degree of saturation. In addition, the distribution of pores creates a capillary suction which increases with height above the water table up to some threshold value. This capillary suction is a function of the degree of saturation, as is the hydraulic conductivity (Figure 5.5) so that flow depends not only on gravitational head.

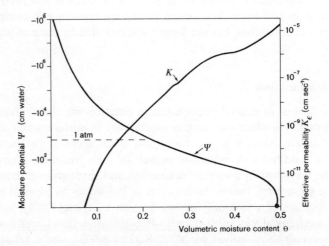

Figure 5.5. Typical curves showing the relationships between effective permeability and moisture potential as functions of volumetric soil moisture content

In a hillslope during rain, there may be lateral movement (throughflow) under saturated conditions. This can continue after the rainfall has ceased. A sharp decrease in permeability with depth encourages this kind of flow, which may contribute to the growth of a saturated wedge at the foot of the slope as envisaged by Kirkby and Chorley (1967) and demonstrated by Weyman (1973) and Harr (1977) amongst others. In addition isolated upslope saturated zones may occur. These lead to important spatial variations in the production of excess surface water.

5.2.4 Channel flow

Water flowing in rills and gullies, where the depth is usually appreciably greater than the coarsest roughness elements, occurs under conditions very similar to conventional open channel flow, the processes of which are well known (Chow, 1959). The main difference lies in the rapid rise to a peak value and an attenuated, but still rapid, decrease as the flood wave passes. This reflects the very rapid response to rainfall, the relatively weak significance of base flow components in perennial headwater areas and the low or non-existent role played by storage in ephemeral channels. Three channel flow hydrographs for small channels in headwater areas in an area undergoing construction (Gregory, 1974), an ephemeral (Thornes, 1976), and a tropical catchment (Nortcliff and Thornes, 1977) are shown in Figure 5.6. All have shapes typical of those which can cause extensive flood damage.

Another important feature of the short-lived flash flooding of headwater areas is the very rapid response time. In ephemeral areas this may reflect the very immediate translation of the hillslope hydrograph to the channel but in wet areas it is accounted for by the development of saturated overland flow in the immediate channel environment. Once the flow is in the channel it responds, at least in very small catchments, more or less linearly to catchment area and to the network characteristics. In ephemeral headwater channels, however, complexities arise from the loss of water into the channel by bed infiltration. Keppel and Renard (1962) reported that such losses were related in a power function to average peak discharges. Basically, the depth of flow controls hydraulic head, the width of flow controls infiltrating area and the speed of flow determines whether or not the appropriate thresholds have been passed for the removal of fines from the bed. All these factors are concerned with rate. The time over which the infiltration occurs may be regarded essentially as that required for the passage of the hydrograph and as such is a kind of 'opportunity time'. The problem is comparable to the advancing wave front in ponding irrigation (Thornes, 1977; Butcher and Thornes, 1978).

Channel flow in gullies and rills is normally rapid and accompanied by high sediment yields. Slopes are often in excess of 10 per cent in rills and as much as 3–5 per cent in headwater channels. Flows as high as 3 m/sec are recorded and suspended sediment concentrations may not infrequently reach 300–500,000 p.p.m., i.e. verging on mudflows. Under these hyperconcentrations viscosity is markedly increased (Beverage and Culbertson, 1964), the capacity of streams to carry coarser sediments is appreciably enhanced and the friction factor is frequently diminished (Simons *et al.*, 1963). These conditions help to explain the capacity of flash ephemeral flows to carry, occasionally, particles several hundred times larger than the mean bed material size.

The conventional conditions of channel flow and the mechanics governing them also tend to break down when flow occurs over lip- or headcuts in gullies

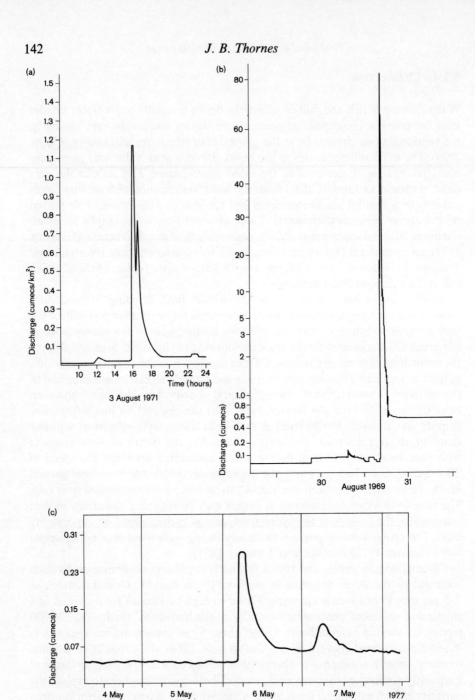

Figure 5.6. Strongly peaked hydrographs representing (a) headwater areas in a catchment undergoing urban construction (Gregory, 1974), (b) a semi-arid channel catchment and (c) a tropical catchment

and rills, and in scourholes. Unfortunately most observations of these phenomena are restricted to relating gully growth at the headcut to total volume of
flow.

5.2.5 Processes within the soil

In the relatively long term, the mineral composition of the soil is controlled by
weathering and this, in turn, is related to the relative stability of the various
minerals in the parent rock, the timing and nature of waterflow through the soil
and the effects of supply and withdrawal of minerals from the soil by organic
processes including leaf fall and the uptake of nutrients. In the medium term,
the soil characteristics may to a large extent be controlled by the vegetation
which occurs on them, at least in the horizons most sensitive to soil erosion.
For example White and Riecken (1955) contrast forests and grasslands in the
North American continent. The forest soils commonly show greater leaching
of cations, correspondingly lower pH and greater clay translocation than do
adjacent grassland sites. In addition organic matter content is higher at greater
depths in the grassland sites, presumably due to the more rapid breakdown of
root matter. Contrasts are also demonstrable beneath individual trees, partly
due to the chemical composition of stem flow, partly to the nature and type of
litterfall, as illustrated by Gersper and Holowaychuk (1971) and Zinke (1962).
In the short-term the soil characteristics are influenced by the permeating solutions through the cation exchange process, by the pH and CO_2 status of the soil
and by the role played by the breakdown of organic materials. These conditions
largely determine the shear strength properties of the soil, their capacity for
change in volume through swelling, the relative ease with which they can
flocculate and deflocculate and the water repellency of the soil. Each of these
properties significantly affects the erosional processes.

The propensity of the individual soil particles to aggregate has long been
recognized to be important in soil erosion (Middleton, 1930; Bryan, 1976).
Some of this tendency can be accounted for by the readiness of small colloidal
particles to be attracted by fluctuating dipole bonds called van der Waals
forces. The attractive forces are given as

$$F = \frac{C}{d_\mu^4} \text{ dynes cm}^{-2} \tag{5.19}$$

in which d_μ is the separation distance and C is a constant, which is a function of
the dielectric constant of the medium separating the particles. This theory,
due to Lifshitz, is for parallel plates which is not a probable arrangement of
particles in a clay soil where particle corners, edges, and projections are likely
to be in contact with adjacent particles. For the interaction of a flat plate and
one or two spheres the attractive force decays as $1/d^3$ instead of $1/d^4$ (Ingles,
1962).

The soil particles, especially clay particles, also produce an important repelling force generated in the double electric layer. This phenomenon is extremely important in cation exchange. At the boundary of negatively charged clay particles, in the presence of a cation rich solution, the positively charged cations are attracted to the surface of the clay and because the adsorbed cation provide a much higher concentration at the surfaces of particles, there is a tendency for equalization of concentration by diffusion away from the boundary. The extent of this outward movement is restricted however by the negative electric field originating in the particle surface. The result is to produce a negative exponential decay in the concentration of cations away from the surface, described by the Boltzmann equation. The anions are repelled from the surface, on the other hand, so that their concentration tends to *increase* away from the surface. In general, the thicker the double layer the lesser the tendency of particles to flocculate and the higher the swelling pressure in cohesive soils (Mitchell, 1976). Any increase in the electrolyte concentration tends to reduce the surface potential and increase the rate of decay of potential away from the surface, so that swelling tendency depends on the electrolyte concentration. The effect of valency is the same, an increase in valency surpresses the potential between interacting plates and therefore leads to a decrease in the interparticle repulsion.

The net combined effect of the van der Waals force and the double layer effect is then a function of distance as shown schematically in Figure 5.7, after Mitchell (1976). The energy of repulsion is sensitive to change in electrolyte concentration, cation valence, dielectric constant and pH; whereas the attractive energy is sensitive only to changes in the dielectric constant and temperature.

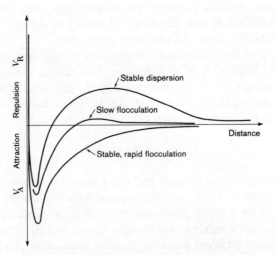

Figure 5.7. Combined effects of van der Waals and
double electrical layer (from Mitchell, 1976)

The electrolyte solution of the soil is heavily dependent on the concentration of unadsorbed cations it contains, and this in turn will depend on external factors such as soil drying and wetting, or plant consumption. Exchange of ions between those in solution and those adsorbed to the colloidal surfaces (mineral and organic), particularly the exchange of hydrogen ions in the solution for cations adsorbed to the surface, fundamentally influences the soil properties and hence their susceptibility to erosion. The smaller cations tend to replace larger ones under average solution conditions, so that the cation have relative degrees of mobility. For example potassium tends to be replaced by calcium so the former is more mobile. The relative rates of exchange are important in relation to residence time of water. In general kaolinite minerals show an almost instantaneous reaction, illites may need a few hours, while smectites (montmorillonite) require much longer because most of the exchange occurs in the interlayer region.

From double layer theory and from empirical relations the Gapon equation is used to express the relative proportions of monovalent and divalent ions in the soil. Following Mitchell (1976) if the subscript s refers to the equilibrium solution, M and N are monovalent concentrations and P refers to the concentration of divalent ions, then

$$\left(\frac{M^+}{N^+}\right)_s = k_1 \left(\frac{M^+}{N^+}\right)_e \tag{5.20}$$

$$\left(\frac{M^+}{P^{2+}}\right)_s = k_2 \left[\frac{M^+}{(P^{2+})^{1/2}}\right]_e \tag{5.21}$$

where k_1 and k_2 are selectivity constants. As will be seen the equations predict the activity ratio of ions in the solution from the proportion of ions on the exchanger in terms of a single constant that may be determined experimentally. As shown later (Section 5.2.6) the proportion of sodium in the adsorbed layer is critically important in the stability of clay soils expecially in relation to the development of pipes. This is expressed by the exchangeable sodium percentage which is related by a form of the Gapon equation to the sodium adsorption ratio. The latter can be determined from chemical analysis of a pore water sample. The equation is

$$\left(\frac{Na^+}{Ca^{2+} + Mg^{2+}}\right)_s = k \left[\frac{Na^+}{[(Ca^{2+} + Mg^{2+})/2]^{1/2}}\right]_e \tag{5.22}$$

in which

$$\left[\frac{Na^+}{[(Ca^{2+} + Mg^{2+})/2]^{1/2}}\right]_e = SAR(meq/litre)^{1/2} \tag{5.23}$$

In addition to the changes in soil properties by the leaching process and by changes in the mineral ions in solution, the organic matter in the soil plays an important role in the physical and chemical processes. The organic matter content correlates with such properties as plasticity, shrinkage, compressibility, permeability, and strength. These effects result from the role of organic matter as a colloid onto which cations may be adsorbed, the adsorption of organic particles onto mineral particles and its effect in coating particles and providing bonding between them. It forms the basis of some soil conditions which are applied to enhance the strength of the soil against erosion. There is also some evidence that organic matter induces water repellency. In desert communities the soil beneath the canopy of shrubs was found by Adams *et al.* (1970) to be water repellent whereas the area beyond the canopy was completely wettable. This results from the activity of fungi and basidomycetes beneath the canopy, presumably due to slightly higher moisture contents. Savage (1975) asserts that year-round water repellency is associated with woody vegetation whilst seasonal variations in water repellency relate to moss-type plants. Several workers have indicated that extracts of various plants are capable of causing water repellency and other have indicated its association with high levels of fresh decomposing organic matter. Certainly the effect of fire seems to induce high water repellency when there are high levels of organic litter and low colloid content in the soil. Finally we note that the effect of water repellency on water movement is complex. Water repellency may actually decrease with time during the infiltration process partly due to solution of the substances which lead to repellency; partly due to an incomplete covering so that some water may be adsorbed on the wettable sites. Notwithstanding this, the movement of the wetting front during horizontal infiltration into soil columns packed with wettable soils can be as much as 25 times faster than in columns packed with a similar water-repellent soil. On the other hand water repellency increases in soils with an increase in the soil suction (Scholl, 1971). It was concluded by De Bano (1975) that the slow uptake of water at the beginning of infiltration is probably quite important from the point of view of soil erosion but only of marginal significance for total runoff volumes.

5.2.6   Sediment entrainment—erodibility

Sediment entrainment into flowing water is the most important aspect of soil erosion. It proceeds by splash, detachment by running water, mass failure and creep and by miscellaneous processes such as suffusion and the burrowing activity of animals. The processes act on a substratum which may comprise either weakly consolidated bedrock or soil produced by weathering. Soil erodibility is dynamic in that it may change through a storm, from season to season, with the effects of tillage and manure application and so on. It is related to the properties of materials, vegetational, and climatic conditions.

The erodibility of a soil is essentially related to the stability of soil aggregates rather than its textural composition, though the two are not unrelated, and to the cohesive forces holding the aggregates together. This is partly because many aggregates are too large to be removed by low velocity runoff so that removal of material will be confined to smaller aggregates and separates. However aggregate density and shape is as important, if not more important, than size. This means that entrainment of aggregates of a given size may occur at velocities lower than those expected by standard empirical curves, such as the Hjulström (1935) curve. This has been experimentally observed by Bryan (1976) though some of this difference may be partially ascribed to difference in the flow conditions of deep channels and hillsides. In the same paper Bryan observed that subsurface aggregates performed more akin to comparably sized quartz particles.

With large stable aggregates infiltration can take place more readily reducing the net volume of surface runoff. The resistance offered by an aggregate to the impact of falling water drops (a common test of stability) will reflect its response to wetting. Aggregates which disperse or slake in water will offer little stability. Slaking involves the destruction of aggregates due to compression of air in the dry soil ahead of the wetting front. Bryan (1969) found this process to be of appreciable significance early in a rainfall cycle and to be followed by the development of a compact and relatively stable crust. Similarly aggregates containing important particles of clay may suffer from swelling of the oriented clays resulting in breakdown and a negative correlation occurs between the degree of clay orientation and aggregate stability (Imeson and Jungerius, 1977). Organic compounds influence aggregate stability leading to more stable aggregates. It is found, for example, that a greater biological activity of forest soils is reflected by the relatively large numbers of primary aggregates when compared with farmland soils.

The formation of surface crusting is important in conditioning the entrainment processes. Three types of surface sealing processes appear to be important: raindrop impact, interstitial straining due to evaporation, and chemical crusting. The first has been described in detail by Farres (1978). He defined crusts as areas with a continuous particle cover without any definable aggregate boundaries. Initially the rate of areal development of crusts in his experiments was found to be slow, the rate then became very rapid and finally slowed again until 100 per cent cover was reached (Figure 5.8). These stages were related to the reduction of the internal aggregate strength of the particles during the first period to a value below the stress applied by the falling raindrops. The second phase involved rapid break up of the aggregates into the fundamental parent particles followed by organization of the particles in a surface cover of closely packed individuals that give the crust. The thickness of the crust, once developed, remains fairly constant, presumably as a result of the sealing off of lower horizons.

The effect of crusting is not only on the entrainment of particles but also to some extent on infiltration. McIntyre (1958) found two layers in the surface seal, a washed-in zone of 1.5 mm overlain by a clay seal 0.1 mm thick. The latter was found to be 10 times more effective than the former in reducing infiltration. Hillel and Gardner (1970) have also shown that the thickness of the crust through time is vital to the study of infiltration processes. When clay seals are generated they may encourage high rates of surface runoff unless the soils become so dry that surface dessication cracks occur, at which point the cracks then dominate the infiltration process.

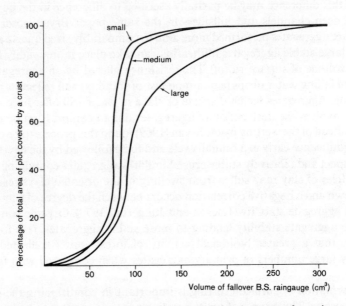

Figure 5.8. Changes in the degree of areal development of crusting through time for large (mean diameter 7.7 mm), medium (3.4 mm) and small (1.7 mm) aggregate sizes (from Farres, 1978)

The mechanism of clay-failure by surface forces is different from that prevailing in non-cohesive granular soils. In particular it is controlled by electrochemical as well as mechanical forces as described earlier (Section 5.2.2). The mechanical effects stem from the structural organization of the particles. The clay particles have a thin needle or plate shape with one or two of their dimensions much longer than the other two. The edges of each particle are attached to some low spot of the surface irregularities of some other particle and the two linked by electrochemical forces and adsorbed water. It is the weakest of these links which breaks first, leading to a shearing of aggregates of otherwise tightly-held particles. Electrochemical changes may lead to significant changes in aggregate resistance to erosion through time. Grissinger and Asmussen (1963) found that the erosion resistance of clay soils increased with the time the clay was kept in a wet state as a result of hydration processes. This may occur in four ways (Low, 1961): hydrogen bonding, hydration of exchangeable cations, attraction by osmosis to the highly concentrated cations in the electrical double layer and by charged-surface dipole attraction. When the clay is first wetted the free water releases the bonds between the particles but as free water is adsorbed and clay minerals hydrate, the bond is strengthened. Because hydration is a slow process a clay sediment can be eroded for some time before the process reaches its full strength. By merely leaching salt from a clay, on the other hand, the strength may be reduced to less than 1 per cent of its original value, even though it may create little change in the fabric. This is a result of decrease of the electrolyte concentration and increase in the double layer thickness as described in Section 5.2.5. The identical treatment of the different types of clay, e.g. illite and smectite clays, may have however totally different effects on their strengths.

5.2.7 Rainsplash

Rainsplash occurs essentially as a result of water drop impact forces. A raindrop falling on wetted soil produces a splash-crown, each drop of which is made up of one or two particles of soil encased in a film of water. The drop describes a parabolic curve, moving laterally about four times its height. The total mass carried on a flat surface decays exponentially with distance from the point of impact. Splash in a downslope direction travels further than in the upslope direction, about three times as far on a 10 per cent slope. With soil containing coarse fragments, armouring of the slope may eventually result from downslope movement of the saltating fines. In addition, particles are ejected by impact of particles arriving from upslope. The main sources of energy for this process are the kinetic energy of falling rain and gravitational (potential) energy.

In a large tropical storm a deadweight of 350,000 kg of water might fall in a thirty minute period over one hectare. Under the usual intensity–energy relationships the energy would be equivalent to approximately 10 million joules per hectare (Stocking, 1977). The computation of energy of rainfall depends upon a knowledge of distribution of drop sizes, terminal velocity and intensity of rainfall. Wischmeier and Smith (1958) and Hudson (1965) have both expressed kinetic energy as an empirical function of intensity. This relationship has been widely used to model splash and has been empirically calibrated and validated.

Where erosivity is assumed constant the net flux of soil due to splash is a function of slope since mean distances moved increase with slope gradient. This effect is shown by data from Mosley (1973) given in Figure 5.9. However, for slopes steeper than 33 per cent, Foster and Martin (1969) found that, depending on soil's bulk density, detachment by raindrop impact rose to a maximum and then decreased again on steeper slopes. Moreover, at constant erosivity, the rate of splash detachment appears to be a time-dependent function. In some cases the rate increases rapidly at the beginning, reaches a peak and then decays exponentially to a steady rate. Occasionally it decreases through time.

The significance of splash under conditions of overland flow depends on the depth of flow. Where this exceeds about three drop diameters the flow essentially protects the soil from raindrop impact. However, the raindrops increase the transporting potential of the flowing water by lifting detached particles into the flow, so that erosion is greatest where a thin film of water is present.

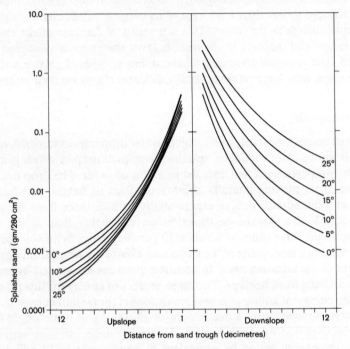

Figure 5.9. Weight of splashed sand as a function of distance from an experimental sand trough for angles of 0°, 5°, 10°, 15°, 20°, and 25° (from Mosley, 1973)

Moeyersons and De Ploey (1976) describe another type of splash process called splash creep which they observed on weathered granite slopes. After the finer components had been splashed out by saltation a coarser fraction remained which underwent a creep-like movement under raindrop impact. Unlike saltation splash rate (proportional to steepness), creep-wash showed an effect due to slope length, implying a runoff discharge effect while still maintaining discontinuous movement of individual particles. Vegetation effects operate primarily in the role of protector. Both Imeson (1977) and Imeson and Kwaad (1976) found a significant seasonal effect in the amount of rainsplash under deciduous forest which results from the litter-fall and decay cycles. The former and Yair (1973) both found the effects of animal burrowing to be important to splash amounts through their effect on bare soil exposed.

5.2.8 Entrainment by flow

The shearing detachment of sediments by flowing water is probably the most basic process apart from transport itself. Where flow is concentrated into rills total sediment removed is several orders of magnitude greater than unrilled flow. For the latter the term sheetwash is used, though in fact flow is rarely, if ever, in the form of a continuous sheet over the surface. Rather it occurs as surges or subdivided flow, checked by small obstacles and continually shifting in location (Schumm, 1956; Emmett, 1970). Detachment of non-cohesive sediments is generally assumed to take place as a result of the drag exerted by differential stress on the upstream and downstream faces of the particle, by Bernouilli lift from differential stress in the horizontal plane and by vertical turbulence resulting in sharp bursts of lift at the boundary. These effects are treated in the regular texts on sediment transport (e.g. Raudkivi, 1967) and form the basis of the standard equations for sediment transport. In the latter, emphasis is laid on the tractive shear stress developed by the velocity of the flowing water and expressed in terms of the depth of flow and slope. This usually proceeds through the adoption of one of the velocity equations, such as the Manning equation, in which velocity is expressed as a function of overland flow generation. The Horton model adopts this procedure and a development of it is discussed by Carson and Kirkby (1972). Estimated shear stresses for channel flow in rills are several times higher than typical critical shear stresses for average agricultural soils. For clay soils Partheniades (1965) found that the erosion rate depended strongly on the increase of average bed shear stress past a threshold value. He derived a detachment capacity equation based on the assumption that the bed shear varied as a normal distribution with time, yielding a square increase of detachment with average shear stress.

Practically, there is a need to incorporate effects other than straightforward tractive shear stress. One important effect is that of sediment concentration. Meyer and Monke (1965) and Willis (1971) noted empirically that the rate of detachment at the head of a non-cohesive bed is a linear function of transporting capacity deficit, i.e. they assume that

$$\frac{dG_f}{dx} = C(T_c - G_f) \tag{5.24}$$

in which G_f is the sediment being transported, T_c is the total capacity for sediment transport and C is a constant. This model is considered in further detail in Chapter 6.

In unrilled areas conventional wash produces shear stresses so small that flow can, in theory, detach only a small percentage of the particles when compared with rainsplash. However, the practical conditions are such that the particles are usually protruding through the flow. In subcritical flow a coherent laminar sublayer develops and the flow remains rough even in the absence of rainfall. Under these conditions the Bernouilli pressure force acts on the particles suspended in the boundary layers and flow separation and wake forces are characteristic. If the flow is supercritical then drag forces become relatively more important. As drag is preferentially imposed, the larger particles are preferentially entrained. These conditions have been experimentally verified by Savat and Poeson (1977) in a study of entrainment of sand materials from the Kalahari Sands. At high rainfall intensites, each particle may be covered with a thin film of water over its upper surface, held by surface tension. Under these conditions the development of scour downstream of coarse particles may induce movement and this process has been given the name run-off creep. It may be that entrainment of fine particles behind vegetation stems occurs as a result of similar 'wake' effects. Empirical evidence (De Ploey, Savat, and Moeyersons, 1976) suggests that detachment in shallow flows increases non-linearly with slope angle up to a relatively low angle after which the concentrations remain relatively constant. In such shallow flows, where there is a constant detachment and deposition of particles and where infiltration losses lead to complex hydrology there is no reason to expect that transport rate is dependent on distance from divide. This is not, of course, true in relatively deep flows.

Although very restricted in occurrence and minor in total effect, the flow of water in gully plunge pools is quite important in detaching material and causing recession of the headcut. When water is flowing over the lip, in addition to free fall and the associated scour, part of the flow clings to the wall by surface tension and leads to severe undercutting. Ireland et al. (1939) found this to be a very important process in headcut migration.

5.2.9 Mass movement

Although Leopold, Emmett, and Myrick (1966) indicate that in the semi-arid environment they studied, mass movement is relatively unimportant, elsewhere it may be very important, at least in specific locations and at specific times. In so far as they tend to be regarded, spatially as well as temporally, as extreme events, they have received relatively little attention in the soil erosion literature. In gully development they may be particularly important. In non-cohesive sediments the rapid incision by gullies may be constrained by influxes of coarse material so that the slopes are maintained at the angle of residual shear associated with the material (Thornes, 1975). In cohesive sediments gullies tend to build steep-sided gorges where slab failure may be common. Not unusually these are either undercut by the stream or fail due to seepage along the base of the cliff, both instances leading to eventual collapse. In areas of very high relief the process may lead to deep seated failures, involving large rock masses and whole hillsides.

On the hillslopes, shallow planar slides are also commonly associated with soil erosion. These are normally initiated during extreme storms and result from high pore water pressures arising from heavy rains. For example Temple and Rapp (1972) observed more than 1000 fresh slide scars in the Mgeta area of Tanzania following the February 1970 extreme rainfall event. The shallow slides were typically 5–20 m wide, 10–50 m long and up to 2 m deep. These shallow failures are usually accounted for by the Terzaghi (1950) model in which high pore pressures lead to a reduction in the shear strength of the material. Water movement below the soil surface creates a seepage force which acts in the direction of flow and is proportional to the hydraulic gradient. To this is added the unit weight of the water in the sloping mass and the combined effect produces failure usually along the boundary between the regolith and the un-weathered material. The general model used to describe this process is outlined in detail in Carson and Kirkby (1972) and need not be repeated here. The importance of these shallow planar slides in soil erosion arises partly from the fact that they leave bare areas which may be the site of gully initiation or rapid removal of material. Temple and Rapp (1972) in the example cited, estimated that it would take 25–50 years of 'normal' processes (i.e. under the present land use) to achieve the same volume of denudation as achieved by slides resulting from the single storm. Other workers have reported similar results (Wentworth, 1943; Thomas, 1977).

Soil creep operates at rates which are too slow to be significant, but rapid flowage such as debris flows may contribute significantly to soil erosion problems. Repeated flow activity may be concentrated along a line to form a gully with a low-angled debris cone of accumulation at its base. Again, the debris flows are triggered by heavy rainfall (Statham, 1976). The process of debris flowage is quite different from that of water flow since the flow is laminar rather than turbulent. A small amount of suspended clay and silt is adequate to dampen out eddies and reduce the tendency towards turbulent flow. Under these conditions dispersive stresses of the type described by Bagnold (1954) prevail. Where flowing granular solids are bounded on one side by a rigid boundary, grains can move more readily parallel to the boundary or away from it rather than towards it. The attempts of grains to move towards the boundary is resisted by forces exerted on the grains by the boundary. This leads to the sorting and transport of very large particles.

5.2.10 Piping

The process of piping ought logically to be thought of as an extension of the weathering processes under special conditions of soil chemistry and hydraulic head. Pipes are essentially large subsurface open channels sometimes several metres in diameter. They occur in a wide variety of environments from semi-arid (Stocking, 1977) to temperate (Knapp, 1970) though they appear to be most common in dry-lands because they are most clearly seen. Besides effecting very rapid drainage of the slope, the pipes are responsible for erosion and evacuation of entrained material and, as they break to the surface, the initiation of gullies. It is not uncommon to observe that pipeflow and rill flow are intimately related and flow in the pipes and rills may occur sometimes while the surrounding bedrock is quite dry (Bryan, Yair, and Hodges, 1978). The pipes may occur beneath crusts and commonly appear on the sides of gullies with outlets in the walls.

Similar features occur in dam structures and Wood *et al.* (1964) have accounted for them in terms of clay deflocculation. Clay is flocculated at densities determined by the cation exchange status of the clay (see Section 5.2.6) and the composition of the electrolyte surrounding each clay particle. Deflocculation results from changes in the cation balance and the concentration of the pore fluid surrounding the clay particles. The deflocculated particles are fine enough to move in suspension through the pore space within the soil and then escape from the body of the dam. These circumstances require a suitable clay, an appropriate chemical environment and a high rate of movement of pore water.

One of these conditions, if well developed, may offset lack of another. Lee (1968) for example, noted the onset of piping in a soil having high exchangeable sodium percentage with permeabilities as low as 10^{-5} cm/sec. Non-dispersable soils would require permeabilities as high as 10^{-2} cm/sec or greater for the soil particles to remain in suspension. Exchangeable sodium percentage is the degree of saturation of the soil exchange complex with sodium. Heede (1971) found that all exchangeable sodium percentages for nonpiping gully side slopes were less than 1.0 whereas those for piping slopes were greater than 1.0 and averaged 12.0. Stocking (1977) found values in the range 4–68 per cent for piping slopes and Sherard, Dunnigan, and Decker (1975), after the study of piping in many dam failures, have produced a graph showing the relationship between dispersibility and dissolved pore water salts (Figure 5.10) which is valid when the eroding water is relatively pure.

The requirement of a strong hydraulic gradient, encouraging subsurface flow and transport, is easily generated where gullies are developed, but they may also occur in steep slopes when the pipe outlets may be found on the surface of the same slope segment as the inlet (Jones, 1971). The process may also be assisted by decreasing permeability with depth in the soil.

That pipes produce and transport sediment is clear from the alluvial fans built up at their exits. The sediment production rates are then related to the flows through the pipes in an almost linear fashion. The high moisture levels required are probably produced by prolonged infiltration or saturated conditions due to snowmelt rather than a sudden incidence of large amounts of precipitation. Once pipes have been formed however, the latter produce dramatic effects.

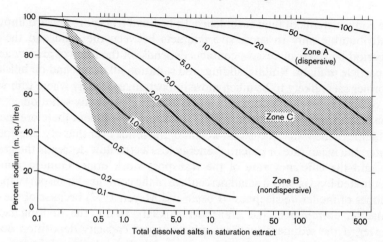

Figure 5.10. Relationship between susceptibility to colloidal erosion and dissolved pore water salts based on pinhole tests and experience with erosion in nature. The solid lines represent SAR values obtained from equation 5.23. The diagram is based on work by Sherard, Dunnigan, and Decker (1976) and published in Mitchell (1976)

Once a pipe is started it provides a new base for the local hydraulic gradient. The water flow lines are then concentrated towards the pipe and the rate of piping accelerates. Eventually the roof becomes unstable and may collapse leading to gully formation.

5.2.11 Sediment transport and gully formation

Once material has been entrained it moves as suspended or bedload transport. The loose particles which make up the load are of greater density than the water and hence differ in essential characteristics from materials which are chemically combined and therefore move as the solute load. The essential features of granular flow have been described by Bagnold (1966). A shearing motion occurs in which successive layers are moved over one another. The motion has to be maintained by an impelling or tractive force. For transport to be maintained the forces acting on every layer of solids must be in statistical equilibrium. The carpet-like layer of immersed solids must be supported by a stress equal to the immersed weight of the solids. This stress arises from the transfer of momentum from particle to particle by continuous or intermittent contact and by transfer of momentum from one mass of fluid to another. The solid transmitted stress arises from the shearing of solids over one another. The fluid transmitted stress arises from the shearing of the fluid in the form of turbulence. In suspension the whole weight is supported by a random succession of upward pulses attributable to the eddy currents of fluid turbulence. In bedload transport upward impulses are attributable only to successive contacts between the solid and the bed. The period between successive contacts depends on the force of gravity acting on a particle in flight. Rolling, sliding, and saltation all occur, and on hillslopes the former seem likely to be much more important, especially where thin water films occur. In the channels mass transport occurs largely by saltation.

The basic properties that control the transport of sediment therefore include the volume of the water, the excess weight of the solids over that of a like volume of water, the mean rate of potential-energy loss as the flow descends the gravity slope and the efficiency rate of the transport work done. Quite commonly the estimated load is only a small percentage of the computed transport capacity for slopes of moderate steepness (Foster and Meyer, 1972) because of the constraints on entrainment (i.e. transport capacity is not the limiting factor). However, if the sediment load exceeds transport capacity deposition occurs. This most frequently results from flattening of slopes widening of the stream or rill channel, ponding (as in reservoir sedimentation), flow retardation by vegetation and mulches and by transmission losses to the bed. All these are spatially variable. In addition, recession of flow in the declining stages of the hydrograph is perhaps one of the most important effects on hillslopes. The nature of the deposition is important for erosion in subsequent high flows. The actual processes of deposition are conditioned mainly by the character of the flow and the environment in which deposition takes place. There are three: deposition on open hillslopes and on fans, deposition in stream channels, and deposition in lakes and reservoirs.

Deposition on open hillslopes usually involves the terminal portion of discontinuous gullies, the bottom of fields along hedgerows where velocity is checked or between plough lines at the late stages of flow. Deposits at the base of eroding slopes are usually heterogenous comprising rock and soil that have not been transported through any well-defined channel system. Occasionally, however, the very high water content of material at the foot of the slope leads to a delta-like effect. On alluvial fans, large and small, the processes of debris movement are highly varied, ranging from debris flows to stream flows. The deposits of debris flows are poorly sorted, with coarse particles embedded in fine materials; they are unstratified and the boundaries of recent flows are sharply defined (Hooke, 1967). Mudflow deposits largely follow defined stream channels, but may overtop banks and spread out as a sheet and are also characterized by abrupt margins, a high proportion of clay and poor sorting, and some large 'rafted' boulders (Cooke and Warren, 1973).

Extensive and catastrophic upland erosion is usually accompanied by deposition along the channels of the stream systems draining in the area. This may occur simply and in one event or in a complex fashion over a period of years. Examples of the former have been investigated by Scott (1973), Williams and Guy (1973), and Thornes (1976), amongst many others. Deposition in channels resulted in these cases from decrease in the water volume during recession flow. Longer term deposition in the form of either extensive sand-bed aggradation or braiding is a more complex process and occurs mainly by the growth and migration of bars in the river. A shift from meandering to braiding channels is related to a combination of discharge, slope, and sediment size (Leopold and Wolman, 1957; Henderson, 1963). The association with soil erosion is most probably related to all three, but particularly to slope and sediment size. Braiding may represent an equilibrium pattern of transport to adjust to the change in slope resulting from aggradation, even while discharge remains constant. With an already steep channel, a relatively small change in aggradation may lead to instability and braiding if the channel is at or near a threshold condition of stability (Schumm, 1973). The shift of channel pattern with increase in slope has been studied in the laboratory by Schumm and Khan (1972) and a typical result in terms of channel morphology is shown in Figure 5.11.

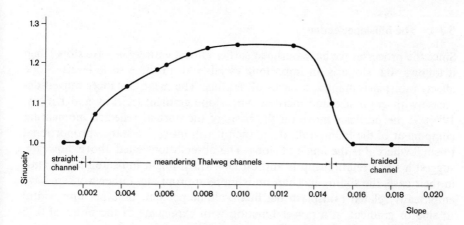

Figure 5.11. Relationship between sinuosity and valley slope based on experimental studies by Schumm and Khan (1972)

Deposition in reservoirs is conditioned by the normal processes of deltaic formation where water is present. As the stream enters a reservoir the flow depth increases checking velocity and causing a drop in transporting capacity. Deposition generally begins with coarser sediments dropping in the reservoir area whereas the finer sediments may be transported by density currents down to the dam. Virtually all kinds of reservoirs and check dams lead to a depletion of total load and result in some degradation of the channel downstream of the dam. In the development of erosion control studies these phenomena play an important role.

5.3 SPATIAL VARIATIONS OF PROCESSES

Soil erosion as a problem varies enormously in space both in terms of the total magnitude of erosion and in the type of erosion occurring. This is because of variations in the processes involved and in the relative magnitudes of the individual processes. It is also a result of variations in relative resistance of the soil and the surface cover. The problem is generally dealt with at one of three scales: the field, where the main contrast is between upslope and downslope and between rilled and non-rilled areas; the catchment, where generation of flow and of sediment lead to strong contrast in the type and amount of sediment and where the channel phase of erosion and deposition contributes significantly to the problem; and finally the region, where variations are largely attributable to differences in erosivity, erodibility and land-use practices.

5.3.1 The hillslope section

Since the processes we have outlined earlier almost entirely involve slope, then it follows that slope is an important variable at all three scale levels. Slope affects rainsplash mainly in terms of net flux. The mean distances moved decrease with grain size and increase with slope gradient (Kirkby and Kirkby, 1974). If the particles move on the basis of the normal reflected momentum component of the falling rain drop, then it will move a distance proportional to $\sin \beta$ where β is the angle of slope. The observations cited above, however, suggest a $\sin^2 \beta$ relationship is appropriate which might indicate that, at least in their case, the sliding mechanism is more appropriate. Other workers have empirically related rainsplash net flux (volume per unit time per unit width) to sine of gradient in a power function with exponents of the order of 0.75 (Moeyersons and De Ploey, 1976).

Gradient is also important in controlling surface wash, both concentrated and unconcentrated. An increase in slope should increase the velocity of flow and through it the tractive shear stress, leading to greater capacity for both entrainment and transportation.

Slope wash is related to both the shearing force acting on the particle and the transporting capacity of flowing water. In the upper part of a slope the former may be most important, in the lower part, the latter. If the tractive stress of flowing water exceeds a critical tractive stress for the sediment, erosion occurs. Tractive force is again proportional to $\cos \beta \sin \beta(s)$ and velocity proportional to $s^{1/2}$, so that the effects of sheet wash discharge should be approximately proportional to the 1.5 power of slope. This is for the deep-channel Manning equation condition and variations in the exponents of slope in the overland flow equation (Carson and Kirkby, 1972) may be expected to yield corresponding variations in the relative role of slope. Observed relations of erosion as a function of slope suggest relationships in whoch slope tangent is raised to the power 1.4 (Musgrave. 1947) when taken in conjunction with slope length, but these empirical results usually represent a combination of circumstances. A quadratic function of slope is also commonly used in the universal soil loss equation (Mitchell and Bubenzer, Chapter 2).

Recent work has qualified relations between gradient and soil erosion (as net flux). The relationship relies on a particular form of runoff, and the amount of runoff itself is thought to be a function of slope angle. Horton (1945) and Meyer and Monke (1965), for example, attempted to relate runoff coefficients to slope angle. There is here, however, a complex interaction with other slope properties, notably materials. For example Yair and Klein (1973) found no clear relation between runoff and slope angle, but under the desert conditions they studied, the relationship between runoff and slope angle seems to suggest an increase in the amount of runoff as gradient decreased. This was attributed to the greater roughness and higher infiltration capacities on steeper slopes. They found that the gentlest slopes of smallest area and shortest length yielded the greatest amount of material.

Where rainsplash is concerned there appears to be a relatively poor relationship with slope steepness in practice and above 20 per cent the effect of steepness seems to level off (Foster and Martin, 1969). This implies that the rate of soil detachment by splash changes only slowly as steepness increases and that this largely governs the rate of loss. This was shown empirically by Meyer, Foster and Romkens (fig. 5.12).

The length of the slope should have a negligible effect on rain splash but may be expected to be important in both detachment and transport by wash. Assuming a steady state rainfall and runoff, the total discharge will increase with slope length for a given gradient. Horton (1945) used this argument to suggest that there would be a 'belt of no erosion' near the crest, in a zone over which the developed shear stress of the flow was less than the resistance to shear of the soils forming the surface. Erosion is frequently expressed (per unit area) as proportional to L^n where n ranges from 0–0.9 (Zingg, 1940; Wischmeier and Smith, 1965) and values near $n = 0.5$ have been used for most purposes. In fact where erosion is predominantly by concentrated flow the exponent should approach unity and with predominant rainsplash should approach zero. Under conditions of unsteady flow, the role of shape is likely to be more important than steepness (Evans, Chapter 4). The failure to identify the different roles of splash and wash may be the reason for some of the apparently conflicting results obtained in field experiments (e.g. Lam, 1977).

Finally the slope shape, as the interaction of angle and distance, has important effects on the total magnitude of erosion. Where slope length increases but steepness decreases the effects should offset each other. On convex slopes with increasing length and slope the rates should be at a maximum. Figure 5.13

shows results of a simulation of erosion by Li, Simons, and Carder (1976) in which erosion rates on convex slopes appear to be five times those on uniform slopes. Young and Mutchler (1969) examined empirically the erosion rates on slopes of different shape and found that downslope the depth of erosion from concave slopes dropped to almost zero and slope foot deposition occurred, while on convex slopes the depth of erosion increased linearly with distance. Slope shape is also important in terms of the relative contributions of overland flow and interflow and for the development of basal seepage. Freeze (1972) noted that a necessary condition for the dominance of the subsurface flow

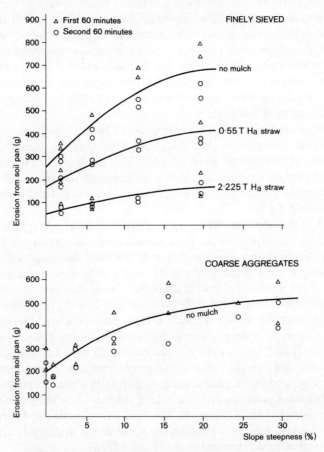

Figure 5.12. Relationship of soil loss to slope steepness on short interrill areas, largely attributable to splash processes. The data are for soil pans 0.61 × 0.61 m with a rainfall intensity of 6.35 cm/hr T/H_a = tonnes/hectare (after Meyer, Foster, and Romkens, 1975)

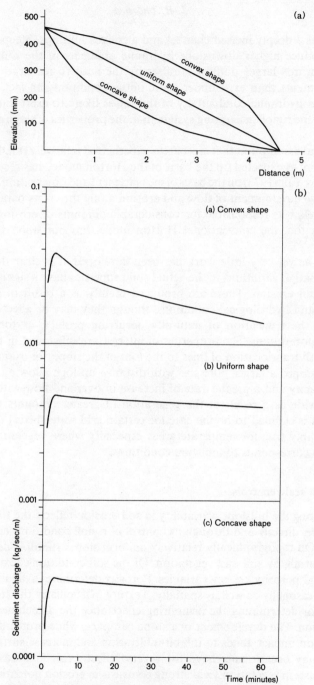

Figure 5.13. The effect of slope shape on temporal variations in sediment yield according to the model of Li,
Simons, and Carder (1976)

mechanism is a deeply incised channel and a convex hillslope. Steeper sloping hillsides produce higher downslope hydraulic gradients in the soil, but this tendency towards larger outflow is offset by the need to recharge the lower moisture contents than exist under static initial conditions. In fact, however, the saturated hydraulic conductivity of the soil is likely to exert a far greater influence on the runoff generating system than the properties of slope configuration.

The general assumption that crest areas experienced relatively small amounts of soil erosion, as expected on the basis of the Horton model, has recently been questioned by Yair (1973) on the basis of measurement and observation. He noted that extensive development of flow and erosion along the crests parallel to the main channels was responsible for considerable amounts of erosion. In temperate areas too, the conventional Horton model has not stood the test of time.

Although as yet very little work has been developed, it is clear that profile (as well as spatial) variations in the actual soil properties have a significant role to play in soil erosion. These are produced mainly as a result of long-term weathering and hydrological conditions, though they may be affected by soil erosion, by the truncation of naturally occurring profiles, as for example following deforestation. The properties of interest are subsumed in the catena concept. With translocation of fines to the foot of the slope, for example, there may be developed a larger pore size with distance upslope allowing for lower storage capacity and a greater rate of increase in overland flow with distance from the divide as a result of the generation of greater amounts of rainfall excess. Such is claimed to be the case for certain arid watersheds (Yair, 1973) and is probably true for temperate areas especially where vegetation zoning on the slope corresponds to moisture conditions.

5.3.2 Basin scale controls

Laterally, along the hillslope, variability in soil erosion reflects the topography of the hillside, directly and through its control of runoff conditions and vegetation cover. On topographically relatively uniform slopes runoff is determined, if it occurs at all, by soil and vegetation. Of the soil conditions, texture, gross structure, and porosity are key variables. Porosity determines available storage and varies seasonally as well as spatially. Texture, particularly the stoniness, is important for determining the armouring effect once the detachment process is in operation. The development of a stone carapace, while protecting the soil from raindrop impact, tends to inhibit infiltration and increase surface runoff, though it may occasionally have the reverse effect (Yair and Lavee, 1977). Local contrasts in vegetation yield strong contrasts in erosion potential through the development of macropores inducing higher permeabilities (see for example, Arnett, 1976) and the contribution of organic leaf litter which maintains a better aggregation of soil particles (Section 5.2.6). These show a strong feedback relationship with soil, since maturer soils may be expected to be more stable (Beckett, 1968).

When hillslopes have well developed drainage lines in the form of rills significant changes in sediment yield occur from rilled to non-rilled slopes, the latter having much lower yields than the former. This appears largely to result from the fact that channelled runoff has a much greater detachment capacity than inter-rill flow. Moreover, as the rill density increases the sediment yield increases. This effect is likely to be produced by the greater probability that inter-rill splashed soil fragments will reach a channel. Empirical investigations of this contrast have been carried out by Meyer, Foster, and Romkens (1975) though they fail to establish the initial causes of rilling.

In recent years it has been widely recognized that within a river basin (and even on a hillslope) overland flow is generated from precipitation falling on very small partial areas that quickly become saturated during rainfall events. The partial areas are often wetlands whose location is controlled by the topographic and hydrological conditions of the basin. These areas expand and contract both seasonally and through a storm, and are partly responsible for generating the hillslope and channel hydrograph shape. In so far as sediment transfer to the channel from the hillslope is dependent almost exclusively on the overland flow mechanism, then similar partial contributing areas for sediment are to be expected. In dry times and dry areas sediment transport on the hillslopes is a much more complex phenomenon since the flows often dry up before reaching the channel and so 'arrivals' at the channel are controlled rather more by the storm dynamics than by subsurface moisture conditions. Bryan, Yair, and Hodges (1978) have also indicated that partial contributing areas in semi-arid badlands in Alberta are controlled by variations in material properties rather than variations in moisture regime on shales. On the sandstone units they examined erosion takes place over the complete surface (although not necessarily uniformly) regardless of slope angle, and runoff does not conform to the partial area contribution model.

In addition to the common observation that some parts of river basins are intensively gullied whilst adjacent areas are relatively untouched, there are more systematic trends which support the proposition of partial contributing areas for sediment in all channel systems. It is observed, for example that with increasing basin size the sediment yield per unit area diminishes (Figure 5.14) reflecting the smaller specific yield of sediment of lowland in comparison with upland parts of the basin. Although we know little of the relative contributions of material from gullied slopes and channel erosion, evidence suggests (e.g. Anderson, 1954) that channel bank erosion may provide up to one-third of the total sediment load and in rilled and gully areas this is much larger because of the high level of coupling between the channels and slopes in these areas. Accordingly areas of high drainage density may be expected to be losing more sediment. This problem is generally supposed to be most acute and most easily recognized in intensely gullied areas.

Gullies are dynamically similar to small stream channels except for their ephemeral flow and the direct supply of material from the slopes to the actual channel by collapse and by entrainment. They are characterized by high levels of dynamic activity and rapid change of form. Gullies are perturbations which grow and enlarge by positive feedback until they become inhibited either through spatial competition or excess sediment production. The causes of the initial perturbation may be random as for example the trampling of vegetation by cattle, the survival of a rill through consecutive seasons, the collapse of a sub-surface hollow, or the cutting of forest for road construction. However particular

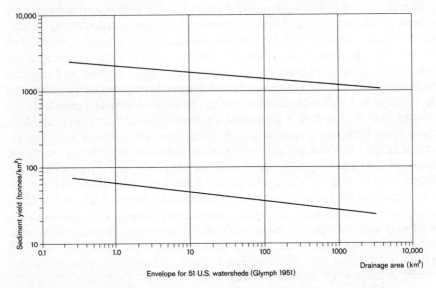

Envelope for 51 U.S. watersheds (Glymph 1951)

Figure 5.14. Outer bounds of the relationship between sediment yield and catchment areas for 51 watersheds in the United States according to Glymph (1951)

conditions seem conducive to their growth and expansion across otherwise uneroded terrain, the most important of which is very high runoff rates from infrequent storms. Thus in temperate areas gullies appear to be largely confined to lithologies producing high runoff rates such as clays though the assumption that the causes of such 'badlands' is only related to high volumes of runoff is clearly at variance with the facts. In arid and semi-arid areas, however, gullies are notably absent from areas producing coarse-textured regolith with coarse debris. The general instability conditions may also relate in part to slope and Patton and Schumm (1975) argued that the distribution of gullies within the semi-arid valleys they studied is largely controlled by geomorphic thresholds.

According to them sediments accumulate on the floors of valleys until a threshold gradient is achieved above which entrenchment occurs.

Heede (1975) describes how gullies may begin in either rills, in the case of a continuous gully, or in an abrupt headcut in the case of discontinuous gullies. Discontinuous gullies form on slopes and as soon as a headcut is initiated it begins to retreat upslope. Concentration of water in the trench reduces 'channel' storage and increases the flow depth so further erosion can occur. As the channel floor of the discontinuous gully 'emerges' onto the slope deposition occurs. The salient feature of the discontinuous gully is the relatively small slope of its gradient or bed. It is this flat slope that makes the gully discontinuous, for the bed profile must intersect the profile of the original valley. At that point the gully depth has diminished to zero. Runoff over the headcut contributes to gully growth by exerting forces on the channel boundary, by removing accumulated soil debris from the channel and by eroding gully banks by undercutting them and by gravity loading to a value above the critical shear strength (Piest *et al.*, 1975). This upstream extension is frequently associated with piping and eventually discontinuous gullies may become continuous, though sometimes they become atrophied. Gullies have been studied intensively in attempts to obtain predictive empirical equations and most results indicate that growth rates are a function of drainage area (Mitchell and Bubenzer, Chapter 2). In terms of modelling soil erosion what is more critically needed is a general model for the stability–instability conditions of the growth of initial perturbations along the hillslope comparable to that suggested for channels. Work along these lines has been initiated by Kirkby (1978b).

The efficiency with which the material produced from the inter-rill areas, the rills, and the gullies is transported is largely related to the performance of the axial stream of the system. Eventually if extensive deposition occurs along the main valley this may actually inhibit erosion in the upstream areas; this is the principle involved in the construction of check dams. Progressive sedimentation leads to infiltration losses and further deposition so that an alluvial fill may grow up-valley filling the trench of a former gully. The hydrograph characteristics are determined in part by the network configuration of the drainage system and the magnitude and frequency of flows in the gully-channel system determines the character of channel erosion and deposition. In ephemeral channels as well as in alluvial fans, sediment wedges build up throughout the channel in complex patterns (Thornes, 1977). It is only in extreme events or as a result of the exceedence of the slope thresholds that this process is arrested.

When intensive scour occurs in channels, perhaps as a result of a large storm, or a change in the dynamic conditions relating to the basin as a whole, downcutting may also be triggered in adjacent areas. This problem is discussed further in the final section of this chapter. Within the basin as a whole the spatial distribution of erosional processes is critically related to the relative concentration of water and sediment. In upstream areas this is largely controlled

by hillslope hydrology and hence by soil and slope; further downstream the
network characteristics, contributing area, and vegetation characteristics
assume a progressively more important role.

5.3.3   Regional spatial controls

The most important spatial variant on the regional scale is climate which controls natural vegetation as well as soil erosion. Its significance lies principally in
determining the hydrological balance at the surface through the operation of
precipitation and evaporation. These climatic controls have largely been
inferred through the sediment yield of rivers and a useful review is contained
in Gregory and Walling (1973). Langbein and Schumm (1958) graphed the
relationship between sediment yield and effective precipitation which is the
rainfall equivalent of the observed runoff and hence expresses the overall
hydrological budget. This graph, which is based on a very small number of
observations, peaks at an effective precipitation of about 300 mm, trailing off
at lower values because of lack of runoff and at higher values because of the
encroachment of vegetation (Kirkby, Chapter 1). Regional patterns of suspended
sediment production appear to confirm this overall proposition (e.g. Rainwater,
1962).

In decomposing the climatic controls two elements are especially important
for soil erosion. These are rainfall intensity and energy, which affect the splash
process and the seasonality of rainfall which, with evaporation, controls the
overall runoff and vegetation characteristics. Rainfall intensity is also important
in channel erosional processes because of its effects on the hydrograph. Intensity is usually expressed in volume/unit time for a given return period and
several workers have expressed the erosivity of rainfall as a function of the
maximum intensity of rainfall for a 30 minute interval with a two-year return
period. More commonly the distribution of intensity is compounded with rainfall energy to provide a better base for splash erosion and Stocking (1972),
for example, found a good correlation between measured erosion amounts
and rainfall energy.

The seasonality effect is important for its control on natural vegetation and
soil moisture depletion. There are many measures of this effect, such as those
of Thornthwaite (1948) and Blaney and Criddle (1947), both of which attempt
to incorporate the consumptive use of water by existing plant cover. These
indicate the differences for example between areas having precipitation mainly
during the winter season and those which have rain mainly during the summer
season, a very important distinction in terms of the generation of hillslope runoff.
Any attempt to model regional variability of water-produced soil erosion will
have to attempt to determine the available soil moisture storage at any time as
well as rainfall erosivity. Carson and Kirkby (1972) used the exponential
distribution of daily rainfall amounts in conjunction with a value r_c, which is

the amount able to infiltrate and evaporate during rainfall, to produce estimates of variations in annual slope wash transport in terms of mean annual precipitation. Like the Fournier (1960) empirical correlation of world wide sediment yield in rivers this model is strongly sensitive to seasonal effects. The r_c term presents most difficulty and is defined in terms of cover land use. In this sense it attempts to incorporate the other elements of the water balance mentioned earlier, and the pattern eventually produced is very similar, for high temperatures, to that derived by Langbein and Schumm (1958) and Douglas (1967).

The other major control of regional significance is geology. This operates in three important ways. First it controls relief through the long-term susceptibility of the rock petrology to weathering processes. Second, the weathering products and soils are quite profoundly affected by rock type. The soil capacity for holding water is also critically affected by the vegetation cover of course. Thirdly, the macrostructure of the rock itself determines in a coarse way the relative contributions of groundwater and surface water flow. Although the potential water storage capacity of soil is of crucial importance so far there have been few attempts to characterize this parameter on a regional basis. Schumm (1956) attempted to differentiate between the Brule and Chadron formations in South Dakota on the basis of the time to runoff and many studies since have attempted to differentiate erosion-susceptible soils on the basis of infiltration (e.g. Selby, 1972). A parameter leading to the definition of the r_c value based on general bedrock and vegetation cover would be of more general application on a regional basis.

Finally, the effect of human activity on a regional basis is probably most important of all within a framework of overall natural susceptibility to soil erosion. Very strong contrasts over short distances reflect, all too often, the mismanagement of soil and vegetation. These effects are further discussed in Morgan (Chapter 8).

5.4 TEMPORAL VARIATIONS IN PROCESS

5.4.1 Introduction

Through time the type and rate of operation of process may vary considerably leading to changes in the pattern and magnitude of soil erosion. These variations may be short term such as the changes in available sediment for erosion which take place in a storm lasting a few hours. Seasonal effects lasting a few months are essentially cyclical and stem largely from climatic controls. Long term changes in soil erosion mainly reflect the effects of human activity, but there are some natural processes, such as runs of drought years, or progressive weathering of materials, which may also give rise to such changes.

Short term changes are usually expressed in terms of the frequency of occurrence of events of a given magnitude. Magnitude may be expressed in absolute terms, for example the volume of rain in a given storm, in terms of exceedence of a particular threshold, such as the shearing resistance of the soil, or in terms of the rate of operation of a process such as sediment transport. The area under the frequency curve indicates the total amount of time (duration) over which the process is acting. The total volume of rainfall or the total amount of debris moved is equal to the area under the curve formed by the product of the frequency curve by magnitude. There tends to be a dominant size of event which is most 'effective', for example, which produces the largest rainfall volume, or represents the dominant rate of debris transport. Wolman and Miller (1960) argued that this event is larger than the most frequent event (that corresponding to the peak of the frequency curve) and it is the size of event which does more work than any other. Carson and Kirkby (1972) draw attention to two difficulties associated with the concept. First, the distribution may be multimodal, resulting in a number of 'dominant' events and second, the larger events may have effects which are different in kind as well as in degree. Thornes and Brunsden (1977) draw attention to two further difficulties. The first is that the events are regarded as independent of each other whereas in fact they are serially and sometimes cyclically related. Thus, for example, storms occur in clusters so that an event of a given magnitude may be much less effective if it follows hard on the heels of an event of comparable magnitude than if the two events were separated by an appreciable period of time. Secondly, hysteresis effects due to other controls lead to multivalued functions between process and effect relationships.

In the longer term (over several hundred years) the historical or serial-approach to variations in soil erosion has to be employed because we are dealing with factors largely associated with unique events such as forest clearance or animal grazing changes. These phenomena are less susceptible to modelling although the processes remain essentially the same.

5.4.2   Variations in erosion on a storm basis

Several studies have shown the shift in suspended sediment load during individual storms and the associated hysteresis effect between suspended sediment load and discharge. This results from a phase lag between the peak of sediment concentration, which is usually ahead, and the main flow of water (Figure 5.15). The phase relationships depend on the location of the guaging station, the principal contributing areas of soil erosion, the relative contribution of hillslope to channel sediment and storm intensity. The onset of a storm usually leads to removal of the most readily available material so that there is a depletion of available materials as the storm progresses. Moreover one might expect during the progression of a storm that the textural composition of the eroded material will change, and this has been observed, for example, by Meyer, Foster, and Romkens (1975) on experimental plots. In areas subject to severe rilling they found that during the progression of a storm the textural composition of the sediments produced approached those of the original soil more and more quickly. Progressive rain beat adds to the depletion of materials to some extent by compaction of the soil and development of a more stable resistant crust. This effect is the short term equivalent of armouring.

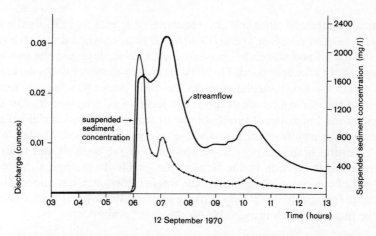

Figure 5.15. Lagged responses of streamflow peak relative to peak in suspended sediment concentration which give rise to typical hysteresis curves between sediment concentration and discharge (Walling, 1974)

In the more general situation, in ephemeral headwater channels and gullies, as well as on the slopes, the antecedent conditions play an important role in determining the volume of material removed. The rate of sediment production during an individual event depends not only on the stage of gully development but also on the relative stability of the materials. Heede (1975) found that sediment loads are more often related to the time and duration of flow rather than to total volume, as the material already prepared is removed quickly. Likewise Piest, Bradford, and Wyatt (1975) observed in Iowa that one third of the average annual rainfall causes four-fifths of the gully sediment movement. This largely occurs in May–June and is attributed to the fact that accumulation of gully soil debris is greatest before the spring runoff season. Concentrations of gully materials decrease with successive spring rainstorms, although some debris is produced throughout the year.

In the longer term the reverse effect, the provision of more rather than less material, takes place with the incidence of large, infrequent flood-producing storms. These storms transport large volumes of material to the channel which are available for the immediately succeeding lower flows. This source may then be progressively depleted. Anderson (1970) described the effects of the 1964 floods in Oregon as producing substantial increases in the sediment yields of succeeding years which gradually diminished in size until they reached the pre-flood levels. Schumm and Lichty (1963) investigating the scouring processes along the Cimarron River also found that there was a well-developed sequence of channel widening and narrowing associated with individual extreme events.

In small ephemeral channels the sequence of events in individual storms appears to be quite complex. Wertz (1966) found three separate stages of development, the main flood stage, the immediate post-flood stage and the prolonged dry stage prior to the next flood. The flood stage thoroughly reworks the channel debris within the valley. At the moment the flood begins to wane, the heaviest elements of the whole load are dropped in a series of primary patterns arranged in steps up the longitudinal profile of the stream, depending on the nature of the channel, the volume of the flow and the velocity of the flood. The late waters of the recession limb readjust the morphology of the channel and resort the finer materials in relation to the steps in the profile. In the period before the next flood of comparable magnitude, the smaller flows etch out minor channel features, progressively but slowly moving material downstream. This is essentially the material which is available for the next flood.

5.4.3   Seasonal variations in soil erosion processes

Seasonal variations in the relative efficacy of the various processes are related to changes in the soil and vegetation characteristics as well as to changes on the overall hydrologic budget. Areas of strong seasonal contrasts, especially in rainfall, such as the Mediterranean areas of the Old World, are areas of high susceptibility to soil erosion.

Seasonal changes in soil character result from changes in volume of swelling soils, the activities of animals, fluctuations in organic matter production by leaf fall and the effects of winter frost. These affect the infiltration and storage capacity of the soils and the type and nature of materials at the surface. In an interesting study of the Mancos shales in Western Colorado, Schumm and Lusby (1963) found that greater volume changes occurred in soils with higher sodium content and those having vermiculite. Figures for the free swell test indicated volumetric increases of between 20 and 30 per cent and a combination of swelling and frost action produced observable seasonal variations in the infiltration rates. They hypothesized seasonal changes in infiltration according to the rainfall regime of a particular year. In a normal year infiltration rates would be highest in winter and early spring before the rains begin, and then decrease in late spring and summer due to compaction and surface sealing. With a normal winter but a dry spring and summer, infiltration rates could be sustained at a high level throughout the spring and summer. With a dry winter the frost action is inhibited and so infiltration capacities remain relatively low throughout the year. Similar variations in infiltration capacities throughout the year have been observed in other semi-arid areas.

In temperate areas too there are seasonal changes in the permeability of soils. Arnett (1976) showed strong variations in soil permeabilities seasonally due to swelling of the clay fraction and consequent changes in the pore size distribution. Topsoil permeabilities are much higher in summer than in winter. These variations in porosity indicate the dangers of using texture as a surrogate for infiltration capacity, since texture is constant throughout the year. In cold areas infiltration may be retarded in spring as a result of frost penetration into soils although frozen soils may have significant, though declining, hydraulic conductivities well below 0°C (Burt and Williams, 1976).

Although the seasonal activity of burrowing animals is obviously quite important, the most significant seasonal controls of soil erosional processes probably relate to variations in vegetation cover, both natural and human. This has been widely recognized by use of a cropping-factor in soil erosion equations. The effect of vegetation cover on a seasonal basis is threefold: principally it controls surface runoff by the total standing crop, it affects the hydrological balance of the stream bed and the slopes by withdrawal of soil moisture and produces seasonal variations in the interception. The significance of cover depends on its relationship to the occurrence of falling rain. Some locations have nearly all their erosive rainfall during the growing season. Continuous corn is much more susceptible to erosion in the first month after seeding than when it is approaching maturity. It may be that a more effective incorporation of seasonal vegetation variations into soil erosion models can be based on their consumptive use of water. For semi-arid plants leaf cover and consumptive use of water operate in the same way with respect to soil erosion. Leaves increase the interception of water and reduce the raindrop impact while increasing the total available storage, at least in the root zone, and so tending to reduce overland flow production. Phreatophytes in ephemeral channels effectively decrease the subsurface storage volume causing increased transmission losses to the channel.

Finally we note that seasonal contrasts in rainfall lead to seasonal contrasts in direct runoff. Usually, in areas with winter rains, the character of the storms is also strongly varying. In the western Mediterranean, for example, winter storms are of the cyclonic type, whereas summer storms are mainly convective. The former tend to provide sustained periods of light rain, the latter short periods (a few hours) of very heavy rain. Thus in summer, despite the high availability of storage, runoff is very swift both on the hillslopes and in the channels. Opportunity time for infiltration is shorter and transmission losses to the channel may be expected to be less. In winter more sustained rains and higher infiltration lead to water storage, so that by spring base-flow is occurring in the channels. The effectiveness of storms is greatest therefore in spring, when antecedent moisture is very high and in autumn when the rainfalls are most intense and infiltration rates are at their lowest.

5.4.4 Long-term variations in processes

Long term changes in the type, magnitude and frequency of soil erosional processes are more difficult to explain and model because they frequently relate to historical events such as episodic land clearance, individual floods and extreme droughts. Moreover they are often succeeded by long periods of transient behaviour. The problem is augmented by the spatially discontinuous character of the responses and the time lags involved.

One the one hand, local effects may lead to the triggering of erosion which is then enhanced or dampened by the regional stability conditions. For example Tuckfield (1964) found that gully growth in the New Forest, England, was largely initiated along footpaths and local clearings and Schumm and Hadley (1957) showed that trenching of valley fills is commonly associated with local steepening of the valley fill as a result of aggradation. On the other hand, regional instability over longer periods must largely be related to either land-use changes, climatic changes or, in the very long term, to changes in the overall energy conditions of a region as a result of diastrophic or eustatic change. These last are beyond the scope of this book.

There are several major studies which have sought to elucidate the historic changes involving soil erosion in terms of controlling variables as reflected by morphological or sedimentological changes. There are also many empirical investigations which discuss the correlation of sediment yields with specific effects such as grazing density, the percentage of the area logged and so on. The best of these studies, such as Cooke and Reeves' (1976) investigation of the arroyo problem in the American South West or the Coweeta Hydrologic Experiment (Dils, 1957) add greatly to our understanding of the complexity of such relationships and their functional structure and they indicate the caution required in attempting to model the processes involved in detail. From such studies it is evident that particular attention needs to be paid to two areas, the effects of vegetation removal and the impact of climatic changes.

Vegetation removal occurs in a wide variety of circumstances, and it is hardly surprising that a variety of different results have been obtained. Rarely is there total devastation of the type carried out, for example, at Hubbard Brook experimental watershed (Pierce *et al.* 1970). Rather forest clearance has involved partial felling or selective cutting with some ground clearance at one, highly mechanized extreme; crude felling with no clearance whatsoever, followed by burning and planting at the other. The effects of clearance involving heavy machinery might be expected to be different from the impact of primitive cultivation, especially, in tropical areas, where regeneration of secondary cover may be very rapid indeed. However there are some minimal effects which might be anticipated. Depletion of standing crop may be expected to lead to an overall reduction in evaporation and decrease in the percentage of the bare area covered, both of these should enhance soil erosion. Moreover, the reduction in total leaf litter production should lead to a fall in the organic content of the soil, and a lower effective strength. Removal of the vegetative roots may reduce binding action, but if the roots are left in the soil, their eventual death and decay leads to larger soil macropores. Added to this is the fact that both clearance of forest and afforestation have commonly been associated with artifical drainage, so that runoff conditions may be quite dramatically changed. As a result of an investigation in Oregon, Anderson (1975) showed that the sediment production increases expected as a result of logging in that area would mainly be associated with road development (80 per cent) rather than logging itself (20 per cent).

Another major cause of change in vegetation has been cattle grazing. Lusby (1970), for example, studied the highly erodible Mancos Shale, near Grand Junction, Colorado and found that over a twelve and a half year period grazed watershed sediment yields averaged 3901 kg/ha/year compared with 2672 kg/ha/year for ungrazed watersheds. No change was noted between the two types in the first year but thereafter the sediment yield in the grazed area averaged 1.8 times those of the ungrazed area. However, as with forest cutting, so with grazing the relationship to soil erosion varies considerably. The effectiveness of range vegetation in inhibiting soil erosion depends on the life form of the plant and particularly the ground cover it offers and its ability to sustain a strong root mat. Their productivity is density dependent, so for example, *Larrea divaricata* (creosote bush) has a primary productivity of about 1000 kg/ha for a relatively young shrub community, but as the age of the plant increases density decreases and productivity increases. The effect of light grazing may be to enhance productivity and cover, at least in the first instance. Ephemeral grasses, on the other hand, may easily be removed without further growth. Besides direct reduction of the surface cover, excess defoliation may lead to reduced root growth in both grasses and shrubs (Goodin and McKell, 1971). Shrub reproduction may also be affected by the degree of use. The direct defoliation depends too on the type of animal. Nord (1965) found that sheep seek out small, tender plants, whereas cattle generally graze the larger and better established plants. Goats can attack species which are by and large unpalatable ot other animals.

Other important effects arise in the grazing-soil erosion complex. One of these is the gradual replacement of palatable with unpalatable species of plant, particularly those that are very spiny, or those that are toxic to animals. These species may be associated with lower cover densities. A third effect of human activity is the introduction of new species which replace the older, perhaps more drought resistant or erosion-resistant species. Burcham (1957) for example has traced the invasion of California by exotic species over the last 200 years.

Unfortunately as Cooke and Reeves (1976) point out the implications of these for runoff (and soil erosion) are not yet fully understood. Besides the height and density (both related to primary productivity), changes in the growing season of the freshly introduced species have also to be evaluated. The authors suggest that the initial changes from bunchgrass to wild oats, mustard, and associated exotics may actually have led to a *decrease* in runoff. Removal of vegetation in the valley floors also changes the erosiveness of flows and the erodibility of the valley floors leading to instability. In early times ploughing and sowing replaced ciengas, willow swamps, and stands of shrubs and trees in the Mediterranean as well as in North America. A contemporary analogue of this process is the clearance of salt cedar (*Tamarix chinensis*) from certain arid and semi-arid channels in the American south-west. In the flood-plain of the Gila River in south eastern Arizona saltcedar and mesquite (*Prosopis juliflora*) have been replaced with short rooted grass. The saltcedar and mesquite increase both resistance to flow and the stability of the flood plain boundary. Burkham (1966) concluded that these changes have significantly affected the peak discharge major flood parameters of stage, mean cross-sectional velocity, channel-boundary roughness and mean cross-sectional depth.

A third major type of vegetation change is produced by burning. In an intensive study in California (Rowe, 1944) it was shown that burning increased surface runoff and erosion rates mainly through reduced infiltration capacities. Although Veihmeyer (1953) found inconclusive effects of burning, others have confirmed the findings of Rowe (e.g. Glendening, Pase, and Ingebo, 1961).

The observation, from prehistoric records, that cutting and filling occurred before the arrival of man in significant numbers indicates that climatic changes can also be responsible for shifts in the prevailing extent of soil erosion. Shifts of climate, even over a relatively short period of time (a few hundred years) may reasonably be expected to induce changes in the mass budget of hillslopes. There is a large literature to indicate the disagreement over what these effects are likely to be, but three viewpoints have been expressed. The first is that prehistoric erosion on the slopes was associated with drier conditions because the vegetation cover would have been reduced and the storms would have been of higher intensity. This in turn would have lead to deposition in the higher order valleys (Huntington, 1914). Bryan (1928) on the other hand considered that alluviation in the valley floors represented more humid conditions. This might be expected because wash erosion is related to rainfall intensity and wash erosion makes up the bulk of material being supplied to channels in semi-arid regions (Leopold, Emmett, and Myrick, 1966). Rainfall intensity, however, has a very variable correlation with total volume and so neither of these speculative explanations can claim much confidence. A third type of explanation based on rainfall intensity or, even more realistically, rainfall in excess of storage, is likely to be more profitable. In addition, the earlier arguments seem to take only a modest account of the spatial characteristics of the channel network system. Schumm (1965) has shown how responses should be different on hill-slopes and channels; evidently we need to consider different parts of the channel system. In this way some of the past discrepancies might be resolved.

Much of the recent work in erosional systems has highlighted the importance of extreme events in erosional systems when soil erosion appears dispropor-tionately large, sediment yields remarkably high, and channel changes dramatic and persistent (Schumm and Lichty, 1963; Burkham, 1976; Thornes, 1976). Bull (1964) related channel entrenchment in the American Southwest to periods when rainfall was very high and there was an increased frequency of large daily falls. The disproportional effects of large floods may relate as much to the large inter-arrival time, as to the volume of the discharge in the individual event, because this reflects a sustained period of accumulation of material on the slopes and in the channels.

Finally, we note with respect to changing climatic control the widely varying response in terms of lag time. Most historic explanations are obliged, to some extent, to seek for 'causes' of the events of interest in the period in which the response occurs. However in modelling soil erosion controls the historic causes must be treated with extreme caution. The temporal lack of coincidence between say a run of wet years and valley alluviation need not indicate that one is independent of the other. The diffusion of erosional pulses across space needs time. One effect may beget another, apparently related to the original cause. In this way the 'complex responses' described by Schumm (1973, 1977) may be such that the historical approach to investigating soil erosion phenomena is relatively unproductive. All the material reviewed in this chapter indicates that climatic controls are vitally important in the essential processes producing soil erosion, so we may expect them to have an important effect over time, as and when variations in these parameters occur.

REFERENCES

Adams, S., Strain, B. R., and Adams, M. S. (1970). Water repellent soils, fire and annual plant cover in a desert scrub community of south eastern California, *Ecology*, **51**(4), 696–700.

Anderson, H. W. (1954). Suspended sediment discharge as related to stream flow, topo-graphy, soil, and land use. *Trans. Am. Geophys. Union*, **35**(2), 268–281.

Anderson, H. W. (1970). Principal components analysis of watershed variables affecting suspended sediment discharge after a major flood. *Int. Assoc. Scient. Hydrol. Pub.*, **96**, 404–416.

Anderson, H. W. (1975). Relative contributions of sediment from source areas and trans-port processes. In: *Present and Prospective Technology for Predicting Sediment Yields and Sources*—Proceedings of Sediment-Yield Workshop, United States Department of Agriculture Sedimentation Laboratory, Oxford, Mississippi, November 1972 (Agricultural Research Service Report ARS-S-40), United States Department of Agriculture, Washington, D.C., pp. 66–73.

Arnett, R. R. (1976). Some pedological features affecting the permeability of hillside soils in Caydale, Yorkshire. *Earth Surface Proc.*, **1**(1), 3–16.

Bagnold, R. A. (1966). An approach to the sediment transport from general physics. *U.S. Geol. Surv. Prof. Paper*, **422-I**, 37.

Bagnold, R. A. (1954). Experiments on the gravity-free dispersion of spheres in a Newtonian fluid under shear. *Royal Soc. (London) Proc.*, **A**, 225. 49 pp.

Beckett, P. H. T. (1968). Soil formation and slope development. *Z. für Geom*. Neue Folge, Bd. **12**(1), 1–24.

Beverage, J. P., and Culbertson J. K. (1964). Hyperconcentrations of suspended sediment. *J. Hydraulics Div. Am. Soc. Civil Engrs*, **90** (HY6), 117–128.

Blaney, H. F., and Criddle, W. D. (1947). *A method of estimating water requirements in irrigated areas from climatological data*, Soil Conservation Service (Mimeo), United States Department of Agriculture, Washington, D.C.

Bryan, K. (1928). Historic evidence on changes in the channel of the Rio Puerco; a tributary of the Rio Grande in New Mexico. *J. Geology*, **36**, 265–282.

Bryan, R. B. (1969). The relative erodibility of soils developed in the Peak District of Derbyshire. *Geogr. Ann.*, **51A**, 145–159.

Bryan, R. B. (1976). Considerations of soil erodibility indices and sheetwash. *Catena*, **3**, 99–111.

Bryan, R. B., Yair, A., and Hodges, W. K. (1978). Factors controlling the initiation of runoff and piping in Dinosaur Provincial Park Badlands, Alberta, Canada. *Z. für Geomorph*. Suppl. Bd. **29**, 151–168.

Bull, W. B. (1964). History and causes of channel trenching in the western Fresno County, California. *Am. J. Sci.*, **262**, 249–58.

Burcham, L. T. (1957). *California Range Land: An historical study of the range resources of California*, California Division of Forestry, Sacramento.

Burkham, D. E. (1976). Hydraulic effects of changes in bottom-land vegetation on three major floods, Gila River in Southeastern Arizona. *US Geol. Surv. Prof. Paper*, **655-J**, 14.

Burt, T. P., and Williams, P. J. (1976). Hydraulic conductivity in frozen soils. *Earth Surface Proc.*, **1**(4), 349–360.

Butcher, G. C., and Thornes, J. B. (1978). Spatial variability of runoff processes in an ephemeral channel. *Z. für Geomorph.*, Suppl. Bd. **29**, 83–92.

Carson, M. A., and Kirkby, M. J. (1972). *Hillslope Form and Process*, Cambridge University Press. 475 pp.

Childs, E. C., Collis-George, N., and Holmes, J. W. (1957). Permeability measurements in the field as an assessment of anisotropy and structure development. *J. Soil Sci.*, **8**, 27–38.

Chow, V. T. (1959). *Open-Channel Hydraulics*, McGraw-Hill, Koga Kusha, Tokyo. 680 pp.

Cooke, R. U., and Reeves, R. W. (1976). *Arroyos and Environmental Change in the American South-West*, Oxford Research Studies in Geography, Oxford University Press. 213 pp.

Cooke, R. U., and Warren, A. (1973). *Geomorphology in Deserts*, Batsford, London, 413 pp.

De Bano, L. F. (1975). Infiltration, evaporation and water movement as related to water repellency. In: *Soil Conditioners* (Moldenauer, W. C. Ed.), Soil Science Society of America, Special Publication No. 7, pp. 25–33.

De Ploey, J., Savat, J., and Moeyersons, J. (1976). The differential impact of some soil loss factors on flow, runoff, creep, and wash. *Earth Surface Proc.*, **1**(2), 151–162.

Douglas, I. (1967). Man, vegetation and sediment yield of rivers. *Nature*, **215**, 925–928.

Dixon, R. M. (1975). Infiltration control through soil surface management. *Proceedings of Symposium on Watershed Management*, American Society Civil Engineers, Logan, Utah, pp. 543–567.

Dils, R. E. (1957). *A Guide to the Coweeta Hydrologic Laboratory, SE Forest Experimental Station*, Forest Service, Washington, D.C., US Dept of agriculture, 40 pp.

Dunin, F. X. (1976). Infiltration: its simulation for field conditions. In: *Facets of Hydrology* (J. C. Rodda, Ed.), Wiley, Chichester, pp. 199–229.

Emmett, W. W. (1970). The hydraulics of overland flow on hillslopes. *U.S. Geol. Survey Prof. Paper*, **662-A**, 46.

Farres, P. (1978). The role of time and aggregate size in the crusting process. *Earth Surface Proc.*, **3**, 243–254.

Foster, R. L. and Martin, G. L. (1969). Effects of unit weight and slope on erosion. *Irrigation and Drainage Div., Proc. Am. Soc. Civil Engrs*, **95** (IR4), 551–561.

Foster, G. R., and Meyer, L. D. (1972). A closed-form soil erosion equation for upland areas. In: *Sedimentation: Symposium to Honour H. A. Einstein* (H. W. Shen, Ed.), Colorado State University, Vol. 12, pp. 1–9.

Fournier, F. (1960). *Climat et érosion: la relation entre l'érosion du sol par l'eau et les précipitations atmospheriques*, Presses Universitaires de France, Paris. 201 pp.

Freeze, R. A. (1972). Role of subsurface flow in generating subsurface runoff upstream source areas. *Water Res. Res.*, **8**, 1272–1283.

Gersper, P. L., and Holowaychuk, N. (1971). Some effects of stem flow from forest canopy trees on the chemical properties of soils. *Ecology*, **52**, 691–702.

Glendening, C. E., Pase, C. P., and Ingebo, P. (1961). Preliminary hydrologic effects of wildfire in Chaparral. *Proceedings, Fifth Annual Arizona Watershed Symposium*, Sept. pp. 146–153.

Glymph, L. M. (1951). *Relation of Sedimentation to Accelerated Erosion in the Missouri River Basin*, Soil Conservation Service, United States Department of Agriculture, Washington, D.C.

Gregory, K. J. (1974). Stream flow and building activity. In: *Fluvial Processes in Instrumental Watersheds* (K. J. Gregory and D. E. Walling, Eds), Institute of British Geographers, Special Publication. No. 6 pp. 107–122.

Gregory, K. J., and Walling, D. E. (1973). *Drainage Basin Form and Process*, Arnold. 456 pp.

Goddin, J. R.. and McKell, C. M. (1971). Shrub productivity: A reappraisal of arid lands. In: *Food, Fiber and the Arid Lands* (McGinnies, W. G., Goldman, B. J., and Paylove, P., Eds), University of Arizona Press, Tusson, Arizona, 235–246.

Grissinger, E. H. and Asmussen, L. E. (1963). Discussion of 'channel stability in undisturbed cohesive soils, by E. M. Flaxman. *J. Hydraulics Div., Proc. Am. Soc. Civil Engrs*, **89** (HY6), 259–64.

Harr, R. D. (1977). Water Flux in soil and subsoil on a steep forested slope. *J. Hydrology*, **33**, 37–58.

Heede, B. H. (1971). *Characteristics and Processes of Soil Piping in Gullies*, Forest Service Service Research Paper RM-58, Department of Agriculture, United States, Washington, D.C. 15 pp.

Heede, B. H. (1975). Stages in the Development of Gullies in the West. In: *Present and Prospective Technology for Predicting Sediment Yields and Sources*—Proceedings of Sediment-Yield Workshop, United States Department of Agriculture Sedimentation Laboratory, Oxford, Mississippi, November 1972 (Agricultural Research Service Report ARS-S-40), United States Department of Agriculture, Washington, D.C., pp. 115–161.

Henderson, F. M. (1963). Stability of alluvial channels. *Trans. Am. Soc. Civil Engrs*, **128**(1), 657–686.

Hillel, D. I., and Gardener, W. R. (1970). Transient infiltration into crust-topped profiles. *Soil Sci.*, **109**, 69–76.

Hjulström, F. (1935). Studies of the morphological activity of rivers as illustrated by the River Fyris. *Bulletin of the Geological Institute, University of Uppasla*, **25**, 221–527.

Hooke, R. Le B. (1967). Steady-state relationships on arid-region alluvial fans in closed basins, *Am. J. Sci.*, **266**, 609–629.

Horton, R. E., Leach, H. R. and Van Vliet, R. (1934). Laminar sheet flow. *Trans. Am. Geophys. Union*, Hydrology, **15**(2), 393–404.

Horton, R. E. (1933). The role of infiltration in the hydrological cycle. *Trans. Am. Geophys. Union*, **14**, 446–460.

Horton, R. E. (1945). Erosional development of streams and their drainage basins: hydrophysical approach to quantitative morphology. *Bull. Geol. Soc. Am.*, **56**, 275–370.

Hudson, M. W. (1965). *The Influence of Rainfall on the Mechanics of Soil Erosion*, M.Sc. Thesis, University of Cape Town.

Huntington, E. (1914). The climatic factor as illustrated in arid America. *Carnegie Institution Washington, Publication No. 192*.

Imeson, A. C. (1977). Splash erosion, animal activity and sediment supply in a small forested Luxembourg Catchment. *Earth Surface Proc.* **2**(2–3), 153–160.

Imeson, A. C., and Jungerius, P. D. (1977). The widening of valley incisions by soil fall in a forested Keuper Area, Luxembourg. *Earth Surface Proc.*, **2**, 2–3, 141–152.

Imeson, A. C., and Kwadd, F. J. (1976). Some effects of burrowing animals and slope processes in the Luxembourg Ardennes. *Geogr. Ann.*, **58A**, 317–28.

Ingles, O. G. (1962). Bonding forces in soils, part 3: A theory of tensile strength for stabilized and naturally coherent soils. *Proc. First Conference of the Australian Road Research Board*, Vol. **1**, 1025–1047.

Ireland, H. A., Sharp, C. P. and Eargle, D. H. (1939). *Principles of Gully Erosion in the Piedment of South Carolina*, Technical Bulletin No. 633, United States Department of Agriculture, Washington, D.C.

Izzard, C. F. (1944). The surface profile of overland flow. *Trans. Am. Geophys. Union*, **25**(6), 959–968.

Jones, A. (1971). Soil piping and stream channel initiation. *Water Res. Res.*, **7**, 602–610.

Keppel, R. V., and Renard, K. G. (1962). Transmission losses in ephemeral stream beds. *J. Hydraulics Div., Proc. Am. Soc. Civil Engrs*, **88**(HY3), 59–68.

Keulegan, G. H. (1944). Laws of turbulent flow in open channels. *Journal of Research, National Bureau of Standards*, **21**, 707–741.

Knapp, B. J. (1970). A note on throughflow and overland flow in steep mountain watersheds. *Reading Geographer*, **1**, 40–43.

Kirkby, M. J. (1969). Infiltration, throughflow, and overland flow. In: *Water, Earth, and Man*, (R. J. Chorley, Ed.), Methuen, London.

Kirkby, M. J. (1978a). Implications for sediment transport. In: *Hillslope Hydrology* (M. J. Kirkby, Ed.), Wiley, Chichester, pp. 325–364.

Kirkby, M. J. (1978b). The stream head as a significant geomorphic threshold. School of Geography, *University of Leeds Working Paper 216*, 17 pp.

Kirkby, M. J., and Chorley, R. J. (1967). Throughflow, overland flow, and erosion. *Bull. Int. Assoc. Sci. Hydrology*, **12**, 5–21.

Kirkby, A. V., and Kirkby, M. J. (1974). Surface wash at the semi-arid break in slope. *Z. fur Geomorph.*, Neue Folge, Suppl. Bd. **21**, 151–76.

Kirkby, M. J. (Ed.) (1978). *Hillslope Hydrology*, Wiley, Chichester.

Lam, K. C. (1977). Patterns and rates of slopewash on badlands of Hong Kong. *Earth Surface Proc.* **2**(4), 319–32.

Langbein, W. B., and Schumm, S. A. (1958). Yield of sediment in relation to mean annual precipitation. *Trans. Am. Geophys. Union,* **39**, 1076–84.

Lee, I. K. (Ed.) (1978). *Soil Mechanics—Selected Topics,* Butterworths, London.

Leopold, L. B., Emmett, W. W. and Myrick, R. M. (1966). Channel and hillslope processes in a semi-arid area. *U.S. Geol. Surv. Prof. Papers,* **352-G**, 193–253.

Leopold, L. B., and Wolman, M. G. (1957). River channel patterns—braided, meandering, and straight. *US Geol. Surv. Prof. Paper,* **282-B**, 39–85.

Li, R. M., Simons, D. B. and Carder, D. R. (1976). Mathematical modelling of overland flow for soil erosion. *National Soil Erosion Conference,* Purdue University, Lafayette, May 25–26.

Linden, D. R., Dixon, R. M., and Guitjens, J. C. (1977). Soil air pressure under successive border irrigations and simulated rain. *Soil Sci.,* **124**, 3, 135–138.

Low, P. F. (1961). Physical chemistry of clay-water interaction. *Adv. in Agron.,* **13**, 269–327.

Lusby, G. C. (1970). Hydrologic and biotic effects of grazing versus non-grazing near Grand Junction, Colorado. *US Geol. Surv. Prof. Paper,* **700-B**, 232–236.

McIntyre, D. S. (1958). Permeability measurements of soil crusts formed by raindrop impact. *Soil Sci.,* **85**, 185–189.

Melton, M. A. (1965). The geomorphic and palaeoclimatic significance of alluvial deposits in Southern Arizona. *J. Geology,* **66**, 177–194.

Meyer, L. D., Foster, G. R. and Romkens, J. M. (1975). In: *Present and Prospective Technology for Predicting Sediment Yields and Sources*—Proceedings of Sediment-Yield Workshop, United States Department of Agriculture Sedimentation Laboratory, Oxford, Mississippi, November 1972 (Agricultural Research Service Report ARS-S-40), United States Department of Agriculture, Washington, D.C., pp. 177–189.

Meyer, L. D., and Monke, E. J. (1965). Mechanics of soil erosion by rainfall and overland flow. *Trans. Am. Soc. Agric. Engrs,* **8**(4), 572–577 and 580.

Middleton, H. E. (1930). *Properties of soils which influence soil erosion.* Technical Bulletin 178, 1–16. United States Department of Agriculture, Washington, D.C. 16 pp.

Mitchell, J. K. (1976). *Fundamentals of Soil Behaviour,* Wiley, New York. 422 pp.

Moeyersons, J., and de Ploey, J. (1976). Quantitative determination of splash erosion, simulated on unvegetated slopes. *Z. für Geomorph.* suppl. Bd. **25**, 120–131.

Mosley, M. P. (1973). Rainsplash and the convexity of badland divides. *Z. für Geomorph.,* Suppl. Bd., **18**, 10–25.

Musgrave, G. W. (1947). The quantitative evaluation of factors in water erosion, *J. Soil and Water Conserv.,* **2**(3), 133–138.

Muzik, I. (1974). Laboratory experiments with surface runoff. *J. Hydraulics Div., Am. Soc. Civil Engrs,* **100**(HY4), 501–513.

Nord, E. C. (1965). Autecology of Bitterbush in California. *Ecological Monographs,* **35**(3), 309–334.

Nortcliff, S., and Thornes, J. B. (1977). Water and cation movement in a tropical rain-forest environment: I. Objectives, experimental design, and preliminary results. London School of Economics, *Graduate School of Geography, Discussion Paper,* 62.

Nortcliff, S., Thornes, J. B., and Waylen, M. J. (1979). Tropical weathering: a hydrological approach. *Amazoniana,* **6**, 4, 557–568.

Palmer, V. J. (1948). Retardance coefficients for low flow in channels lined with vegetation. *Trans. Am. Geophys. Union,* **27**, 35–73.

Parthenaides, E. (1965). Erosion and deposition of cohesive soils. *Proc. Am. Soc. Civil Engrs., J. Hydraulics Div.,* **91**(HYi), 105–139.

Parsons, D. A. (1949). *Depths of overland flow*, Soil Conservation Service Technical Paper No. 82, United States Department of Agriculture, Washington D.C. 33 pp.

Patton, P. C., and Schumm, S. A., (1975). Gully erosion, Northwestern Colorado: A threshold phenomena. *Geology*, 1, 88–90.

Petryk, S., and Bosmajian, G. (1975). Analysis of flow through vegetation, *J. Hydraulics Div., Proc. Am. Soc. Civil Engrs*, 101(HY7), 1105–1120.

Phillip, J. R. (1969). Theory of infiltration. In: *Advances in Hydroscience*, (V. T. Chew, Ed.), Vol. 5, pp. 214–296.

Pierce, R. S., Hornbeck, J. W., Likens, G. E. and Borman, F. H. (1970). Effect of elimination of vegetation on stream water quantity and quality. *Int. Assoc. Scient. Hydro. Pub.*, 96, 311–328.

Piest, R. F., Bradford, J. M. and Spomer, R. G. (1975). Mechanisms of erosion and sediment movement from gullies. In: *Present and Prospective Technology for Predicting Sediment Yields and Sources*—Proceedings of Sediment-Yield Workshop, United States Department of Agriculture Sedimentation Laboratory, Oxford, Mississippi, November 1972 (Agricultural Research Service Report ARS-S-40), United States Department of Agriculture, Washington, D.C., pp. 162–176.

Piest, R. F., Bradford, J. M., and Wyatt, G. M. (1975). Soil erosion and sediment transport from gullies. *J. Hydraulics Div., Proc. Am. Soc. Civil Engrs.*, 101(HY1), 65–80.

Ploey, J. de, and Moeyersons, J. (1976). Runoff creep of coarse debris: experimental data in some field observations. *Catena*, 2, 275–288.

Ploey, J. de., Savat, J., and Moeyersons, J. (1976). The differential impact of some soil loss factors on flow, runoff, creep, and rainwash. *Earth Surface Proc.*, 1(2), 151–162.

Rainwater, F. H. (1962). United States Geological Survey. *Atlas*, HA-61.

Raudkivi, A. J., (1967). *Loose Boundary Hydraulics*, Pergamon, Oxford.

Ree, W. O. (1958). Retardation coefficients for row crops in diversion terraces. *Trans. Am. Soc. Agric. Engrs*, 1, 78–80.

Rowe, B. B. (1944). Soil moisture records from burned and unburned plots in certain grazing areas of California. *Trans. Am. Geophys. Union*, 25, 84–86.

Rubin, J. (1966). Theory of rainfall uptake by soils initially drier than their field capacity and its applications. *Water Res. Res.*, 2(4), 739–749.

Savage, S. M. (1975). Natural and fire induced water repellency. In: *Soil Conditioners* (W. C. Moldenauer, Ed.), Soil Science Society of America, Special Publication No. 7, pp. 165–172.

Savat, J. (1977). The hydraulics of sheet flow on a smooth surface and the effect of simulated rainfall. *Earth Surface Proc.*, 2(2–3), 125–140.

Savat, J., and Poeson, J. (1977). Splash and discontinuous runoff as creators of fine sandy lag deposits with Kalahari Sands. *Catena*, 4, 321, 332.

Scholl, D. G. (1971). Soil wettability in Utah Juniper Stands. *Proc. Soil Sci. Soc. Am.*, 35, 344–345.

Schumm, S. A. (1956). The role of creep and rainwash on the retreat of badland slopes. *Am. J. Sci.*, 254, 693–706.

Schumm, S. A. (1965). Quaternary palaeohydrology. In: *The Quaternary of the United States* H. E. Wright and D. G. Frey, Eds), Princeton, pp. 783–794.

Schumm, S. A. (1973). Geomorphic thresholds and complex response of drainage systems. In: *Fluvial Geomorphology* (M. Morisawa, Ed.), Publications in Geomorphology, State University of New York, Binghamton, New York.

Schumm, S. A. (1977). *The Fluvial System*, Wiley Interscience. 338 pp.

Schumm, S. A., and Hadley, R. F. (1957). Arroyos and the semi-arid cycle of erosion. *Am. J. Sci.*, 225, 161–174.

Schumm, S. A., and Khan, H. R. (1972). Experimental study of channel patterns. *Bull. Geol. Soc. Am.*, **85**, 1755–1770.

Schumm, S. A., and Lichty, R. W. (1963). Channel widening and flood plain construction along Cimarron River in Southwestern Kansas. *US Geol. Surv. Prof. Paper*, **352-H**.

Schumm, S. A., and Lusby, G. C. (1967). Seasonal variations of infiltration capacity and run-off on hillslopes in Western Colorado. *J. Geophys. Res.*, **68**, 3655–66.

Scoging, H. (1978). The relevance of time and space in predicting sheet erosion from semi-arid fields. *Workshop on Assessment of Erosion*, Ghent, Belgium. March.

Scoging, H., and Thornes, J. B. (1980). Infiltration characteristics in a semiarid environment. *Int. Assoc. Scient. Hydrol.*, *Pub.* **128**, 159–168.

Scott, K. M. (1973). Scour and fill in Tujunga Wash—A fanhead valley in Southern California—1969. *US Geol. Surv. Prof. Paper*, **658-B**.

Selby, M. J. (1972). The relationship between land use and erosion in the Central North Island, New Zealand. *J. Hydrology* (N.Z.), **11**, 73–87.

Sherard, J. L., Dunnigan, K. P., and Decker, R. S. (1976). Identification and nature of dispersive soils. *J. Geotech. Div.*, *Proc. Am. Soc. Civil Engrs*, **102**(GT4), 287–301.

Simons, D. B., Richardson, E. V., and Haushild, W. L. (1963). Some effects of fine sediments on flow phenomena. *Water Supply Paper 1498G*, United States Geological Survey, Washington, D.C.

Smith, R. E. (1972). The infiltration envelope: results from a theoretical infiltrometer. *J. Hydrology*, **17**, 1–21.

Statham, I. (1976). Debris flows on vegetated screes in the Black Mountain, Carmarthanshire, *Earth Surface Proc.*, **1**(2), 173–180.

Stocking, M. (1972). *A Geographical Analysis of the Factors in the Erosion of Soils in Rhodesia*, M.Phil. Thesis, University of London.

Stocking, M. (1977). *Erosional Soils in Central Rhodesia*, Unpublished Ph.D. Thesis, University of London.

Temple, P. H., and Rapp, A. (1972). Landslides in the Mgeta Area, Western Uluguru Mountains, Tanzania. In: *Studies of Soil Erosion and Sedimentation in Tanzania.* (A. Rapp, L. Berry, and P. H. Temple, Eds.), University of Uppsala, pp. 157–193.

Terzaghi, K., (1950). Mechanism of Landslides. *Engineering Geology* (Berkley Vol.), pp. 83–123.

Thomas, M. F. (1977). Contemporary denudation systems and the effects of climatic change in the humid tropics: Some problems in Sierra Leone. *Proceedings of the VI Anglo-Polish Seminar*, Sheffield University.

Thornes, J. B. (1975). Lithological control of hillslope erosion in the Soria Area, Dueroalto, Spain. *Boletin Geologico Minero*, **85**, 11–19.

Thornes, J. B. (1976). *Semi-arid Erosional Systems*, London School of Economics, Department of Geography, Occasional Papers, No. 7.

Thornes, J. B. (1977). Channel changes in ephemeral streams: observations. problems and models. In: *River Channel Changes*, (K. J. Gregory, Ed.), pp. 317–335.

Thornes, J. B., and Brunsden, D. (1977). *Geomorphology and Time*, Methuen, London. 208 pp.

Thornwaite, C. W. (1948). An approach towards the rational classification of climate. *Geographical Review*, **38**. 85–94.

Tuckfield, G. G. (1964). Gully erosion in the New Forest, Hampshire. *Am. J. Sci.*, **262** 795–807.

Veihmeyer, F. J. (1953). Use of water by native vegetation versus grasses and forbs on Watersheds. *Trans. Am. Geophys. Union*, **34**, 201–212.

Walling, D. E. (1974). Suspended sediment and solute yields from a small catchment prior to urbanization. In: *Fluvial Processes in Instrumented Catchments*. (K. J. Gregory and D. E. Walling, Eds), Institute British Geographers, Special Publication No. 3, pp. 33–45.

Wentworth, C. K. (1943). Soil avalanches on Oahu, Hawaii. *Bull. Geol. Soc. Am.*, **54**, 53–64.

Wertz, J. B. (1966). The flood cycle of ephemeral mountain streams in the Southwestern United States. *Annals. Association of American Geographers*, **56**, 589–633.

Weyman, D. R. (1973). Runoff process, contributing area and streamflow in a small upland catchment. In: *Fluvial Processes in Instrumented Watresheds* (K. J. Gregory, and D. E. Walling, Eds), Institute of British Geographers, Special Publication No. 6, pp. 33–45.

White, E. M., and Riecken, F. F. (1955). Brunizem gray-brown podsolic soil biosequences. *Proc. Soil. Sci. Soc. Am.*, **19**, 504–509.

Williams, G. P., and Guy, H. P. (1973). Erosional and depositional aspects of Hurricane Camille in Virginia, 1969. *US Geol. Survey. Prof. Paper*, **804**, 80 pp.

Willis, J. C. (1971). *Erosion by Concentrated Flow*, Agricultural Research Service Report ARS-41-179, United States Department of Agriculture, Washington, D.C. 15 pp.

Wischmeier, W. H., and Smith, D. D. (1958). Rainfall energy and its relationship to soil loss. *Trans. Am. Geophys. Union*, **39**, 285–91.

Wischmeier, W. H., and Smith D. D. (1965). *Rainfall-erosion Losses from Cropland East of the Rocky Mountains*, Agricultural Handbook No. 282, United States Department of Agriculture, Washington, D.C.

Wolman, M. G. (1955). The natural channel of Brandywine Creek, Pennsylvania. *US Geol. Surv. Prof. Paper*, **371**.

Wolman, M. G., and Miller, J. P. (1960). Magnitude and frequency of forces in geomorphic processes. *J. Geol.*, **68**, 54–74.

Wood, C. C., Aitchison, G. D. and Ingles, O. G. (1964). Physics-chemical and engineering aspects of piping failures in the small earth dams. *Proceedings of the coloquium on Failure of Small Earth Dams*, Soil Mechanics Section, CSIRO, Melbourne.

Woolhiser, D. A. (1975). Simulation of unsteady overland flow. In: *Unsteady Flow in Open Channels* (K. Mahmood and V. Yevjevich, Eds), Water Resources Publications, Fort collins, Colorado. Vol. 2, pp. 485–508.

Woolhiser, D. A., Holland, M. E., Smith, G. L., and Smith, R. E. (1971). Experimental investigation of converging overland flow. *Trans. Am. Soc. Agric. Engrs*, **14** (4), 684–687.

Yair, A., (1973). Theoretical considerations on the evolution of convex hillslopes, *Z. für Geomorph.*, Neue Folge, Suppl. Bd. **18**, 1–9.

Yair, A., and Klein, M. (1973). The influence of surface properties on flow and erosion processes on debris covered slopes in an arid area. *Catena*, **1**, 1–18.

Yair, A., and Lavee, H. (1976). Runoff generation processes and runoff yield from arid talus mantled slopes. *Earth Surface Proc.* **1**, 285–247.

Young, R. A., and Mutchler, C. K. (1969). Soil and water movement in small tillage channels. *Trans. Am. Soc. Agric. Engrs*, **12**, 543–5.

Youngs, E. G. (1964). Water movement in soils. *The State and Movement of Water in Living Organisms*, Cambridge University Press, pp. 89–112.

Zingg, A. W. (1940). Degree and length of landslope as it affects soil loss in runoff, *Agric. Engng*, **21**, 59–64.

Soil Erosion
Edited by M. J. Kirkby and R. P. C. Morgan
© 1980 John Wiley and Sons Ltd.

CHAPTER 6

Modelling water erosion processes

M. J. Kirkby

School of Geography, University of Leeds, UK

6.1 STRATEGIES FOR MODEL BUILDING

The Universal Soil Loss Equation (Wischmeier and Smith, 1960, 1965) has been the most widely applied erosion model since the late 1960s. Its value and degree of adoption, particularly in the USA can be judged by the rate at which it is still being extended to incorporate new regions (e.g. The Pacific NW: McCool, Papendick, and Brooks, 1975). new environments (e.g. undisturbed areas: Wischmeier, 1975; roadsides: Meyer, Schoenberger, and Huddleston, 1975) and new scales (e.g. drainage basins: Williams, 1975). Despite its widespread use and the breadth of experience which it incorporates, the equation suffers from the conceptual defect that rainfall and soil factors (among others) cannot simply be multiplied together because of the subtractive effect of soil infiltration capacity in generating erosive runoff from a given rainfall. A fuller understanding of soil erosion must therefore go closer to first principles, and this chapter is essentially concerned with the development of a physically based component model. Research on soil erosion processes, described elsewhere in this book, allows at least some progress towards a physically based model, which in turn suggests directions for future research.

Alternative approaches to erosion modelling have been reviewed by Renard (1977) who distinguishes component models, mystery or black-box models, and stochastic models. Black-box models most commonly rely on multiple regression methods, usually with logarithms of sediment yield and other variables (Flaxman, 1972; Meeuwig, 1970) to produce a multiplicative form of forecasting equation similar to that of the Universal Soil Loss Equation (USLE) though with some differences in variables or factors. The greatest variety is seen in surface and soil variables, perhaps indicating the complexity of the physical influence of soil properties on erosion rates (Thornes, Chapter 5).

Stochastic models are needed to generate a sample sequence of events drawn from a known underlying distribution. As a forecasting tool they can be used

183

for generating sediment yield directly (Woolhiser and Todorovic, 1971). Alternatively rainfall or runoff can be generated stochastically and linked with a sediment equation or rating curve to forecast sediment yield (Woolhiser and Blinco, 1975). In principle stochastic rainfall inputs may also be linked to many physically based or black box models to generate the time-distribution of sediment yield and extrapolate from actual measurement periods to 'average' years. Following established hydrology, stochastic sediment models have tended to forecast yields for drainage basins rather than hillslope plots. Their range of application is therefore complementary to that for physically based models, which have greater potential at the plot scale.

In working towards an appropriate physically based component model for soil erosion, it will be helpful to state some basic modelling principles from which to work. These may be summarized as efficiency, range of validity, and constraints. For a model to be efficient those processes which have the greatest influence on the overall behaviour of the model must be simulated in the greatest detail. Quantitatively less important processes can in practise be ignored if never near-dominant or dominant. If a minor process is locally dominant, then it should be modelled with a minimum of parameters and complexity, and with reasonable accuracy necessary only for its zone of dominance. No practicable model is valid for all conditions, but a model can also gain in efficiency in another sense by maximizing its range of validity for a given number of parameters. If five given parameters, say, are thought to describe the main sources of variation in sediment yield over the *whole* of an area, then a good model is one which is valid throughout the area and not just for a part of it. In maximizing the range of a model, one technique is to make use of functions which take physically reasonable values at zero and infinity, so that responses to extreme conditions remain inherently plausible. The main constraints for an erosion model are thought to relate to conservation of mass for sediment and water, and to meeting boundary conditions at plot boundaries, divide, and slope base as relevant. Mass balance models of the type advocated by Ahnert (1964) and Foster and Meyer (1975) therefore provide an appropriate model structure.

The literature on soil erosion processes shows a wide range of mechanisms, with the greatest variety in discussions of rainsplash and inter-rill transport and detachment. Sediment yields directly from rainsplash are however rather low even under maximal conditions. Splash yields are dominant only near the upper end of erosion plots (or near natural divides) and where rilling is absent. In other words high sediment yields are associated with rill or gully erosion. It is therefore argued that the construction of an efficient erosion model should concentrate on the accurate representation of channelled erosion, and that a relatively crude model for inter-rill processes is sufficient. Given the complexity of inter-rill mechanisms it may therefore be most practicable, at the present state of the art, to fall back on a largely empirical model for rainsplash and inter-rill wash rather than to insist on a fully physical representation of these processes.

Soil erosion as a problem, and soil erosion models in consequence, have been most concerned with cultivated areas. Although efficient model construction argues for greatest physical understanding for these problem areas, it is important to be able to forecast sediment losses from natural areas as well. The two main reasons for needing reliable models for natural areas are first because they provide the base-line against which attempts to control accelerated erosion must be judged (Smith and Stamey, 1965); and second because natural soil erosion, although slow, is nevertheless one of the dominant processes of landscape lowering. An appropriate sediment yield model is therefore of importance to geomorphologists in understanding landscape evolution. These considerations suggest an appropriate range of validity for an erosion model, in that it should cover natural and disturbed surfaces, bare and vegetated surfaces and areas near divides as well as downslope. If the evolution of natural landscapes is to be modelled as well as the erosion of field plots, then a range of time spans from a few minutes up to at least thousands of years must be encompassed. In principle it may be possible to disaggregate a model based on a long unit time span, but physical understanding is greatest for short time spans, so that aggregation for long spans is likely to be preferred. The aggregation must be explicit if the same underlying process model is to be applied throughout.

These criteria for a sound model which satisfies the condition of efficiency, particularly with regard to splash erosion, and which is valid for natural as well as cultivated hillsides, set a standard against which existing physically-based models can be assessed. In this way suggestions for improving or replacing existing models can be approached rationally.

6.2 EXISTING PHYSICALLY-BASED MODELS

6.2.1 The main problems

Recent work has concentrated on the modelling of three main aspects of hillslope soil erosion. They are first the rates of soil transporting capacity and detachment capacity by raindrop impact; second the rates of transport and detachment capacities by overland flow, either rilled or unrilled; and third the interaction between transport and detachment capacities in determining actual transport and erosion rates. A fourth area of concern which is now also receiving some attention is the interaction between rill and inter-rill processes. In reviewing these topics, repeated reference to the seminal paper by Meyer and Wischmeier (1969) indicates the continuity of soil erosion research which seems likely to lead ultimately to the abandonment or radical modification of the USLE (Wischmeier and Smith, 1960).

All recent erosion modelling implicitly or explicitly works within the constraints of a mass balance framework, which forms a common link between short and long term erosion models. Thus for a laterally uniform hillslope:

$$\frac{\partial S}{\partial x} + \frac{\partial s}{\partial t} = A \qquad (6.1)$$

where

S = rate of actual sediment transport (averaged across slope),
x = horizontal distance from divide or plot head,
z = elevation of soil surface (averaged across slope),
t = time elapsed, and
A = rate of external addition of sediment (averaged across slope).

This equation assumes no change in soil density. It can be applied for individual processes provided that rate of addition (which may be plus or minus) includes all inter-process transfers. If Equation (6.1) is applied for all processes together the rate of addition is normally zero. This equation has been widely used (Meyer and Wischmeier, 1969; Culling, 1963; Young, 1963; and many others) in various forms, and is not in dispute.

6.2.2 Transport and detachment interaction

The interaction of transport and detachment (topic 3 above) is also a necessary part of the modelling framework, and will therefore be discussed next. Two

SOIL EROSION PROCESS

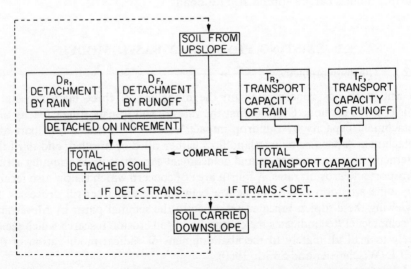

Figure 6.1. Possible interaction between transport and detachment processes (from Meyer and Wischmeier, 1969)

basic approaches have been proposed, although neither has been tested exhaustively and both yield plausible results. Meyer and Wischmeier (1969) have proposed that the detachment *capacity* at each point should be compared with the transporting capacity and actual transport rate at that point. The *actual* increment in transport rate is then taken as the lesser. This concept is illustrated by the flow diagram in Figure 6.1. Rowlison and Martin (1971) have illustrated

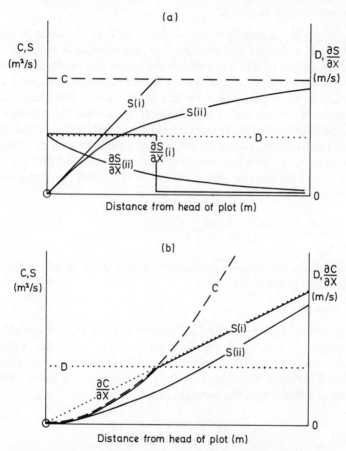

Figure 6.2. Transport/detachment interaction models (a) Initially clean flow of constant transport (*C*) and detachment (*D*) capacities, flowing down a plot and eroding soil as it flows. (b) Rainfall on a plot with increasing flow and transport capacity, but with constant detachment capacity Actual sediment transported (*S*) predicted by two models:
  (i) Transport/detachment comparison (Meyer and Wischmeier, Wischmeier, 1969)
  (ii) Transport deficit model (Foster and Meyer, 1972)

a variant of this concept qualitatively and compared it with their experimental data, but with inconclusive results. In the notation of Equation (6.1) above;

$$\frac{\partial S}{\partial x} = D \text{ if either } \frac{\partial C}{\partial x} \geqslant D \text{ or } S < C \tag{6.2}$$

where

C is the transporting capacity, and
D is the detachment capacity.

If neither condition is met then $S = C$ until the first condition is met. The predicted actual sediment transport for two simple cases is shown in Figure 6.2(a) and (b), model (i). The main disadvantages of this model is that discontinuities in erosion rate, $\partial S/\partial x$, arise as in Figure 6.2(a) where actual transport rate increases to the capacity rate, but not as in Figure 6.2(b) where transport rate begins to fall below the capacity rate. If the surface morphology of the slope is allowed to change then these discontinuities can be shown to diffuse out with only a small movement of sediment, so that adjustment will be rather rapid.

An alternative approach has been proposed by Foster and Meyer (1972), in which the rate of surface lowering is related to the deficit between actual and capacity transport rates. In the notation above:

$$\frac{\partial z}{\partial t} = \frac{C - S}{h} \tag{6.3}$$

where h, which has the dimensions of length, may depend on other variables. In Foster and Meyer's analysis;

$$h = C/D \tag{6.4}$$

so that the rate of erosion, $-\partial S/\partial x$, approaches the detachment capacity when actual transport is very much smaller than its capacity. In their subsequent analysis the ratio C/D is derived as a constant. Equation (6.3) has also been proposed from a different standpoint (Kirkby, 1971) in which the distance h is interpreted as a function of the constant ratio:

$$h = f \left(\frac{\text{cohesive resistance}}{\text{frictional resistance}} \right) \tag{6.5}$$

This equation implies a very rapid pick-up of material for non-cohesive deposits, and a slower response where cohesion is significant. In both cases, equation (6.3) forecasts a gradual rather than a sudden transition from detachment-limited to transport-limited removal, as is illustrated by model (ii) in Figure 6.2, based on Equations (6.3) and (6.4) above. For conditions of deposition ($C < S$), the parameter h is usually considered to take lower values than for erosion. Foster and Meyer (1975) argue that sediment removal is, in many cases, far below transporting capacity, so that Equation (6.3) may then be simplified to

$$-\partial z/\partial t = C/h \qquad (6.6)$$

This form may be a useful approximation under some erosive conditions, but is plainly inappropriate for deposition. Both direct evidence and intuition suggest some interaction between transport and detachment processes to be appropriate, but it is not clear how strong the link really is. Two relevant views on the transport deficit model of Equation (6.3) are first that it is the simplest first-order reaction model for the interaction (Bennett, 1974, p. 487); and second that it is a convenient mathematical device for combining both the transport-limited ($h \to 0$) and deficit limited (h large) cases in a single expression.

Existing models for transport and detachment by raindrop impact apply mainly to erosion plots with prepared surfaces. Selective removal of fines leads to progressive armouring by a lag deposit of the coarse material on many natural unvegetated surfaces. This provides a negative feedback which complicates the simplest view. Other complications arise from a variety of experimental conditions, and from some confusion in the literature concerning the difference between transport and detachment.

6.2.3 Rain splash process models

In an idealized detachment experiment, soil material within a very small (horizontal projection) area is labelled, and the net loss from the area is measured over a very short period of rainfall. No account is taken of material moving into the area. If a water layer covers the surface, then labelled material which is suspended in the water at the end of the rainfall period is also considered to be detached. In an idealized transport experiment, all materials above a line is labelled. After a very short period of rainfall (with or without a water layer), the net transport in the direction at right angles to the line is given as (total labelled material below line) minus (total unlabelled material above line). Because real experiments cannot generally reproduce these ideal conditions, there are serious problems of interpretation. Thus loss from a pan of soil tends to measure detachment for very small pans, and to measure transport if the pan is large relative to the mean travel distance of a soil grain. On low gradients, it is also important that material moving out of the top of the pan is subtracted from that moving out at the bottom, to obtain an estimate of net transport. As pans become large, flow processes also tend to complicate the measurement of rainsplash in isolation.

Rainsplash detachment rates have generally been related to storm energy and intensity (Free, 1960; Bubenzer and Jones, 1971) or momentum (Elwell and Stocking, 1973), which in turn are commonly correlated with intensity, with some concensus (Foster and Meyer, 1975) at a splash detachment rate,

$$D_s \propto i^2 \qquad (6.7)$$

where i is the instantaneous rainfall intensity. Palmer (1963) has measured the impact of raindrops beneath a water film, and found first an increasing impact until the water layer thickness was about 85 per cent of the drop diameter, and then a decline. This type of curve has been incorporated into the conceptual model of Rawlison and Martin (1971), but it is not clear whether the increased impact is effective in detaching additional soil *pro rata* or whether the impact is more diffused and so less effective. Other authors have assumed independence of water depth, or a decay function (David and Beer, 1975). The influence of soil type or grain size is also unclear, and Foster and Meyer (1975) suggest use of the USLE soil erodibility factor as an indicator. Surface sealing, the interaction of different grain sizes at the surface and changing surface properties both during and between rainstorms are among the factors which complicate the influence of soil type.

Rainsplash transport in the absence of a water layer is generally considered to be directly proportional to slope (Ekern, 1953) although there is some evidence that the exponent ranges between about 0.7 and 2.0 according to the way in which grain sizes alter on natural slopes (Kirkby and Kirkby, 1974, p. 173). Expressions which include a non-zero transport rate on zero slope (Meyer, Foster, and Romkens, 1975) are thought to show a misinterpretation of plot experiments, particularly with regard to plot head as well as plot foot losses, as described above. The effect of a thin water layer on rainsplash transport is confused by the difficulty of distinguishing transport generated by the flow itself. For this reason the two are sometimes combined as 'inter-rill' transport (Meyer, Foster, and Romkens, 1975), but this distinction blurs the process distinction, valuable though it is in other ways. There is evidence that large (20 mm) material is transported by rainsplash in a creeping motion, and that this motion is increased by flow which does not submerge the grains (Moeyersons, 1975), and is not rapid enough to roll or otherwise entrain them. There is no comparable evidence for finer material, but most authors implicitly assume that strict rainsplash transport capacity is unaffected or suppressed (along with detachment capacity) by moderate water depths.

6.2.4 Wash process models

The capacity of a water flow to detach soil under maximal conditions (i.e. initially clear water) has been related by most authors to the tractive stress at the bed. Given that even fine soils are carried mainly as aggregates, bedload equations are thought to be appropriate models for transport and detachment (Foster and Meyer, 1975, p. 195). Since bedload is commonly modelled as a balance between detachment and deposition, the detachment capacity is arguably more basic than the transporting capacity. Thus the rate of net sediment pick up

$$\partial S/\partial x = D - S/h \tag{6.8}$$

where h is a critical distance. The detachment capacity D will take different values according to the strength of the sediment, being greatest (D_0) under conditions of deposition, and least for clear water $(S = 0)$ flowing over an intact cohesive surface. This approach leads readily to Equations (6.3) and (6.4) above if it is assumed that detachment rate changes linearly between extreme values in proportion to the rate of settling sediment, S/h (see Section 6.3.1 below). The parameter h in Equations (6.3) and (6.8) is interpreted as the mean travel distance for an entrained sediment particle, and the capacity rate of transport is given by the maximum steady state value

$$C = hD_0 \qquad (6.9)$$

Meyer and Wischmeier (1969, p. 755) assume detachment rates directly proportional to tractive stress. Foster and Meyer (1975, p. 195) who quote exponents of tractive stress from 1.5 to 1.9 prefer the expression:

$$D \propto (\tau - \tau_c)^n \qquad (6.10)$$

where

τ is the shear stress, at the bed of the flow,
τ_c is its threshold value, and
n is an exponent in the range 1–2.

For low thresholds, that is fine material, then a simple exponent of shear stress is an adequate approximation, with a somewhat higher value for the exponent. Foster and Meyer (1975) use the exponent 1.5. By applying a Manning or Darcy-Weisbach equation for velocity;

$$v^2 = 2grs/f \qquad (6.11)$$

and the continuity equation;

$$q = vr \qquad (6.12)$$

where

v = flow velocity,
r = flow depth,
s = energy slope,
q = discharge per unit width, and
f = Darcy-Weisbach drag coefficient,

the expression for bed shear becomes:

$$\tau = \rho grs = 2^{-1/3} \rho^{2/3} f^{1/3} q^{2/3} s^{2/3} \qquad (6.13)$$

where

τ is the shear stress, and
ρ is the fluid density

Thus for detachment proportional to (shear)$^{1.5}$, and for constant drag coefficient, f, the rate of detachment:

$$D \propto qs \tag{6.14}$$

This expression is convenient and simple, and provides a link to expressions for bed material movement based on stream power per unit area (Bagnold, 1966), which is, in this notation, simply:

$$\tau v = \rho g q s \tag{6.15}$$

Flow transporting capacity has been partly dealt with in discussing detachment above. Most expressions for the transport capacity use the same exponents for discharge, slope, and velocity as for detachment, and this may be justified by assuming that the mean travel distance, h in Equation (6.9), is characteristic of the sediment rather than the flow. This assumption is far from self evident, in that for a particle lifted into the flow, the travel *time* depends largely on settling velocity but the travel *distance* in this time is also related to the flow velocity. An alternative approach to transporting capacity is represented in the recent erosion literature by Tödten (1976). He examines the concentration gradient of grains in a turbulent flow, following Vanoni (1946), to obtain:

$$\frac{c}{c_1} = \left(\frac{r-z}{z} \cdot \frac{z_1}{r-z_1} \right)^{\omega/\kappa(grs)^{1/2}} \tag{6.16}$$

where

c = concentration of grains at elevation z above the bed,
c_1 = concentration of grains at edge of boundary layer, at elevation z_1 above the bed,
w = settling velocity of grains, and
κ = von Kármán constant.

Following this approach, the generalized analogue of equation (6.8) above is:

$$V \frac{\partial c}{\partial x} + W \frac{\partial c}{\partial z} = \epsilon \frac{\partial^2 c}{\partial z^2} \tag{6.17}$$

where

x is the downstream direction coordinate, and
ϵ is the coefficient of sediment diffusion $[\simeq 0.25 \, \kappa r \, (rgs)^{1/2}]$.

Equation (6.17) can be solved for constant conditions, and the solution applied to gradually varied conditions. Thus:

$$\frac{c}{c_0} = \tfrac{1}{2} \text{erfc}\left(\frac{z}{2r} \sqrt{\frac{vr^2}{\epsilon x}} + \frac{wr}{2\epsilon} \sqrt{\frac{\epsilon x}{vr^2}} \right) + \tfrac{1}{2}[\exp(-wz/\epsilon)].$$

$$\text{erfc}\left(\frac{z}{2r} \sqrt{\frac{vr^2}{\epsilon x}} - \frac{wr}{2\epsilon} \sqrt{\frac{\epsilon x}{vr^2}} \right) \tag{6.18}$$

where

c_0 is the concentration at the bed ($z = 0$), and
erfc(x) is the error function complement $2/\sqrt{\pi}\int_x^\infty e^{-t^2}\, dt$.

This expression shows a progressive pick-up of material by an initially clean flow, and the process is essentially complete within a downslope distance of 100 flow depths, which for most hillslope flows is less than one metre. Thereafter, the concentration follows the equilibrium distribution:

$$c = c_0 \exp(-wz/\epsilon) \tag{6.19}$$

In this approach, Tödten (1976, p. 98) considers a zone of no erosion in which turbulent eddy velocities are insufficient to overcome the resistance of the grains, and this is a possibility which has not been discussed by recent authors, although it is implicit in Horton's (1945) paper. The possibility of a pick-up process which is limited by the detachment rate is however not considered, so that this approach is thought to offer no real advantage over that described by Equations (6.8) and (6.9) above.

6.2.5 Process interactions

The final area of concern in current modelling is the interaction between rill and inter-rill processes. This topic has been made more explicit in the papers of Foster and Meyer (1975) and Meyer, Foster, and Romkens (1975). The underlying issue is the relationship between each of the subprocesses involved in soil erosion, but the discussion to date has concentrated on the transfer of material eroded between rills to a (given) rill system. It is clear that inter-rill water flows do not build up indefinitely where rills are present, but that the water and any sediment it carries is eventually removed (or re-deposited) via the rill system. It might be expected therefore, that an increasing proportion of the rill sediment capacity would be taken up by inter-rill sediment contributions, but the field evidence (Meyer *et al.*, 1975, pp. 180–81) and the analysis of it below suggests that the inter-rill sediment has a negligible effect on the course of rill erosion. Instead inter-rill erosion at a cross-section increases to a constant value, while rill erosion starts slowly and then *increases* at a uniform rate (Figure 6.3). The conclusion to be drawn is either that rill erosion is essentially detachment-limited, with only a small proportion of its transport capacity filled (Foster and Meyer, 1975, p. 201) or that the inter-rill erosion is of such fine-grained material (silt-clay) that it is carried in suspension by the rillflow and so interferes minimally with the essentially bed load transport processes involved in rill erosion, which are dominated by sand material (Meyer *et al.*, 1975, pp. 186–187).

Related problems which have received less attention in the literature are the conditions for rill initiation, and the relationship between splash and both inter-

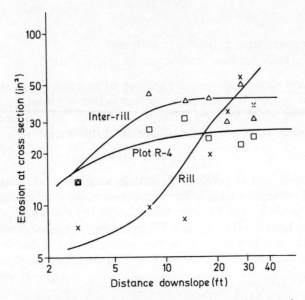

Figure 6.3. Local erosion rates from two plot experiments. For Plot R-4 (squares) rilling was negligible. For Plot R-3, rill (crosses) and inter-rill (triangles) erosion are separated (from Meyer, Foster, and Romkens, 1975, Figures 1 and 2)

rill wash. It is clear from Figure 6.3 that the presence of rills has little influence on inter-rill erosion, but that rills increase the total erosion, with increasing effect downslope. It is also clear that both inter-rill and rill wash are greatly reduced by placing net screens above the soil to reduce raindrop impact. In some experiments (Hudson and Jackson, 1959) this reduction is clearly associated with a corresponding reduction in runoff, although in other experiments the effect on runoff is not discussed (Meyer *et al.*, 1975, pp. 184–185). Conditions for rill initiation are not however analysed in either case, and the relationship between splash and inter-rill wash is blurred by their combination as inter-rill erosion or transport.

6.3 STEPS TOWARDS IMPROVED EROSION MODELS

6.3.1 Foundations

The most secure basis for erosion modelling remains the mass balance equation (6.1) above. Although energy or momentum balances may be locally appropriate, the overall conservation of these quantities forms an impractical basis for budgetting because of the large frictional losses in the mechanical system.

Mass balance must be maintained for each grain size, making allowance for aggregation, or aggregate breakdown.

A second basis for modelling which has a sound basis in sediment transport theory is the relationship between detachment and deposition, given by Equation (6.8):

$$\frac{\partial S}{\partial x} = D - S/h \qquad (6.8)$$

where h is interpreted as the mean free path of the sediment particles in transit. For very unsteady conditions it may be necessary to specify the distribution of mean free paths: Equation (6.8) is then only strictly correct for an exponential distribution of travel distances, $h_2 : p(h_2 > h_1) = \exp(-h_1/h)$. For steady transport at capacity, there will be no net pick up, and detachment will be at a maximum rate, D_0, as the material picked up is the newly deposited material. Thus:

$$S = C = D_0 h \qquad (6.9)$$

At transport rates in excess of capacity, detachment remains at the capacity rate, D_0, so that the (negative) rate of pick up is:

$$\frac{\partial S}{\partial x} = D_0 - S/h = \frac{C - S}{h} = D_0(1 - S/C) \text{ for } S \geqslant C \qquad (6.20)$$

At transport rates below capacity, it may be argued that the newly deposited material (at rate S/h) can be detached at the capacity rate, D_0, but that additional detachment is of intact bed material, which has a lower rate of detachment D_1. If this changeover is linear, then:

$$D = D_1 + (D_0 - D_1)S/C \quad \text{for } S \leqslant C \qquad (6.21)$$

and:

$$\frac{\mathrm{d}S}{\mathrm{d}x} = D_1(1 - S/C) \quad \text{for } S \leqslant C \qquad (6.22)$$

This approach thus supports that of Foster and Meyer (1972), and provides a physical interpretation for the parameter, h. Where the actual transport rate is always much less than the capacity, then Equation (6.22) may be approximated by the detachment limited case:

$$\frac{\partial S}{\partial x} = D_1 \qquad (6.23)$$

At the other extreme S may always be approximately replaced by the capacity C in Equation (6.3), the transport-limited case.

To convert Equations (6.29)–(6.23) into an operational model, explicit functions are needed for D_1 plus two out of h, D_0, and C. From this point on, the evidence becomes considerably less secure. Most models however share a reliance on hydraulic parameters of rainfall and flow as direct predictors of sediment process rates. It therefore follows that a soundly based erosion model should rest on a model for runoff generation. Such a hydrological sub-model is taken as axiomatic in what follows, and is then treated as a foundation on which to build splash and wash process models.

6.3.2 Runoff models

Most erosion modelling assumes an areally uniform rate of runoff generation. There is clear evidence that in humid vegetated areas soil moisture levels tend to build up downslope, especially close to streams, and that near-saturated areas generate a disproportionate amount of overland flow runoff. These topics, and some of their erosional implications, are explored more fully elsewhere (Kirkby, 1978; Thornes, Chapter 5). It is recognized however that humid vegetated areas are not those most sensitive to soil erosion in practice, so that the present account will concentrate on unvegetated soils, where runoff generation is more nearly uniform. In this case total runoff discharge increases linearly downslope, and the relevant hydrological factors are (i) the total runoff generated with its distribution through time; and (ii) the partition of runoff between rill and inter-rill areas.

In modelling the rate of runoff generation per unit area, either infiltration or storage models are appropriate. In the former, the infiltration capacity should be varied with time or accumulated storage, and runoff estimated from rainfall intensities in excess of the current infiltration capacity. In storage models, the rainfall at the start of a storm or a day is considered to begin filling a fixed soil storage capacity. When this capacity is satisfied, subsequent rainfall in the storm, or a fixed proportion of it, is considered to provide runoff, whatever its intensity. Where detailed rainfall data are absent, a daily storage model provides a good estimate for the distribution of daily runoffs. Where detailed rainfall intensity records exist, then an infiltration model, or a combination of the two, is perhaps to be preferred. In each case, one alternative is to assume that evaporation between storms proceeds at a given rate, in order to provide initial conditions for the next storm. Storm sequences can be recorded or simulated, and the sequence used to predict runoff. The simpler alternative, most valid for arid areas, is to assume complete evaporation between storms, so that individual storms can be independently drawn from an appropriate distribution.

The simplest infiltration model is perhaps that based on the Green and Ampt (1911) equation:

$$f = f_0 + B/H \tag{6.24}$$

where

f is the infiltration capacity,
f_0 is its minimum rate,

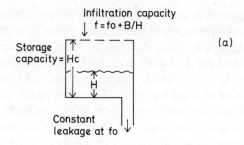

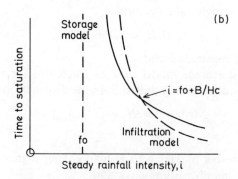

Figure 6.4. (a) Conceptual soil water store, defining f_0 and H in Equation (6.13) and H_c in Equations (6.24) and (6.26). (b) Time to saturation for simple storage and infiltration models.

H is the soil storage level as shown in Figure 6.4(a), and B is a soil constant.

For rainfall at a steady intensity i, saturation is reached after a time:

$$t = B/(i - f_0)^2 \qquad (6.25)$$

The simplest corresponding storage model is one where overland flow runoff occurs when the store shown in Figure 6.4(a) exceeds its capacity, H_c. In this case saturation is reached after a time:

$$t = H_c/(i - f_0) \qquad (6.26)$$

It may thus be seen (Figure 6.4(b) that the storage model is the more appropriate for low intensities and a high ratio B/H_c; and the infiltration model is better for high intensities and a low ratio B/H_c.

The importance of a runoff generation model is strongly related to the problem of soil surface sealing by rainsplash, and it is thought that conditions favouring rainsplash detachment favour sealing and vice versa. Sealing is essentially prevented by physical protection of the surface, either by a vegetation crown cover (if not more than 10 m above the surface), by mulches or by large stones. Surface seals consist of silt and clay particles in oriented films (Tackett and Pearson, 1965; Farres, 1978) which reduce effective infiltration or storage capacities by factors of at least ten times. Their formation is less where soils contain sand grains and/or stable aggregates which will not break down to silt or clay particles. Although tough, the surface seal does not appear to prevent continued detachment which coexists with a stable crust thickness. In the absence of direct evidence it is suggested that appropriate soil storage and infiltration parameters might take the form:

$$f_0, B, H_c \propto (D + b)^{-n} \qquad (6.27)$$

for suitable positive constants b and n.

The very simplest storage model may be calculated for zero leakage (f_0) and daily rainfalls independently drawn from the distribution:

$$N(r > r_1) = N_0 \exp(-r_1/r_0) \qquad (6.28)$$

where

$N(r > r_1)$ is the number of days with rainfall greater than r_1,
$N_0 = N(r > 0) =$ number of raindays, and
$r_0 =$ mean rain per rain day,

The total annual runoff generated per unit area is then:

$$-\int_{H_c}^{\infty} \frac{dN}{dr_1}(r_1 - H_c)\, dr_1 = R \exp(-H_c/r_0), \qquad (6.29)$$

where $R = N_0 r_0 =$ total annual rainfall.

This model estimates annual runoff from a soil plot with fair reliability if the storage capacity, H_c assumes values ranging from about 10 mm for bare areas to 100 mm for fully vegetated or otherwise protected soils (Carson and Kirkby, 1972, pp. 212–216). Thus, for example, an area with 1000 mm of annual rainfall, falling on 70 days per year is estimated to produce from 1 mm to 500 mm of runoff. Such large variations in runoff clearly produce corresponding differences in sediment yield which may outweigh all other factors, and certainly must not be ignored in model construction.

The second important factor in runoff production is its distribution between rill and inter-rill areas. If the contours (and flow lines) on a rilled slope are considered as a series of identical curves translated successively in a downslope direction (Figure 6.5(a)), and rills have a spacing of $2u$, then at distance x from the divide, the area drained by rills rises asymptotically to $(2xu - \lambda u^2)$, and the inter-rill area to λu^2 where λ is a shape factor depending on the exact contour shape. Thus the average areas drained per unit slope width tend respectively

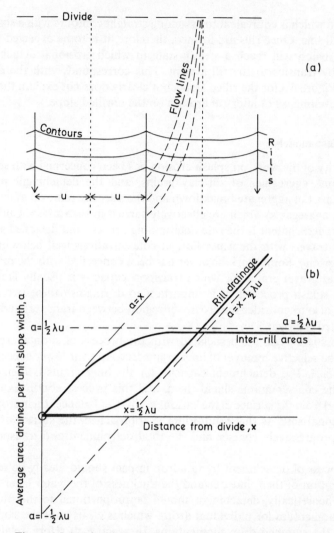

Figure 6.5. (a) Idealised rill pattern showing translated flow and contour lines. (b) Rill and inter-rill drainage areas for idealized pattern in (a)

to $(x - \frac{1}{2}\lambda u)$ and $\frac{1}{2}\lambda u$ (Figure 6.5(b)). Very close to the divide there are no rills, so that inter-rill area per unit width is equal to the total distance, x. The distance $\frac{1}{2}\lambda u$ may be associated with the average position of rill heads, and so wtih Horton's (1945) critical erosion distance. For many purposes the curves joining the straight-line portions of the relationships in Figure 6.5(b) which depend on the exact shape of the rill contours, may perhaps be ignored in modelling, an assumption which is equivalent to assuming straight contours with a sharp nick for each rill line. Once rills are formed, therefore, it is to be expected that the inter-rill erosion will reach a steady state in which erosional detachment is balanced by transfer to the rill system. This corresponds with the evidence shown in Figure 6.3 for the rilled slope, but clearly does not explain the almost identical levelling off of inter-rill erosion on the unrilled slope.

6.3.3   Splash models

The majority of the work on splash erosion has been concerned with soils containing some aggregates of silt/clay grains, and the detachment processes studied consist of aggregate breakdown and the selective removal of fine grains (and small aggregates) which are effectively carried in suspension. Under these conditions, detachment is the rate-determining process and detached sediment essentially travels with the water flow at concentrations well below its transporting capacity. Some work however has been concerned with the movement of sand and gravel grains for which transport capacity is usually limiting. In modelling splash processes, it is important to distinguish these two extreme cases, and also to consider the crossover region between transport and detachment limited removal.

The crossover region is associated with the process of surface armouring, in which the selective removal of finer grains leads to their lower concentration at the surface. The detachment capacity for the finer grains is then reduced because the coarser grains shield them, and this process continues until the net removal from the surface of the various grain sizes balances their proportions in the original soil. As this equilibrium is approached, the surface of the soil becomes progressively coarser and the total detachment and transport rates less.

The process of detachment by raindrop impact should ideally be related to the momentum of the raindrops and the thickness of the water layer over the surface. Theoretically, detachment should be proportional to the sum of the squared momentum for individual drops, which is clearly not the same as the square of the summed drop momentums. In practise therefore, relationships with total rainfall kinetic energy or intensity provide at least as good a fit to observed data, though it is clear that a calibration constant may be needed to compare areas of different rainfall characteristics. The concurrence among authors on Equation (6.7) above however provides a convenient model with adequate empirical backing.

Splash losses from the downslope side of 2 ft × 2 ft (0.6 m) pans of fine-grained soils follow the pattern of Figure 6.6. Detachment can be interpreted as the loss from both upslope and downslope sides of the pan, and is essentially constant as slope varies. Net transport rate may also be interpreted from small pan experiments provided that the mean travel distance, h, is short compared to the

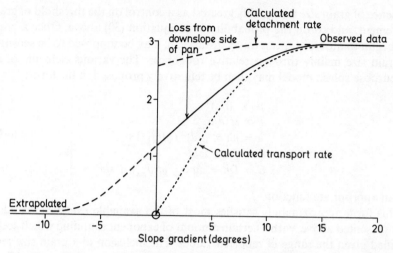

Figure 6.6. Typical soil pan data for splash losses (e.g. Lattanzi, 1973) Detachment at slope θ = loss at θ plus loss at $(-\theta)$ Net transport at slope θ = loss at θ minus loss at $(-\theta)$

length of the pan. The near-linear increase with slope shown at less than about 10° slope is therefore thought to indicate a linear response over the full range of relevant slopes. Since mean travel distance, h, is given by the ratio of transport to detachment, h is therefore also considered to increase linearly with slope.

Splash tests on coarse-grained material have generally measured the total travel distances for marked grains, giving an average velocity, u, which is related to the distance travelled in a single step by:

$$C = Dh = dph = du \qquad (6.30)$$

where

d is the grain diameter, and

p is the proportion of exposed surface grains moving in unit time.

Thus Kirkby and Kirkby (1974, p. 173) obtained the empirical relationship for movement in the field over a series of storms, for 2–10 mm gravel:

$$u \propto (s/d)^{1.6}(d/d_{84})^{0.4} \qquad (6.31)$$

where d_{84} is the grain-size than which 84 per cent of the soil is finer. This result clearly contains some influence of inter-rill wash as well as splash action. Given the normal reduction of grain size on gentler slopes, the d/d_{84} term leads to a somewhat lower exponent of slope in practice. Moeyersons (1975, p. 295) found a direct proportionality between the travel velocity, u and the tangent slope, s, in a laboratory experiment with 20 mm gravel, on slopes of 3° to 30°.

It is concluded that a *linear* increase of splash travel distance, h, with slope gradient provides an adequate model for both fine and coarse material. The influence of grain size is arguably greatest as a control on the threshold of grain movement, and so on the probability, p in equation (30) above. Once a grain has begun to move, its mean travel distance, h, may be supposed to be sensitive to grain size mainly through relative roughness. The various elements of an all-purpose splash model may then be tentatively proposed in the form:

$$\left.\begin{array}{l} p \propto (i/d)^2 \\ h \propto sf(d/d_{84}) \\ u = ph \propto (i/d)^2 f(d/d_{84})\, s \\ D = pd \propto i^2/d \\ C = Dh = du \propto f(d/d_{84})\cdot i^2 s/d \end{array}\right\} \qquad (6.32)$$

for an appropriate function, f.

This form approximately satisfies most of the available evidence including that presented above, with a certain amount of exponent rounding which seems justified given the range of reported values. The inclusion of a grain size term in the detachment capacity, D, can only represent simple granular soils. Allowances must be made both for the effective grain size made available from aggregates, for inter-aggregate bonding, and for an almost proportional reduction to allow for surface protection by vegetation crowns, mulches or large stones. Thus for example, protection by layers of window screen (Meyer et al., 1975, pp. 184–85) reduced detachment to about 10 per cent of its unprotected value.

At the start of a storm, a surface is likely to undergo a period of adjustment as fines are selectively removed and an armouring layer of coarse debris is left behind. The extent of this effect depends on the degree of disturbance of the surface since the last storm; and the period of adjustment is shortest for soils of heterogeneous composition. Most cultivations provide an extreme form of disturbance, but natural processes associated with biological activity can also be significant in both arid (Yair and Rutin, in press) and humid (Van Zon, 1978) environments, mainly in the form of material brought to the surface by burrowing organisms. Freeze-thaw and wetting-drying cycles can also be significant on bare surfaces. Detachment capacity should therefore be modelled as gradually rising in interstorm periods, and as falling back (perhaps exponentially) to a low steady level during rainfall. Such short-term variations must however be averaged out for a long-term sediment yield model.

6.3.4   Inter-rill wash models

Both observations and measurement concur in giving negligible importance to detachment by thin films of flowing water. The role of a water layer in reducing detachment is a real one, but its influence under field conditions is not clear, because of the constantly varying thickness of a water layer flowing over a rough surface. At this stage therefore, there seems no good reason to depart

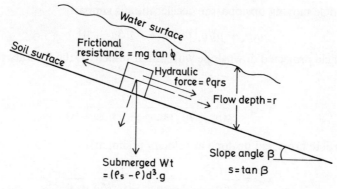

Figure 6.7. Forces acting on a submerged soil particle in a shallow flow

from a modelling assumption that the splash detachment rate remains constant as flow depth is varied.

When a soil particle is struck by a raindrop, its initial velocity and probability of movement is largely determined by its mass and the raindrop momentum. Once in motion, it is slowed by frictional forces, and movement is encouraged by its downslope weight component, assisted by hydraulic tractive forces from the flowing water (Figure 6.7). For a particle sliding over the surface, the forces acting parallel to the surface are as follows

(i) Friction opposing direction of motion:
$$(\rho_s - \rho)d^3g \tan \phi$$
where

ϕ is the angle of friction between the particle and the surface,
ρ is the water density, and
ρ_s is the particle density

(ii) Hydraulic tractive force acting downslope
$$\rho grsd^2$$
where r is the depth of flow

(iii) Downslope weight component
$$(\rho_s - \rho)d^3gs$$

In these expressions the cosine of the slope angle has been taken as 1; and $\sin \theta$ and $s = \tan \theta$ used interchangeably. Thus the deceleration acting on a particle moving downslope is:

$$\frac{(\rho_s - \rho)d^3g \tan \phi - \rho grsd^2 - (\rho_s - \rho)d^3gs}{\rho_s d^3}$$

$$= g(\rho_s - \rho)/\rho_s[\tan \phi - (rs/\Delta d + s)]$$

where $\Delta = (\rho_s - \rho)/\rho$

On a particle moving upslope, the deceleration is similarly:

$$g(\rho_s - \rho)/\rho_s[\tan\phi + (rs/\Delta d + s)]$$

For a particle projected downslope with initial velocity v, the distance travelled is then:

$$x_+ = \frac{v^2\rho_s}{2g(\rho_s - \rho)} \cdot \frac{1}{\tan\phi + (rs/\Delta d + s)}$$

For a particle projected upslope at velocity v, similarly:

$$x_- = \frac{v^2\rho_s}{2g(\rho_s - \rho)} \cdot \frac{1}{\tan\phi - (rs/\Delta d + s)}$$

Averaging those values over a probability distribution of velocities, which is assumed symmetrical with respect to direction, the overall mean travel distance is given by:

$$h = \frac{\varepsilon(v^2)\rho_s}{2g(\rho_s - \rho)} \cdot \left[\frac{1}{\tan\phi + (rs/\Delta d + s)} - \frac{1}{\tan\phi + (rs/\Delta d + s)}\right]$$

$$= \frac{\varepsilon(v^2)\rho_s}{g(\rho_s - \rho)} \cdot \left[\frac{rs/\Delta d + s}{\tan^2\phi - (rs/\Delta d + s)^2}\right] \tag{6.33}$$

where $\varepsilon(v^2)$ is the average value of v^2.

The final term in the denominator of this expression can generally be neglected, giving a direct dependence on slope gradients. The expression outside the square brackets may be treated as roughly constant. The angle of friction, ϕ, is related to the relative grain sizes of surface and moving grain, probably in a way similar to that for screes (Kirkby and Statham, 1975), for which:

$$\tan\phi = \tan\phi_0 + \lambda d_{84}/d \tag{6.34}$$

for suitable constants ϕ_0, λ.

Over a moderate range of the ratio d_{84}/d, this is equivalent to the empirical terms for relative roughness in Equation (6.31); and suggests that in a shallow flow:

$$h \propto (s + rs/\Delta d)/(\tan\phi_0 + \lambda d_{84}/d)^2 \tag{6.35}$$

Assuming a constant Darcy-Weisbach drag coefficient, f, the final term in the first bracket may be replaced by:

$$rs/\Delta d = (f/2g)^{1/3}(qs)^2/\Delta d \tag{6.36}$$

where q is the discharge per unit slope width.

The data presented by Moeyersons (1975, Figure 4) for 20 mm gravel shows fair support for this expression, although the mathematical convenience of a linear function of (qs) in place of Equation (6.36) leads to the proposal of a model which is easier to use, though less well supported by theory, namely:

$$h \propto (s + \mu qs/d)/(\tan \phi_0 + \lambda d_{84}/d)^2 \qquad (6.37)$$

for an empirical constant, μ. For uniform runoff production this expression is then linear in both slope and distance, and will be adopted in the analyses below.

Although no direct evidence for mean travel distance is available for fine-grained material, an analysis based on Equation (6.37) may be compared with the data presented by Meyer *et al.* (1975) and illustrated in Figure 6.3 above. The presence or absence of rills is irrelevant because water and sediment are lost to any rills in proportion for the linear process described by Equation (6.37). The experimental conditions apply to a uniform material and slope gradient, and to uniform runoff production, so that Equation (6.37) may be written in the form:

$$h = a + bx \qquad (6.38)$$

where

x is the distance from the top of the plot, and
a, b are appropriate constants.

At the top of the plot splashed material can be transferred across the boundary, but flow cannot, so that:

$$\text{at } x = 0:$$
$$S = C = aD \qquad (6.39)$$

Combining equations (6.1) for $A = 0$ and (6.8):

$$\frac{\mathrm{d}S}{\mathrm{d}x} + \frac{S}{h} = D \qquad (6.40)$$

The relevant solutions are then:

$$S = D\{(a + bx) + ab[a/(a + bx)]^{1/b}\}/(1 + b) \qquad (6.41)$$
$$\frac{\mathrm{d}S}{\mathrm{d}x} = bD\{1 - [a/(a + bx)]^{(1 + 1/b)}\}/(1 + b) \qquad (6.42)$$

Equation (6.42) is a relationship for the local rate of removal which is in qualitative agreement with Figure 6.3 above. The distance needed to approach equilibrium is seen as approximately 2 m in Figure 6.3. It is also given as

$$a/b \sim \Delta vd/\sigma \qquad (6.43)$$

where σ is the runoff intensity from Equations (6.35) and (6.38) above. Substituting $i = \sigma = 60 \text{ mm/hr}$; $v = 10 \text{ cm/s}$; $d = 0.1 \text{ mm}$, and $\Delta = 1.65$, gives a critical distance of 2.0 m for comparison.

When the influence of individual rainfall events is accumulated over a period, then the effects of adding a series of terms like Equation (6.42) is not strictly additive as rainfall intensity (influencing D) and flow rate (influencing b) vary. For most purposes however, especially for detachment of dominantly fine-grained soils, it will be adequate to replace the expression for detachment rate in Equation (6.32) by:

$$D \propto \varepsilon(i^2)/d \tag{6.44}$$

where $\varepsilon(i^2)$ is the mean square rainfall intensity; and to use Equation (6.37) with q representing annual overland flow, and the constant μ adjusted empirically. Aggregation over time is therefore not seen as an acute problem for splash and inter-rill wash modelling.

In the long term, the shape of divides formed by splash and inter-rill wash can also be obtained from the process Equation (6.37) and from the continuity/detachment Equation (6.40). In this case the equilibrium divide corresponds closely to a state of constant downcutting, that is:

$$-\frac{\partial z}{\partial t} = \frac{\partial S}{\partial x} = T \tag{6.45}$$

where T is the (constant) rate of lowering.

It follows that:

$$
\begin{aligned}
S &= Tx \\
h &= S/(D - \partial S/\partial x) = Tx/(D - T) \text{ from Equation} \\
&= s(1 + \mu\sigma x/d)/(\tan \phi_0 + \lambda d_{84}/d)^2 \text{ from Equation (6.37)}
\end{aligned} \tag{6.40}
$$

For equilibrium then, the shape of the divide is given by its slope:

$$s \propto x/(1 + \mu\sigma x/d) \tag{6.46}$$

which is a convexity, gradually straightening out downslope over a width which is related to the distance for equilibration in Figure 6.3 and Equation (6.42) above. Since semi-arid divides are convex, reaching maximum gradients within a few metres, the model proposed appears to have relevance for long as well as short time spans.

6.3.5 Rill wash models

The series of experiments reported by Meyer and Monke (1965) and Kramer and Meyer (1969) provide meaningful measures of transporting capacity under strongly rilled conditions. Sediment was added to a predetermined flow at the head of a 3.6 m slope table until a dynamic equilibrium was obtained for the sediment on the table. The results quoted show a good fit to the relationship:

$$S/s \propto (qs)^n \tag{6.47}$$

for exponents n ranging from 1.7 to 3.5 (Figure 6.8). Lower exponents are associated with coarser debris and surface mulch; and higher exponents with fine debris and bare surfaces. The effect of rainfall in addition to the flow appeared slight in these experiments. Only for the coarsest material used (460 μm) is there any evidence for a threshold of sediment movement. It will be noted that the dependence on slope gradient is much greater than for most published models, including the Universal Soil Loss Equation.

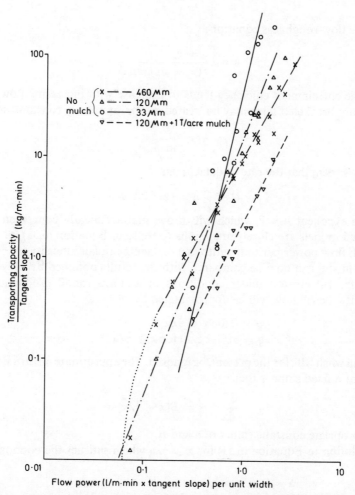

Figure 6.8. The relationship between rill transporting capacity and flow power for uniform grain-size artificial soils. Data from Meyer and Monke (1965) and Kramer and Meyer (1969)

The distance travelled by a moving particle may be analysed as above, starting from Figure 6.7. In this case particles may be considered to travel a distance:

$$h = \epsilon v^2 \rho_s / \{2g(\rho_s - \rho)[\tan \phi - (rs/\Delta d + s)]\}$$

where ϵ is a measure of the efficiency of sediment entrainment relative to the flow velocity, v. Neglecting the final term in the numerator in comparison to $\tan \phi$; and substituting

$$\frac{v^2}{2g} = \frac{rs}{f}$$

from the flow roughness equation:

$$h = \left[\frac{\epsilon r \rho_s}{f(\rho_s - \rho)} \right] \frac{s}{\tan \phi}$$

Since the entrainment efficiency tends to decrease with increasing flow depth, the square bracketed term may be treated as approximately constant, so that:

$$h \propto \frac{s}{\tan \phi} \tag{6.48}$$

It follows then, that the rate of detachment,

$$D = C/h \propto (qs)^n \tan \phi \tag{6.49}$$

For the exponent $n = 2$, which will prove mathematically convenient, these suggested models are closely analogous to those of Equation (6.32) for splash, but with flow power per unit width (qs) replacing rainfall intensity (i).

This model can now be compared with the results obtained by Meyer *et al.* (1975) for rill erosion under conditions of uniform runoff production. For simplicity the rill flow will be approximated by

$$q = 0 \text{ for } x \leqslant \tfrac{1}{2}\lambda u$$
$$q = \sigma(x - \tfrac{1}{2}\lambda u) \text{ for } x \geqslant \tfrac{1}{2}\lambda u \tag{6.50}$$

Inter-rill wash will, for the present, be ignored. The appropriate form of Equation (6.40) for a fixed slope is then:

$$\frac{dS}{dx} + \frac{S}{h} = B(x - \tfrac{1}{2}\lambda u)^2 \tag{6.51}$$

for appropriate constant values of h and B.
The solution to Equation (6.51) for $x \geqslant \tfrac{1}{2}\lambda u$ which satisfies the boundary condition:

$$S = 0 \text{ at } x = \tfrac{1}{2}\lambda u \text{ is:}$$
$$S = h(x - \tfrac{1}{2}\lambda u)^2 - 2h^2(x - \tfrac{1}{2}\lambda u) + 2h^3[1 - \exp(-x/h)]$$
$$\frac{dS}{dx} = 2h(x - \tfrac{1}{2}\lambda u) - 2h^2[1 - \exp(-x/h)] \tag{6.53}$$

The latter equation gives the local rate of erosion at a cross-section, indicating a low initial rate of erosion, and then an approach to a linear increase in erosion rate downslope. The model therefore corresponds well to the rill erosion data of Meyer *et al.* (1975) shown in Figure 6.3 above. The exact way in which rill discharge increases near the divide (Figure 6.5(b)) influences the local erosion rate in this region, but the linear increase downslope is directly related to the choice of exponent *n* as 2 in Equations (6.49) and (6.51) above. This value is thus seen as valid in at least some circumstances. For concave slopes, a similar analysis gives a depositional zone at the slope base with or without a change in the detachment value at the point of transition to deposition.

When the influence of instantaneous flows and individual storms is aggregated in Equation (6.49), each flow must be weighted by its respective frequency density over a year or longer period. For most reasonable distributions, the effect of this aggregation is to lower the exponent *n* rather slightly, if the discharge is now taken as accumulated over a year. In practice a value near $n = 2$ is still seen as appropriate. The influence of rill erosion in the long term cannot normally be considered alone, and this topic is discussed below, after considering the interaction of rill and inter-rill processes.

6.3.6 Rill–inter-rill interactions

Two plausible mechanisms might control the position of rill heads and the consequent importance of rill processes on a slope. The first is that rill enlargement is a balance between infilling by inter-rill processes and erosion by rill processes. This view carries the concept of valley head stability as a balance between slope and channel processes (Smith and Bretherton, 1972) down to the micro-scale. The alternative, and not necessarily contradictory, view is that rills begin as soon as hydraulic tractive forces overcome the resistance threshold for the soil aggregates. Without going into the detailed mathematics, it is argued that because the transfer of sediment into the rills is a roughly linear process, it will have little influence on the onset of instability, which for the more-than-linear rill process will begin as soon as rill processes become active. The critical criterion is therefore seen to be the threshold for hydraulic traction, and this view is supported by observations of rill initiation.

The threshold of hydraulic traction has been well described for splash and flow conditions by Moeyersons (1975, Figure 5) and his data show that the threshold for 20 mm gravel occurs at a flow power of:

$$qs = 10 \, \text{l/m min}$$

with raindrops falling at the same time, and at 13 l/m min without rain. The data of Meyer and Monke (1965) in Figure 6.8 shows a threshold stream power 0.05 l/m min for 0.46 mm (460 µm) sand. These values may be compared with those obtained by entrainment theory from the standard equations for turbulent flow:

$$v^2 = 2grs/f \tag{6.54}$$

$$f^{-1/2} = 1.77 \ln(r/d) + 2.0 \tag{6.55}$$

$$rs/\Delta d = 0.06 \text{ at the threshold} \tag{6.56}$$

By rearrangement, at the threshold:

$$qs = 0.2(1 - 0.88 \ln 10s)^{1/2} d^{3/2} \tag{6.57}$$

The predictions of this equation are compared with the empirical values above in Figure 6.9. The fair agreement suggests that the mechanism is appropriate, and that the initiation of rilling may be related largely to the effective grain size of the soil aggregates for a given flow.

For real soils of mixed particle and aggregate sizes the influence of soil structure on the size of water-stable aggregates is seen as crucial although inter-aggregate cohesion may also be important. Large aggregate size leads to a high value of the critical flow, and to a higher infiltration or storage capacity. Both of these factors contribute to the formation of a broad unrilled area near divides. It has already been argued above (equations (6.32) and (6.37)) that large aggregates also reduce the total inter-rill erosion (Figure 6.3 above) and increase the width of the pure splash zone relative to the inter-rill wash zone. In a heterogeneous soil the hydraulic threshold will be crossed first for the finest aggregates, but the shape of the curves in Figure 6.8 shows that their removal by flow processes will initially be very slow, and will also be limited by the formation of lag deposits. At the other extreme, a few very coarse aggregates can be carried along with finer particles, so that it seems likely that the critical aggregate size is rather coarser than the median, and that d_{84}, the aggregate size than which 84 per cent is finer, may be a suitable measure of effective diameter for rill initiation.

Where the soil grain sizes are uniform enough for the contribution of inter-rill erosion to take up an appreciable part of the rill's transporting capacity, then the solution for rill erosion shown in Equations (6.51) to (6.53) above should be modified to allow for the sediment transferred. In the extreme case of total effective transfer, corresponding to a uniform grained soil, the rate of transfer may be obtained from Equations (6.42) to (6.50). For $x \gg a/b$, Equation (6.51) should be modified to

$$\frac{dS}{dx} + \frac{S}{h} = B(x - \tfrac{1}{2}\lambda u)^2 + bD/(1 + b) \tag{6.58}$$

leading to a slight modification in the form of Equations (6.52) to (6.53).

Turning to a long term view of slope evolution, the equilibrium slope for the divide region $x \leqslant \tfrac{1}{2}\lambda u$ is still appropriately given by Equation (6.46). In the transition region from inter-rill to rill domination, there is no simple analytical

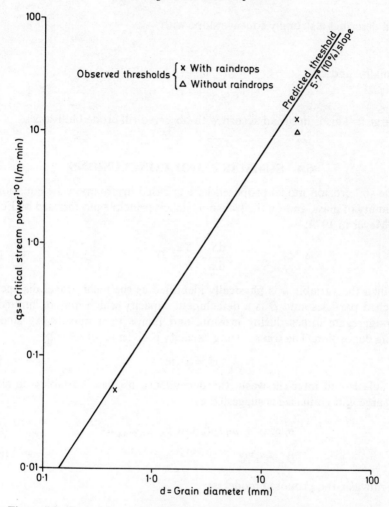

Figure 6.9. Comparison of predicted and observed thresholds for particle traction by overland flow

solution, but at large distances from the divide, Equations (6.48) and (6.49) give:

$$\frac{dS}{dx} + \frac{S}{\alpha s} \simeq \beta(xs)^n \tag{6.59}$$

for appropriate α, β. At equilibrium with a downcutting rate T, $S = Tx$ and:

$$x = (T/\beta)^{1/2n}(x/s)^{1/2}(1 + x/\alpha s)^{1/2n} \tag{6.60}$$

which describes a strongly concave slope with:

$$s \sim x^{-1}$$

for small x and:

$$s \sim x^{-(n-1)/(n+1)}$$

for large x. This is in broad accord with observed rill-eroded hillslopes.

6.4 SUMMARY AND CONCLUSIONS

The soil erosion model proposed here is based firmly upon the equation of continuity of mass, and on the transport deficit principle put forward by Foster and Meyer in 1972:

$$\frac{dS}{dx} + \frac{S}{h} = D \tag{6.8}$$

in which the variable h is physically identified as the mean travel distance of detached particles, and D is a detachment capacity which may be limited by inter-aggregate forces during erosion, and jumps to a maximum value D_0 during deposition. The transporting capacity is given in all cases by:

$$C = D_0 h \tag{6.9}$$

For splash and inter-rill wash, the dependence of these variables on slope, discharge and grain size is suggested as:

$$h_I \propto (s + \mu q_I s/d)/(\tan \phi_0 + \lambda d_{84}/d)^2 \tag{6.37}$$

$$D_{0I} \propto i^2/d \tag{6.32}$$

For rillwash, the proposed model is:

$$h_R \propto s/(\tan \phi_0 + \lambda d_{84}/d) \tag{6.48}$$

$$D_R \propto (q_R s)^n (\tan \phi_0 + \lambda d_{84}/d) \tag{6.49}$$

for $n \simeq 2$.
The respective flows q_I and q_R may be simplified to:

$$\left.\begin{array}{l} q_I = \sigma x \\ q_R = 0 \end{array}\right\} \text{for } 0 \leqslant x \leqslant \tfrac{1}{2}\lambda u$$

$$\left.\begin{array}{l} q_I = \tfrac{1}{2}\sigma\lambda u \\ q_R = \sigma(x - \tfrac{1}{2}\lambda u) \end{array}\right\} \text{for } x \geqslant \tfrac{1}{2}\lambda u \tag{6.61}$$

Sediment is considered as transferring from inter-rill areas to rills once the latter exist $(x \geqslant \frac{1}{2}\lambda u)$ at a rate in proportion to the flow transferred. This form of model is able to accommodate most of the available experimental evidence, including the partition between rill and inter-rill processes documented by Meyer *et al.* (1975), the data on splash erosion for coarse and fine grained materials, and the long-term evolution of water-eroded hillslopes. Its main qualitative points of departure from the model proposed by Foster and Meyer (1975) lie in the higher exponents for both slope and discharge under rill erosion; and in the ability to predict deposition as well as erosion without the introduction of additional assumptions.

Some attempt has been made to bridge the gap between physical processes and empirical relationships, but it is recognized that much needs to be done, both on the theory of shallow flow sediment transport, and on experimental work which effectively controls the relevant causal variables individually. The most important experimental topic is however thought to be the need to distinguish between soil properties which influence hydrological processes and so determine the rate of runoff; and soil properties which determine the resistance to hydraulic forces and so influence the amount of erosion produced by a given runoff.

LIST OF SYMBOLS

A = Rate of external sediment addition
B = Constant
C = Transporting capacity
D = Detachment capacity
F = Frictional force
H = Storage capacity
N = Number of rain days
R = Annual rainfall
S = Actual transport rate
T = Actual rate of lowering

a = Area drained per unit contour length constant
b = Constant
c = Concentration of sediment in water
d = Grain size
f = Infiltration rate, capacity, function
h = Mean travel distance
i = Rainfall intensity
n = Exponent
p = Probability
q = Overland flow discharge per unit width
r = Flow depth, daily rainfall
s = (Tangent) slope
t = Elasped time
u = Grain velocity

v = Flow velocity
w = Settling velocity
x = Horizontal distance from divide/plot head
z = Elevation
α = Constant
β = Constant
$\Delta = (\rho_s - \rho)/\rho$
$\epsilon = \begin{cases} \text{Efficiency of sediment transport} \\ \text{Coefficient of sediment diffusion} \end{cases}$
ε = Expectation
θ = Slope angle
κ = Von Karman's constant (0.4)
$\left. \begin{array}{l} \lambda \\ \mu \end{array} \right\} = \{$ Constants
ρ = Density
σ = runoff intensity
τ = Shear stress
ϕ = Angle of friction

Suffices
F   flow (rill or inter-rill)
R   rill (flow)
I   inter-rill (flow and splash)
s   splash, sediment
c   critical value
0   (zero)   value at $t = 0$
1   (one) etc value at $t = t$
         or $x = x_1$, etc.

REFERENCES

Ahnert, F. (1964). Quantitative models of slope development as a function of waste-cover thickness. *Abstr. of Papers 20th IGU Congress*, London, p. 188.

Bagnold, R. A. (1966). An approach to sediment transport from general physics. *US Geol. Surv. Prof. Paper*, **422-I**. 37 pp.

Bennett, J. P. (1974). Concepts of mathematical modelling of sediment yield. *Water Res. Res.*, **10**(3), 485–492.

Bubenzer, G. D., and Jones, B. A. (1971). Drop size and impact velocity effects on the detachment of soils under simulated rainfall. *Trans. Am. Soc. Agric. Engrs*, **14**(4), 625–628.

Carson, M. A., and Kirkby, M. J. (1972). *Hillslope Form and Process*, Cambridge University Press. 475 pp.

Culling, W. E. H. (1963). Soil creep and the development of hillside slopes. *J. Geology*, **71**, 127–162.

David, W. P., and Beer, C. (1975). Simulation of soil erosion—part 1. Development of a mathematical erosion model. *Trans. Am. Soc. Agric. Engrs*, **18**(1), 126–129.

Ekern, P. C. (1953). Problems of raindrop impact erosion. *Agric. Engng.* **34**(1), 23–25.

Elwell, H. A., and Stocking, M. A. (1973). Rainfall parameters for soil loss estimation in a subtropical climate. *J. Agric. Engng Res.*, **18**, 169–177.

Farres, P. (1978). The role of time and aggregate size in the crusting process. *Earth Surface Proc.*, **3**(3), 279–283.

Flaxman, E. M. (1972). Predicting sediment yield in the western United States. *J. Hydraulics Div., Proc. Am. Soc. Civil Engrs*, **98**(HY2), 2073–2085.

Foster, G. R., and Meyer, L. D. (1972). A closed-form soil erosion equation for upland areas. In: *Sedimentation: Symposium to Honour Professor H. A. Einstein*, (Shen, H. W. Ed.), Fort Collins, Colorado, pp. 12.1–12.19.

Foster, G. R., and Meyer, L. D. (1975). Mathematical simulation of upland erosion by fundamental erosion mechanics. In: *Present and Perspective Technology for Predicting Sediment Yields and Sources*—Proceedings of Sediment-Yield Workshop, United States Department of Agriculture Sedimentation Laboratory, Oxford, Mississippi, November 1972 (Agricultural Research Service Report ARS-S-40), United States Department of Agriculture, Washington, D.C., pp. 190–206.

Free, G. R. (1960). Erosion characteristics of rainfall. *Agric. Engng*, **41**(7), 447–449.

Green, W. H., and Ampt, G. A. (1911). Studies on soil physics. 1. The flow of air and water through soils. *J. Agric. Sci.*, **4**(1), 1–24.

Horton, R. E. (1945). Erosional development by streams and their drainage basins. *Bull. Geol. Soc. Am.*, **56**, 275–370.

Hudson, N. W., and Jackson, D. C. (1959). Erosion research. *Henderson Research Station, Report of Progress, 1958–9*, Federation of Rhodesia and Nyasaland, Ministry of Agriculture.

Kirkby, A. V. T., and Kirkby, M. J. (1974). Surface wash at the semi-arid break in slope. *Z. für Geomorph.*, Suppl. Bd. **21**, 151–176.

Kirkby, M. J. (1971). *Hillslope Process-response Models Based on the Continuity Equation*. Institute of British Geographers, Special Publication No. 3, pp. 15–30.

Kirkby, M. J. (Ed.). (1978). *Hillslope Hydrology*, Wiley, Chichester. 389 pp.

Kirkby, M. J., and Statham, I. (1975). Surface stone movement and scree formation. *J. Geology*, **83**, 349–362.

Kramer, L. A., and Meyer, L. D. (1969). Small amounts of surface mulch reduce soil erosion and runoff velocity. *Trans. Am. Soc. Agric. Engrs*, **12**, 638–645.

Lattanzi, A. (1973). *Influence of Straw Mulch Rate and Slope Steepness on Inter-rill Detachment and Transport of Soil*, MSc. Thesis, Purdue University, W. Lafayette, Indiana. 90 pp.

McCool, D. K., Papendick, R. I., and Brooks, E. L. (1975). The universal soil loss equation as adapted to the Pacific North West. *Proc. 3rd Federal Inter-Agency Sedimentation Conference*, Report SEDCOM-03; **2**, 135–147.

Meeuwig, R. O. (1970). *Sheet Erosion on Intermountain Summer Ranges*, Forest Service, Research Paper, INT-85, United States Department of Agriculture, Washington, D.C. 25 pp.

Meyer, G. J., Schoenburger, P. J., and Huddleston, J. H. (1975). Sediment yields from roadsides: an application of the universal soil loss equation. *J. Soil and Water Conserv.*, **30**(6), 289–21

Meyer, L. D., Foster, G. R., and Romkens, M. J. M. (1975). Source of soil eroded by water from upland slopes. In: *Present and Perspective Technology for Predicting Sediment Yields and Sources*—Proceedings of Sediment-Yield Workshop, United States Department of Agriculture Sedimentation Laboratory, Oxford, Mississippi, November 1972 (Agricultural Research Service Report ARS-S-40), United States Department of Agriculture, Washington, D.C., pp. 177–189.

Meyer, L. D., and Monke, E. J. (1965). Mechanics of soil erosion by rainfall and overland flow. *Trans. Am. Soc. Agric. Engrs*, **8**(4), 572–580.

Meyer, L. D., and Wischmeier, W. H. (1969). Mathematical simulation of the process of soil erosion by water. *Trans. Am. Soc. Agric. Engrs*, **12**(6), 754–758.

Moeyersons, J. (1975). An experimental study of pluvial processes on granite grus. *Catena*, **2**, 289–308.

Palmer, R. S. (1963). Waterdrop impactometer. *Agric. Engng*, **44**(Apr.), 198–199.

Renard, K. G. (1977). Erosion research and mathematical modelling. In: *Erosion: Research Techniques, Erodibility and Sediment Delivery* (T. J. Troy, Ed.), Geo Abstracts Ltd, Norwich, pp. 31–44.

Rowlison, D. L., and Martin, G. L. (1971). Rational model describing soil erosion. *J. Irrigation and Drainage Div., Proc. Am. Soc. Civil Engrs*, **97**(CR1), 39–50.

Smith, R. M., and Stamey, W. L. (1965). Determining the range of tolerable erosion. *Soil Sci.*, **100**(6), 414–424.

Smith, T. R., and Bretherton, F. P. (1972). Stability and conservation of mass in drainage basin evolution. *Water Res. Res.* **8**(6), 1506–1529.

Tackett, J. L., and Pearson, R. W. (1965). Some characteristics of soil crusts formed by simulated rainfalls. *Soil Sci.*, **99**(6), 407–13.

Tödten, H. (1976). A mathematical model to describe surface erosion caused by overland flow. *Z. für Geomorph.*, Suppl. Bd. **25**, 89–105.

Vanoni, V. A. (1946). Transportation of suspended sediment by water. *Trans. Am. Soc. Civil Engrs*, **111**, 67–102.

Williams, J. R. (1975). Sediment yield prediction with universal equation using run-off energy factor. In: *Present and Perspective Technology for Predicing Sediment Yields and Sources*—Proceedings of Sediment-Yield Workshop, United States Department of Agriculture Sedimentation Laboratory, Oxford, Mississippi, November 1972 (Agricultural Research Service Report ARS-S-40), United States Department of Agriculture, Washington, D.C., pp. 244–252.

Wischmeier, W. H. (1975). Estimating the soil loss equations cover and management factor for undisturbed areas. In: *Present and Perspective Technology for Predicing Sediment Yields and Sources*—Proceedings of Sediment-Yield Workshop, United States Department of Agriculture Sedimentation Laboratory, Oxford, Mississippi, November 1972 (Agricultural Research Service Report ARS-S-40), United States Department of Agriculture, Washington, D.C., pp. 118–124.

Wischmeier, W. H., and Smith, D. D. (1960). A universal soil loss equation to guide conservation farm planning. *Trans 7th Int. Cong. of Soil Science*, **1**, 418–425.

Wischmeier, W. H., and Smith, D. D. (1965). *Predicting Rainfall-erosion Losses from Cropland East of the Rocky Mountains*, United States Department of Agriculture, Washington, D.C. Agricultural Handbook No. 282, 47 pp.

Woolhiser, D. A., and Blinco, P. H. (1975). Watershed sediment yield—a stochastic approach. In: *Present and Prospective Technology for Predicting Sediment Yields and Sources*—Proceedings of Sediment-Yield Workshop, United States Department of Agriculture Sedimentation Laboratory, Oxford, Mississippi, November 1972 (Agricultural Research Service Report ARS-S-40), United States Department of Agriculture, Washington, D.C., pp. 264–273.

Woolhiser, D. A., and Todorovic, P. (1971). A stochastic model of sediment yield for ephemeral streams. *Proc. USDA Int. Assoc. for Statistics in the Physical Sciences Symposium on Statistical Hydrology* (Miscellaneous Publication No. 1275), United States Department of Agriculture, Washington D.C.

Yair, A., and Rutin, J. (in press). Some aspects of the regional variation in the amount of available sediment produced by isopods and porcupines, Northern Negev, Israel. *Earth Surf. Proc. & Landforms.* 6(3).

Young, A. (1953). Deductive models of slope evolution. *Nach Akad. Wissen Gottingen*, Ser. II, **5**, 45–66.

Van Zon, H. J. M. (1978). Litter transport as a geomorphic process. *Publicates van het Fysisch–Geografisch en Bodemkundig Laboratorium van de Universiteit van Amsterdam*, No. 24. 135 pp.

Soil Erosion
Edited by M. J. Kirkby and R. P. C. Morgan
© 1980 John Wiley and Sons Ltd.

CHAPTER 7

Wind erosion

S. J. Wilson

NERC Research Student,
Bedford College, London, UK

and

R. U. Cooke

Professor of Physical Geography,
Bedford College, London, UK

7.1 INTRODUCTION

7.1.1 The wind erosion hazard

Wind erosion is the process by which loose surface material is picked up and transported by the wind, and surface material is abraded by windborne particles. The spatial redistribution and resorting of particles by wind erosion may have profound effects on the affected soils, their related microtopography, and any agricultural activity associated with them. The process operates in a variety of natural environments that lack a protective cover of vegetation, and it is particularly significant in both hot and cold deserts, coastal dune areas, and exposed mountain regions. But its human consequences are undoubtedly most serious in those agricultural areas that experience low, variable and unpredictable rainfall, high temperatures and rates of evaporation and high wind velocity, as is the case in semiarid areas, as well as some of the more humid regions that experience periodic droughts. In such areas the natural process of wind erosion may be accelerated by imprudent agricultural practices, and there may follow a number of physical effects, including soil damage, crop damage and related problems, and numerous undesirable economic consequences (Table 7.1).

The record of wind erosion in agricultural areas extends back into Classical antiquity, but the hazard achieved international notoriety with the advent of serious soil erosion and related dust and sand storms during the prolonged droughts of the 1930s in the High Plains of North America. As is so commonly the case with human responses to environmental hazards, it was these extreme,

217

spectacular events, accompanied as they were by serious social disruption and human misery and by much publicity, that provided the major impetus for vigorous research into the nature of the wind erosion system and the development of erosion control methods.

In Britain, where wind erosion is a more localized phenomenon, problems have been reported in recent years in the Breckland (in Norfolk and Suffolk), the Sandlings of east Suffolk, the vales of York and Pickering, the 'Black Fens' and parts of the west Midlands, Lincolnshire, east Nottinghamshire and south Lancashire. The six-day 'blow' of March 1968 in Lincolnshire is a notable

Table 7.1. Some physical and economic effects of wind erosion

Physical Effects	Economic Consequences
Soil damage	*Soil damage*
(1) Fine material, including organic matter, may be removed by sorting, leaving a coarse lag.	(1, 2, 3) Long term losses of fertility give lower returns per hectare.
(2) Soil structures may be degraded	(3) Replacement costs of fertilizers and herbicides.
(3) Fertilizers and herbicides may be lost or redistributed.	
Crop damage	*Crop damage*
(1) The crop may be covered by deposited material.	(1–6) Yield losses give lower returns.
(2) Sandblasting may cut down plants or damage the foliage.	(1–3) Replacement costs, and yield losses due to lost growing season.
(3) Seeds and seedlings may be blown away and deposited in hedges or other fields.	(5) Increased herbicide costs.
(4) Fertilizer redistributed into large concentrations can be harmful.	
(5) Soil borne disease may be spread to other fields.	
(6) Rabbits and other pests may inhabit dunes trapped in hedges and feed on the crops.	
Other damage	*Other damage*
(1) Soil is deposited in ditches, hedges, and on roads.	(1) Costs of removal and redistribution.
(2) Fine material is deposited in houses, on washing and cars, etc.	(2, 3) Cleaning costs.
(3) Farm machinery, windscreens etc. may be abraded, and machinery 'clogged'.	(4) Loss of working hours and hence productivity declines.
(4) Farm work may be held up by the unpleasant conditions during a 'blow'.	

example in which it was estimated that some 9000 hectares were affected and the total damage cost was in the order of £1 million (Pollard and Millar, 1968; Robinson, 1968).

The areas affected by wind erosion in Britain are characterized by light or peaty soils growing a variety of market-garden crops, sugar beet, grains, and potatoes, and they are especially prone to wind erosion in dry periods during spring and early summer when fields are bare following recent cultivation and before a protective crop cover is developed. Many observers have suggested that wind erosion has been increasing in terms of its severity, frequency of

Table 7.2. Increased wind erosion in Britain: some suggested causes and their consequences

---

(1) An increase in arable land and correlative reduction in permanent pasture and the use and length of grass leys, which has reduced the protective effect of vegetation, reduced the time for soil to 'recuperate', and extended both the area susceptible to wind erosion and the periods of susceptibility.
(2) A tendency towards monoculture, and a consequent reduction of the stability provided by crop rotations.
(3) An increase in the practice of stubble and straw burning (and other crop residue removal practices) which reduces the protective effect of such material and reduces the provision of organic matter to the soil of value in maintaining soil structure.
(4) An increase in the use of artificial fertilizers, some of which tend to disaggregate soil clods.
(5) Improved weed control with herbicides, which reduces the protective effect of weeds.
(6) The introduction and rapid extension of sugar beet, a crop requiring a loose, vulnerable seed bed and providing only limited surface protection early in its growth season.
(7) An increase in the use of wide-spaced 'drill-to-stand' techniques which increase the area of ground vulnerable to wind erosion in a field.
(8) A decrease in marling, a practice which can improve soil structure and reduce the erodibility of soil.
(9) Continued removal of hedges and increase in field size, thus reducing the protective effect of field boundaries and increasing wind 'fetch'.

Sources include: Astbury, 1958; Davies and Harrod, 1970; Pollard and Millar, 1968; Radley and Simms, 1967 and 1969; Robinson, 1968; Sheesby, 1953, 1966, and 1968; Williams, 1971; Wilkinson *et al.*, 1969)

occurrence, areal distribution and economic consequences. Such trends have generally been attributed to changes in farm management and husbandry (Table 7.2). But the precise contribution of these changes and their relative importance remains unclear. Indeed, little is known at present, other than by casual observation, of the nature and rates of wind erosion on cultivated land in Britain. The problems of studying the process, its physical damage and economic consequences are considerable. For example, monitoring wind erosion is time consuming, expensive, and suitable equipment needs to be developed,

(De Ploey and Gabriels, Chapter 3), and it is extremely difficult to disentangle the economic consequences of wind erosion from the complex structure of the farm economy. In addition, wind erosion control in Britain has largely been left to the devices of the individual farmer, and little effort has been made to disseminate research findings (especially from the USA) relevant to solving the problems.

7.1.2 Development of wind erosion studies

In the late nineteenth and early twentieth centuries, geomorphological studies of wind were mainly concerned with its relative importance as an erosive agent, and especially with the competence and capacity of the wind to transport fine material and of debris-charged wind to abrade (e.g. Udden, 1894; Gilbert, 1895; Keyes, 1910; Bryan, 1923). Numerous other studies, of potential relevance to the study of wind erosion, considered sand movement in deserts and coastal areas mainly in the context of describing and understanding dune morphology (e.g. Cornish, 1897; King, 1916; Kadar, 1934), whilst some experimental studies were designed to reveal the precise mechanisms of sand movement (e.g. Olsson-Seffer, 1908).

The studies by R. A. Bagnold in the 1930s, published in his outstanding treatise on *The Physics of Blown Sand and Desert Dunes* (1941) marked a fundamental advance in the understanding of the wind-erosion system. Bagnold viewed the problem of wind/sand relationships as one of aerodynamics amenable to direct measurement. Working from a theoretical basis, he tested his ideas by laboratory experiments in the controlled conditions of a wind tunnel and by field observations and measurements in the Libyan Desert. The main assumptions that enabled Bagnold to use the wind tunnel were that large-scale eddies played no appreciable part in keeping sand aloft and that sand movement was a phenomenon taking place within a metre or so of the ground surface. Many of Bagnold's ideas and results provided the basis for subsequent research on the wind erosion problem.

In particular, the work of W. S. Chepil, A. S. Zingg, N. P. Woodruff, and others, carried out largely under the auspices of the Wind Erosion Research Station at Kansas State University in the aftermath of the 'Dust Bowl', built on Bagnold's foundations in the context of wind-erosion control of cultivated lands. This research attempted to identify and quantify the factors influencing the location and rates of soil erosion by wind, and to develop predictors of erosive conditions and soil loss based on a climatic index (e.g. Chepil and Woodruff, 1963). It led to the formulation in the 1960s of a wind erosion equation that predicted potential soil loss from individual fields and facilitated the control of wind erosion by devices designed so to manipulate factors affecting erosion that potential erosion could be reduced to a 'tolerable level' (Woodruff and Siddoway, 1965). This research continues today, concentrating on the tasks of refining the prediction equation, improving control techniques, and determining 'tolerable levels' for different crops.

7.2 THE MECHANICS OF DETACHMENT, TRANSPORT, AND DEPOSITION

7.2.1 Wind forces over a stable surface

As long ago as 1894 Udden noted that, because of the retardation effects of vegetation and other obstacles, wind speeds were lowest near the ground surface and increased with height. Olssen-Seffer (1908) also observed that the wind speed at 3 cm above a rough surface was greater than that over a smooth surface at the same height. These two principles were incorporated in the Prandtl and von Kármán Equation (7.1) (Brunt, 1934) which describes the mean forward velocity profile of a wind in fully turbulent conditions, up to about 150 cm above the ground surface. In the equation

$$u_z = (2.3/k) \ U_* \log(z/z_1) \tag{7.1}$$

where

u_z = velocity at any height z ($z \leqslant 150$ cm) above the mean aerodynamic surface (Figure 7.1),

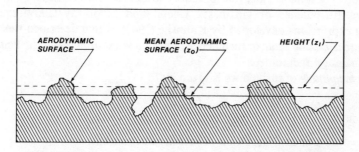

Figure 7.1. Diagram illustrating, for an inflexible ground surface, the aerodynamic surface, the mean aerodynamic surface (z_0) and the height z_1 above z_0 at which the wind velocity is zero (adapted from Chepil and Woodruff, 1963)

z_1 = the height above the mean aerodynamic surface at which the velocity is zero (considered to be 1/30th of the equivalent sand roughness height, k_s),

U_* = the drag (or shear) velocity, and

k = the von Kármán universal constant for turbulent flow, which has a value of approximately 0.4 for clear fluids.

Also, when $k_s \leqslant 1.5$ mm,

$$U_* = (\tau_0/\rho)^{1/2} \tag{7.3}$$

in which,

τ_0 = the shear stress at the bed, and
ρ = the mass density of the fluid.

Equation (7.1) is applicable to hydrodynamically rough boundaries, that is when the Reynolds Number, R_k, has a value greater than 90, i.e.

$$R_k = U_* k_s/v \geqslant 90 \tag{7.3}$$

where v = the kinematic viscosity of the fluid.

When $R_k \leqslant 3.5$–4.0, the boundary is 'smooth', and the roughness elements remain within the laminar (non-turbulent) sub-layer.

Bagnold (1941) found that in using small sand grains his experimental values of z_1 agreed with the value of $k_s/30$, whereas White (1940) found z_1 to be $k_s/9$ in studies using large sand grains. Observations by Zingg (1953a) showed that the value of z_1 varied as the logarithm of the grain diameter, $(d_s = k_s)$, according to the relationship

$$z_1 = 0.81 \log(5.5 \, d_s \text{ ins}) \tag{7.4}$$

Mean wind velocity is normally the only practicable field measurement of wind force, despite the fact that in reality winds are characterized by unsteady motions (turbulence or gustiness). Chepil and Siddoway (1959) examined several expressions developed by Kalinske (1943), which described the *magnitude, intensity,* and *scale* of turbulence in both a wind tunnel and the field using a strain gauge anemometer.

The *magnitude* of turbulence is given by

$$6\sigma_u/\overline{U}_* \tag{7.5}$$

in which

σ_u = the standard deviation of velocity fluctuations, and
$\overline{U}_*$ = mean shear velocity.

This expression is a measure of velocity fluctuations at a particular height in proportion to the shear velocity, and its value was found to increase directly but not proportionately with surface roughness, and to vary with height. The *intensity* of turbulence, a measure of velocity fluctuations at a particular height in proportion to the mean forward velocity ($\bar{u}$) at that height, is given by:

$$\sigma_u/\bar{u} \tag{7.6}$$

and its value decreases with height and varies with each 'wind storm'. The *scale* of turbulence, a measure of eddy size, is given by:

$$\bar{u}/n \tag{7.7}$$

in which n is the number of velocity fluctuation cycles per second and its value increases with the logarithm of height above the surface. Chepil and Siddoway (1959) also investigated the nature of maximum pressure impulses near the surface, as it would seem likely that these would have a significant effect on the initiation of particle movement. Using the expression:

$$(\bar{P} + 3\sigma_p)/\bar{P} \tag{7.8}$$

in which

$\bar{P}$ is the mean pressure, and

σ_p is the standard deviation of pressure,

the maximum pressure to mean pressure ratio (the turbulence factor) was found in the wind tunnel to be greatest (2.7) very close to the surface and to decrease with height (tending towards 1.0). As might be expected, in the field the variation of turbulence was much greater than in the relatively uniform flow of the wind tunnel.

These measures of turbulence illustrate the variety of conditions that may be experienced given the same mean wind velocity, and thus reveal the limitations of using mean velocity, especially in describing the wind condition at the initiation of particle movement.

7.2.2 Wind forces over an eroding surface

Once particle movement has begun, the mobile particles exert a drag effect on the air flow which in turn will alter the velocity profile. It can be seen from Equation (7.1) that over a given stable surface, the height at which velocity is zero remains constant irrespective of any change in shear velocity. In the case of an eroding surface, however, Bagnold (1941) found that the velocity profile was described by:

$$u_z = (2.3/k)\, U'_* \log(z/z_t) + u_t \tag{7.9}$$

in which

$U'_* =$ shear velocity (over an eroding surface),

$u_t =$ the threshold velocity (the velocity at which particle movement is initiated), and

$z_t =$ the height above the mean aerodynamic surface at which the velocity is equal to u_t.

The point z_t remains constant for all shear velocity values, provided surface roughness and threshold velocity remain unchanged. Figure 7.2 gives examples of velocity profiles over an eroding and a non-eroding surface with the same roughness characteristics.

The precise nature of the velocity profile at and below point z_t is of considerable importance in the interpretation of the mechanics of particle entrainment. As indicated above. Bagnold (1941) suggested that the velocity profiles converged at the focal point z_t, and this view was also held by Chepil and Woodruff (1963). They considered that below the focal point velocity decreases as shear velocity increased (Figure 7.2). This may be explained if it is accepted that an increase in the shear velocity gives rise to increased particle movement which in turn

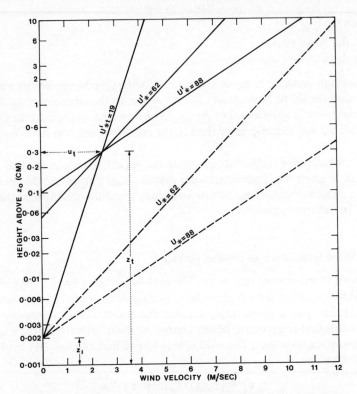

Figure 7.2. Examples of velocity distributions above an inflexible ground surface when eroding (thick lines) and when non-eroding (dashed lines) (after Bagnold, 1941)

increases the drag (or retardation) on the air flow below the focal point. Zingg (1953a), however, thought that the velocity profiles showed a curved convergence when z_t was less than 15.2 mm (Figure 7.3). Bagnold's data (1941, p. 58) could perhaps also be interpreted in this way. The reasoning behind this view is that the energy gained by the particles above the projected focal point is transmitted to the slow-moving air near the surface so that it is speeded up, and the greater the shear velocity the greater will be the increase in velocity near the bed. Zingg (1953a) also calculated the value of k in Equation (7.9)

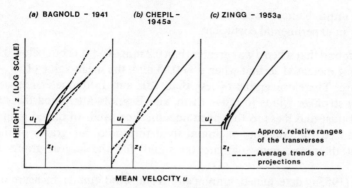

Figure 7.3. Different forms of velocity profiles obtained above drifting 0.25 mm diameter sand surfaces (after *Task Committee on Preparation of Sedimentation Manual*, 1965)

and found it to approximate to 0.375 rather than 0.4, so altering the value of the coefficient $(2.3/k)$ in Equation (7.9) from 5.75 to 6.13. He further suggested that the value of k might possibly vary with rates of particle movement and particle size.

7.2.3 Forces at the threshold of particle movement

The principle that different sizes of material require different velocity thresholds for movement in wind has been recognized for many years. Progressively faster air flow is required to move increasingly large grains, and for a given wind speed there is a maximum movable grain size. Beadnell (1910) and Cressey (1928) showed respectively that the thresholds of sand movement in the Libyan Desert and on the shore of Lake Michigan were 20.91 and 10.94 km/hour (although, of course, such figures are of little value unless details of surface roughness and height of measurement are known.)

Bagnold (1941) approached the threshold problem from a theoretical point of view. By equating the opposing forces of drag on the top-most grains of the surface with the movements about the grains axes of support, he derived the expression:

$$U_{*t} = A\sqrt{\frac{(\sigma - \rho)}{\rho}gd} \qquad (7.10)$$

in which

U_{*t} = the threshold shear velocity,
σ = grain density
ρ = air density,
g = the gravitational constant,

d = grain diameter, and
A is an experimental coefficient.

Bagnold found that when d was greater than 0.2 mm, $A < 0.1$; but below 0.2 mm, the value A increased so that when $d = 0.08$ mm, the mean value of U_{*t} began to increase. Therefore grains of less than 0.08 mm in diameter require progressively stronger winds to move them, and Bagnold attributed the stability of settled dust and loess to this phenomenon. For sands of mixed grain sizes, he introduced the concepts of 'initial fluid threshold' for grains of the predominant diameter, and 'ultimate threshold' for the largest grains in the mixture.

Zingg (1953a) determined graphically from wind tunnel measurements a 'saltation threshold', expressed as follows

$$v_s = 0.007 \, d_s \qquad (7.11)$$

in which

v_s = the saltation threshold shear stress (in lb per ft^2 of bed area) and
d_s = grain diameter (ins). Chepil (1959a) deduced an equation for the mean threshold shear stress on the bed ($\bar{v}_c$), such that:

$$\bar{v}_c = \frac{0.66 \, gd_s \rho' \tan \phi' \eta}{(1 + 0.85 \tan \phi') \tau} \qquad (7.12)$$

where

g = the gravitational constant,
d_s = mean grain size,
ρ' = density difference between the grain and the fluid,
η = the ratio of drag and lift on the whole bed to drag and lift on the topmost grains,
ϕ' = angle of repose of the top grains with respect to the mean level of drag, and
τ = the turbulence factor (given by equation (7.8)) with respect to the mean drag level.

Chepil evaluated the variables and found that $\phi' = 24°$, $\eta = 0.2$, and $\bar{v}_c \simeq 2.5$. In comparing these computed values of $\bar{v}_c$ with experimentally measured values, he found reasonable agreement, although the computed values exceeded the measured values by about 19 per cent. This difference, he suggested, could be remedied by either increasing τ to 3, or slightly decreasing the value of $\tan \phi'$ and η. The coefficients 0.85 in Equation (7.12), representing the lift force as a proportion of drag at the moment of entrainment, was later found to have a value closer to 0.75 (Chepil, 1961).

The effect of the wind forces also depends in part on particle shape, and wind tunnel observations by Williams (1964) suggested that the threshold of movement may increase as the grain sphericity is increased.

Chepil (1945b) found that in a wind tunnel threshold shear velocity varied with the mixture of erodible and non-erodible grains, because the latter tend to dissipate drag. This, and other factors discussed below, mean that for field soils there is no single threshold.

7.2.4 The nature of particle movement

Chepil (1961) measured the ratio of lift (caused by negative pressure at the top of the grain) to drag at various points on a sphere, from the bed of a wind tunnel up to several centimetres above it. At the moment of entrainment the ratio was 0.75, but it decreased rapidly as the particle gained height, until at several grain diameters above the bed, lift became negligible, whereas drag increased in the faster airflow. Having reached its maximum height the particle descends. According to Bagnold (1941) the angle at which particles strike the surface is remarkably constant, between 10° and 16° from the horizontal, irrespective of the height reached. He suggested that this constancy was due to the balance achieved between the force of gravity acting downwards and the maximum forward velocity (both of which increase with height). Chepil's (1961) observations of angle of descent suggested a slightly lower range, from 6° to 12°.

When a particle hits the ground its momentum can be dispersed in several ways. Firstly, the particle may rebound into the air flow, usually with an initial vertical or near vertical component of motion and possibly a transverse deviation from the main downstream direction of air movement (Zingg, 1953b). This is the bouncing motion called *saltation* by Bagnold (1941), who followed Gilbert's (1914) use of the term for similar motion in water. Secondly, the impact of a particle on to a surface of loose material may be sufficient to cause other particles to be thrown into motion which, without this assistance, may not have been entrained by the airflow alone. Entrainment by impact can proceed at shear velocities lower than the threshold velocities required to initiate movement, and Bagnold (1941) called the relevant threshold the 'impact threshold'.

A third way in which momentum may be dispersed is by causing surface disruption such that aggregates may be broken (abraded), and particles struck from behind may be pushed forward by what Bagnold called 'surface creep'. In this way particles with diameters up to six times those of the impacting grains may be moved.

Particles may also be moved in suspension. This can occur when the terminal velocity of fall (determined by grain size, shape, and density) is less than the mean upward eddy currents in the airflow (Bagnold, 1941). Particles moved in this way are usually less than 0.1 mm in diameter (Udden, 1894).

The proportion of material carried by the mechanisms of saltation, surface creep and suspension varies according mainly to wind velocity and the size-distribution of particles. Chepil (1945a) found the proportions varied as follows: from 50–75 per cent in saltation, from 3–40 per cent in suspension, and from

5–25 per cent in surface creep; Bagnold (1941) estimated surface creep at from 20–25 per cent of total movement; and Horikawa and Shen (1960) reported a similar figure of 20 per cent. All agree that saltation is quantitatively the most important process—indeed most creep and suspension would not occur without it.

In the field it is extremely difficult to determine by which process a particle is moving, since most particles continuously change their mode of transport as velocity fluctuates and surface roughness changes.

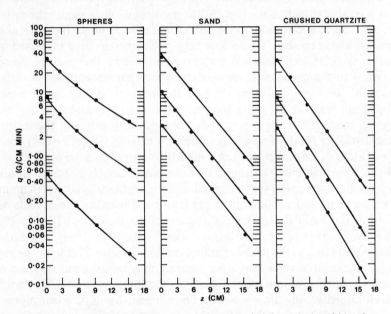

Figure 7.4. Typical sediment concentration-height relationships, for three wind velocities with a symmetrical initial surface distribution (after Williams, 1964)

The mechanisms of movement result in a rapid decrease of load above the surface. Wind-tunnel sampling of mobile material by Zingg (1953a) using sampling tubes at intervals up to 25.4 cm above the bed showed general linear relationships between the logarithm of quantity and the logarithm of height, for both different velocities and different grain sizes. Similar studies by Williams (1964) employed rectangular sediment traps (1 × 10 cm) at five levels up to 16 cm above the wind-tunnel bed, and examined different grain shapes and velocities. His results showed linear relationships between the logarithm of quantity and height (*not* the logarithm of height) (Figure 7.4). It should be noted that for each type of material the gradient of the relationship remains much the same irrespective of wind velocity. The reason for the apparent differences in the results of the two studies is unclear, although a visual inspection of the data suggests that Williams' results show the better linear fit.

Neither of the above studies involved any suspended load and it might be expected that this would have an additional effect on the nature of the concentration/height relationship. The concentration of suspended material should decrease with height more slowly than the rapid decline in saltation load, and since material gradually diffuses upwards, its maximum height would depend on the distance downstream from the point of entrainment.

Olsson-Seffer (1908) investigated sorting in the eight cm zone above a beach sand surface and found that mean particle size in transport decreased with height. Williams (1964) also studied sorting above a wind-tunnel sand bed and he also found that, depending on initial surface grinding, the range of particle size transported decreased with height, and the average height at which grains of a particular size were trapped increased with increased sphericity (a factor also affecting rate of transport).

7.2.5 The downwind progress of erosion

In a situation where there is an upwind limit to the erodible area and the surface comprises a mixture of erodible and non-erodible clods, Bagnold (1941) noted that at a given point in time, sandflow (q) increases downwind until the sandflow is saturated (q_s). No further removal can take place further downwind. As time progresses, and removal continues from the upwind area, the remaining surface debris coarsens and forms a protective pavement, and the area of removal migrates downwind leaving behind an increasingly wide zone of quiescent, stabilized surface. From this account it will be clear that in monitoring debris movement in wind it is important to draw a distinction between the passage of material across a surface which may be suffering no net loss, and the actual loss of material reflected in the depletion of a body of surface sediment.

Chepil (1957) referred to the downwind increase in the quantity of material transported as 'avalanching', and he attributed it to several causes (1959a). Firstly, there is a progressive increase in the number of grain impacts which results in the entrainment of more material by the impact mechanism. Secondly, because of the higher frequency of impact, abrasion increases and thus increases the supply of erodible material; in addition, the erodible material removed upwind supplements the erodible material downwind, making the soil generally more susceptible to erosion. Thirdly, particles dislodged from projections are trapped in depressions so that surface roughness is gradually reduced, which leads to an increase in shear velocity and hence in rate of transport. This process is usually called 'detrusion'. Whilst the saturated flow (q_s) for a given wind velocity was found to be independent of soil type and about the same for all soils, Chepil (1959b) noted that the distance from the point of initiation to q_s varied with soil erodibility. The distance was about 65 m for the most erodible soil and some 1900 m for the least erodible soil. These distances remained approximately the same for all levels of erosive winds.

As erosion progresses, the process of sorting due to differential rates of particle movement becomes more pronounced. Over the whole area erosion and deposition occur contemporaneously, but the net result at a particular site will be determined by the relations between the forces of erosion and deposition. At any given moment the surface grains reflect the associated airflow characteristics, but the ultimate depositional forms represent a sequence of events, probably taking their main characteristics from the peak period of activity.

Where there are no non-erodible grains and wind velocity exceeds the thresholds of movement, no protective layer will be formed and erosion will continue until some other factor causes it to cease (Bagnold, 1941). When soil texture is uniform, non-selective removal may occur in which all particles are moved at a similar rate without producing any deposits notably different from the parent material. Where non-erodible particles are present, the progress of wind erosion may be seriously restricted, for the amount of material removed is limited by their height, number, and distribution. As erosion proceeds, the height and number per unit area of non-erodible particles increases until the non-erodible particles completely shelter erodible material from the wind, creating a 'wind-stable' surface. The final stage can be defined by the 'critical surface barrier ratio' (see below).

7.2.6 The nature of surface deposits and eroded forms

Deposition of some or all of the material in transport will take place if one or more of the following occur: (1) atmospheric flow slackens, reducing or ending its capacity to transport; (2) local wind velocity is reduced by obstructions such as hedges, crops and non-erodible soil material (i.e. increased surface roughness); (3) the surface becomes stabilized by the onset of rain or irrigation; (4) surface compaction may be reduced so that more of the energy of saltating grains is dissipated on the bed rather than in maintaining of forward movement (Bagnold, 1941).

In his discussion of desert sands, Bagnold (1941) recognized three types of deposition: sedimentation, due to a reduction in wind velocity and involving material carried in all three modes of transport; accretion, caused by a reduction in the rate of transport (other than by decreased atmospheric flow) and consisting of creep and saltation loads; and encroachment, caused by a local increase in slope so that surface creep is retarded while saltation continues unhindered.

Chepil and Woodruff (1963) considered that the effects of sorting on soil produced four grades of surface material, arranged in order of increasing erodibility: (1) residual soil materials comprising non-erodible clods and rock fragments (to which semi-erodible clods showing wind etching can be added); (2) lag sands, gravels, and soil aggregates comprising semi-erodible particles moved as surface creep and widely scattered across the surface; (3) sand and clay dunes, together with ripples and possibly sand sheets, composed of erodible material moved primarily by saltation; and (4) loess, material moved in suspension and often deposited a considerable distance from its source.

Bagnold (1941) investigated in a wind tunnel the grading changes that occur to deposits on the sand bed with increasing length of time of exposure to wind action. He found that each deposit has its own peak grade; those parent materials and deposits with a singular peak he called 'regular' sands, and those with two or more peaks he called 'mixed' sands. Such graphical analysis of particle-size distribution is a useful means of comparing parent materials and wind-erosion deposits, as it can reveal the extent of sorting processes within an area.

7.3 THE FACTORS INFLUENCING THE LOCATION AND RATES OF WIND EROSION

7.3.1 Introduction

The rate at which wind erosion occurs, if at all, depends on the erodibility and the erosivity of the wind. Wind-tunnel experiments have been concerned mostly with relatively simple situations in terms of erodibility and erosivity, and this has enabled the influence of various factors to be quantitatively assessed. In the field however, the complexity of soil-erosion systems is much greater; and, in particular, erodibility and erosivity vary both spatially (and often over short distances), and temporally. Compare, for instance, the second by second fluctuations of the wind velocity with the much longer term changes in soil erodibility, determined in part by the previous erosional history at a site. In this section the factors influencing erodibility and erosivity will be examined. The ways in which they can be used in general expressions for actual and potential rates of erosion will be discussed in the following section.

7.3.2 Erodibility factors

It has been shown above that the erodibility of individual grains is dependent upon their diameter, density, and shape. Most soils, however, consist largely of clods comprising individual particles held together by various forces. It is the state and stability (against abrasion) of these structural units which largely determines the erodibility of soil in a field (Chepil and Woodruff, 1963). If a soil is well-structured, the number of soil particles small enough to be moved may be very low and abrasion may be minimal due both to a limited supply of abrasives and to the mechanical strength of the structural units. On the other hand, soils with weak structures and ample initial supplies of erodible material may be rapidly abraded. The state and stability of the structural units are principally determined by water, soil texture, organic cements, and disaggregating processes.

(a) *Water*

Water in the soil tends to bind soil grains together. In the case of sands, water is easily removed by surface drying and the cohesive bond is easily broken, as commonly occurs on sandbanks and sandy beaches that dry out between tides and 'blow'. Where finer materials predominate, as on mudflats, moisture retention properties are much better: water molecules are adsorbed on to grain surfaces by electrostatic forces (Hillel, 1971) and held there despite the high suctions caused by drying: and at the contact between grains the adsorbed water combines with capillary water (held by surface tension) to form a capillary wedge (Hillel, 1971). In a soil that is wetted and then dried, the moisture retention of the fine grains will tend to bind the mass together, and the increased pressure between grains at depth will enhance the strength of the bonding.

Chepil (1956) found that the erodibility of a soil decreased as the square of the soil moisture increased, up to 15 atmosphere percentage (the amount held at a suction equivalent to 15 atmospheres pressure and approximately equal to the permanent wilting point) where no erosion occurred. Belly (1964) similarly showed that threshold shear velocity increases rapidly as moisture content increases, until at a moisture content of two or three per cent by weight the threshold shear velocity is very high indeed.

The other main effect of water in the context of wind erosion is the formation of surface crusts by raindrop impact (Chepil and Woodruff, 1963). The crusts are normally formed of silt and clay, coarser particles being left lying loosely on the surface. The loose particles are easily dried, and may be moved by the wind soon after rainfall has ceased and before any significant drying of the surface has occurred. Once soil movement and abrasion have begun the crust may be fairly rapidly broken down and further drying is enhanced. The crust is part of the general surface smoothing caused by raindrop impact, a trend that increases the avalanching effect.

(b) *Texture*

The influence of texture on erodibility is largely related to the soil moisture retention properties discussed above. Chepil (1955a) found that in general the higher the proportion of silt and clay in a soil, the greater is the production of clods and the lower is soil erodibility. Conversely, a high proportion of sand produces few clods and a highly erodible soil. When mixed with sand, the first 5 per cent of silt or clay produces equal cloddiness, but the clods formed of clay are more resistant to abrasion. From 5–100 per cent, silt forms more clods, but they are softer and more easily abraded than those formed of clay. Chepil also found that clods showing a high degree of stability (i.e. resistance to abrasion) consisted of mixtures containing 20–30 per cent clay, 40–50 per cent silt, and 20–40 per cent sand.

(c) *Cements associated with organic decomposition*

A variety of cements are produced from the breakdown of organic material by micro-organisms. Chepil (1955b) investigated the effects on soil erodibility of increasing organic matter. He found that additions to the soil of between 1 and 6 per cent organic matter during the initial stages of decomposition (less than one year), led to enhanced clod production and decreased erodibility; but that over a period of four years there was a decline in clod production and an increase in erodibility. It was concluded that continuous addition of organic matter is necessary to improve cohesion, and that material left on the surface breaks down at a slower rate and is more useful than material mixed in by ploughing.

(d) *Disaggregating processes*

The effect of adding calcium carbonate to the soil was investigated by Chepil (1954) who showed that in general calcium carbonate weakens soil structure and increases erodibility; sandy soils provide an exception because, as they have little structure in the first place the addition of calcium carbonate is beneficial, for it is mostly silt-sized and acts as a weak cement. Freeze-thaw activity in the surface layer of the soil, particularly in the winter months, may have a marked influence on the breakdown of clods and reduction of their mechanical stability. Chepil and Woodruff (1963) showed that freeze-thaw activity can lead to an increase in erodibility from autumn through to spring. Also, harrowing in the spring causes clods to be broken down and thereby increases soil erodibility.

7.3.3 Erosivity factors

The principal factor affecting erosivity is the force of the wind on the ground surface. The factors affecting this force can be grouped into two main categories: those relating to the nature of atmospheric flow itself and those relating to the main constraint on that flow, surface roughness.

(a) *Atmospheric flow factors*

Wind tunnel tests and field measurements have shown that the rate of soil movement is proportional to the cube of wind velocity (see Equation (7.13) below). Applying this relationship to annual mean velocity records for Dodge City, Kansas, Zingg (1953b) calculated the ratio $(U/\overline{U})^3$ (in which U is the mean wind velocity for a given year and $\overline{U}$ is the mean velocity for all the years in the record). He found that periods of serious wind erosion in the High Plains of the USA were associated in part with periods of years for which the ratio was relatively high.

In a more detailed assessment of wind erosion forces, Skidmore and Woodruff (1968) analysed the wind records for 212 locations in the USA. Only mean hourly windspeeds greater than 5.4 m/s were considered to be erosive and used in the study. This velocity corresponds with the reported lower limits of erosion as measured at the 30.48 cm level; however, since most of the wind records would have been for heights greater than this, perhaps a higher velocity value should have been selected as the lower limit of erosion. For each location three measures were calculated: the *magnitude* of wind erosion forces, the prevailing wind erosion *direction*, and the *preponderance* of wind-erosion forces in the prevailing wind-erosion direction. The magnitude, r_j, was calculated for each month and for each of the sixteen compass points, using the equation

$$r_j = \sum_{i=1}^{n} \overline{U}_i^3 f_i \qquad (7.13)$$

in which

$\overline{U}$ = the mean windspeed in the ith velocity group, and
f_i = the duration, expressed as a percentage of the total observations in the jth direction within the ith velocity group.

The total magnitude for each month is given by the sum of the magnitudes from each direction, and this gives a measure of the relative capacity of the wind to cause soil blowing. The prevailing wind erosion direction (p) calculation is rather more complex, but is based on maximizing the forces parallel to the prevailing direction and minimizing the forces perpendicular to it. The ratio (R) of the 'parallel' forces to the 'perpendicular' is at a maximum (R_m) when the prevailing direction is found, and its value is a measure of the preponderance of the prevailing direction. When R_m has a value of 1 there is no prevailing direction, whereas a value of 2 indicates that wind erosion forces are twice as strong parallel to the prevailing direction as perpendicular to it.

A major deficiency of this technique is that it fails to take account of other weather conditions associated with winds from particular directions. For instance, it may well be that the prevailing wind often brings depressional rain, and that most of the erosion is caused by less frequent but drier winds from another direction.

(b) *Roughness elements*

Although the relative importance of the different factors that contribute to general surface roughness, and surface roughness itself are often difficult to determine in the field, it is convenient for the purposes of description to distinguish five major groups of roughness elements.

(1) *Vegetation.* The most important properties of vegetation cover in the context of surface roughness are its height and density, since these determine the

extent to which air flow contacts the ground surface, and influences the height of the mean aerodynamic surface. The values of height and density will vary with vegetation type and, for a given type, according to the time of year. Time of year is particularly important for annually sown crops, and the length of time taken for each crop to grow to a sufficient size to provide adequate protection against erosion is a fundamental factor. Chepil and Woodruff (1963) suggested that grasses and legumes are the most efficient in establishing a dense cover. Crop residues are also important in protecting the surface. Table 7.3 illustrates some results of wind tunnel tests on the effects that various types and quantities of crop residues have on the amount of material eroded from a sandy loam (Chepil and Woodruff, 1963).

Table 7.3. The average effects of different types of crop residues on the amounts of erosion on a sandy loam in a wind tunnel and at a uniform wind velocity (after Chepil and Woodruff, 1963, p. 276)

Quantity of crop residue above the soil surface (kg/ha)	Quantity of soil eroded in a wind tunnel			
	Covered with wheat residue		Covered with sorghum residue	
	Standing (25.4 cm high) metric tonnes /ha	Flat metric tonnes /ha	Standing (25.4 cm high) metric tonnes /ha	Flat metric tonnes /ha
0	6.47	6.47	6.47	6.47
91.9	1.14	3.49	5.34	5.96
183.8	0.04	1.02	3.33	4.27
367.6	T	0.04	1.60	2.18
551.4	T	T	0.57	0.90
1102.8	T	T	T	0.08

(2) *Clods and non-erodible fractions.* It was mentioned above that erosion continues until a sufficient number of non-erodible elements are uncovered at the surface. At this stage the non-erodible elements provide direct cover, and shelter the remaining erodible grains on the surface. This condition may alter however, with a change in the wind direction. The point at which this cover is just sufficient to prevent movement from continuing or starting is called the *critical surface barrier ratio* (originally called the *critical surface roughness constant* by Chepil (1950a)). The ratio is defined as the distance between the non-erodible barriers divided by the height of the barriers (Chepil and Woodruff, 1963), and was found to vary between 4 and 20 on cultivated soils, depending upon the wind shear velocity at the time and the threshold shear velocity of the erodible fractions. Therefore, the higher the shear velocity and the lower the threshold, the lower is the critical surface barrier ratio.

The use of this concept can be extended to include other roughness elements such as ridges and strip crops.

(3) *Ridges.* Ridges produced by tillage have their greatest effect in reducing erosion by sheltering and trapping when the wind blows at right angles to them, and a decreasing effect as the wind moves parallel to them. Results of portable wind tunnel tests in the field by Woodruff and Siddoway (1965, see Figure 7.5)

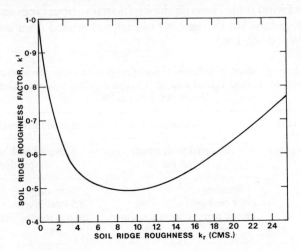

Figure 7.5. Chart for determining soil ridge roughness factor K' from soil ridge roughness K_r (after Woodruff and Siddoway, 1965). Note: K', the soil ridge roughness factor, is expressed in terms of height of standard soil ridges spaced at right angles to the wind and with a height-spacing ratio of 1:4. K_r, a measure of surface roughness, is surface roughness other than that caused by clods or vegetation (i.e. it is the roughness in the form of ridges or small undulations). It can be determined from a linear measure of surface roughness

relate soil ridge roughness factor K' to soil flow. The soil ridge roughness factor refers to a standard height-spacing ratio of 1:4. For example, ridges 2 cm high and 8 cm apart would have an equivalent height of 2 cm, whilst the same ridges 16 cm apart would have an equivalent height of 1 cm. Figure 7.5 shows that equivalent heights of between 3.8 and 7.6 cm are most effective in reducing erosion. Ridges consisting of only erodible elements are of little value because they are easily flattened. Similarly, tall ridges expose the topmost grains to stronger winds, so that in some circumstances ridges may increase rather than decrease the amount of erosion.

(4) *Field shelter belts.* Considerable attention has been paid to the various implications of shelter belt construction by such authors as Jensen (1954), Caborn (1957), and Skidmore and Hagen (1970). Here, remarks will be restricted to the effects of shelter belts on air flow. Usually the term shelter belt is taken to mean a row (or several rows) of trees, and/or a hedge. The principles concerning air flow over shelter belts may be applied to fences, hessian screens, walls, and rows of tall crops grown between more delicate ones.

Measurements made both in wind tunnels and in the field, by the above mentioned authors and others to whom they refer, show that in placing a barrier across the line of the wind there is a significant reduction in the velocity on the leeside of the barrier and a smaller reduction to windward. This reduction is usually expressed as a percentage of the unobstructed flow, and since the reduction is proportional to the height of the barrier, it is possible to express a reduction to a particular percentage in terms of barrier heights (*h*) from the barrier.

The permeability of the barrier has also been found to influence the air flow. For example, the greatest leeward area sheltered near the ground is achieved with a porosity of about 40 per cent (Jensen, 1954; Caborn, 1957; Skidmore and Hagen, 1970). The porosity of many barriers will vary according to the season and the amount of foliage. Chepil and Woodruff (1963) suggested that the maximum amount of protection is provided by a barrier whose cross-section is either triangular or sloping to the windward, rather than vertical to windward; and that little additional protection is gained by increasing the barrier width beyond five rows. They also pointed to the danger of leaving gaps in the belts since these cause accelerated flow and increase the risk of erosion; such velocity increases commonly occur at either end of the barrier.

Figure 7.6 shows the percentage of the unobstructed velocity values to the leeward of a wooden slat barrier with 40 per cent porosity in the field situation

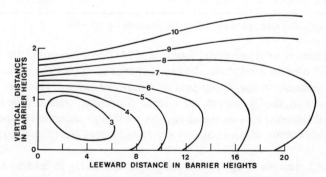

Figure 7.6. Ratio of leeward to windward windspeed in the lee of a 40 per cent open windbreak (after Hagan and Skidmore, 1971)

S. J. Wilson and R. U. Cooke

(Hagen and Skidmore, 1971). This example can be used to illustrate how the area protected from erosion varies with changes in the unobstructed wind velocity and the threshold velocity of the soil. Assuming that the wind is at right angles to the barrier and that all wind velocity measurements are made just above the ground (say 30 cm), then Table 7.4 indicates, for different distances to leeward, all the velocity values at which erosion would occur, given various unobstructed velocity values and threshold values. Clearly the protection is greatest when threshold values are low and the unobstructed velocity is only slightly greater than the threshold; conversely when the threshold is low and the unobstructed velocity is very high, the shelter is insufficient to prevent erosion even at only 2–6 barrier heights to leeward.

Table 7.4. Relations between unobstructed and sheltered wind velocities and threshold velocities leeward of a barrier (after Hagen and Skidmore, 1971)

Distance leeward (in barrier heights)	Percentage of the unobstructed wind velocity	Threshold velocities (km per hour)											
		18				24				30			
		Unobstructed wind velocities (km per hour)											
		24	34	44	54	24	34	44	54	24	34	44	54
		Sheltered wind velocities											
2–6	35	—	—	—	18.9	—	—	—	—	—	—	—	—
8	40	—	—	—	21.6	—	—	—	—	—	—	—	—
10	50	—	—	22.0	27.0	—	—	—	27.0	—	—	—	—
12	60	—	20.4	26.4	32.4	—	—	26.4	32.4	—	—	—	32.4
14	65	—	22.1	28.6	35.1	—	—	28.6	35.1	—	—	—	35.1
16	70	—	23.8	30.8	37.8	—	—	30.8	37.8	—	—	30.8	37.8
18	75	—	25.5	33.0	40.5	—	25.5	35.0	40.5	—	—	33.0	40.5
20	80	19.2	27.2	35.2	43.2	—	27.2	35.2	43.2	—	—	35.2	43.2

Note: only sheltered values exceeding the threshold are shown.

(5) *Local changes in topography.* Bagnold (1941), in examining the nature of the local variation in wind shear above a sequence of ripples, demonstrated that shear was greatest on the upper part of windward slopes. This was found to apply equally to knolly terrain, by Chepil, Siddoway, and Armbrust (1964), who showed that soil loss increases rapidly with both increased slope and distance towards the top of the knoll (for windward slopes less than about 150 m in length).

In the field changes in the microtopography are likely to lead to a complex pattern of erosion, and the presence of knolls and hollows is likely to affect the variables influencing erodibility. For example, lower soil moisture is likely on the higher parts of knolls, due to better drainage.

7.4 THE MEASUREMENT AND PREDICTION OF ACTUAL AND POTENTIAL EROSION

7.4.1 Wind erosion rates

There were several early attempts to measure rates of debris movement by wind. Cornish (1900), for instance, calculated the annual passage of sand 'in the desert' over a point one centimetre square. By estimating visually the velocity of sand grains over the surface and the frequency of erosion, he calculated that the equivalent of a 10 m thick layer was moved each year. O'Brien and Rindlaub (1936) measured sand movement at the mouth of the Columbia River and related sediment discharge to wind velocity. From this they produced the empirical equation

$$G = 0.036u_5^3 \text{ (when } u_5 > 20) \tag{7.14}$$

in which

$G = $ the discharge (in pounds per foot width per day), and
$u_5 = $ wind velocity at five feet above the surface (in feet per second),

the expression being valid for a mean grain size of 0.2 mm. Such an equation is of little general use, and there is clearly a need for a universally applicable relationship.

Bagnold (1941) derived an equation relating the rate of sand flow, q (the weight of sand which moves along a lane of unit width past a fixed point in unit time), to the wind shear velocity over an eroding surface, U'_*, such that

$$q = C \sqrt{\frac{d}{D}} \cdot \frac{\rho}{g} U'^3_* \tag{7.15}$$

In the equation,

$D = $ a standard grain diameter of 0.25 mm,
$d = $ mean grain diameter of the sand in question,
$\rho = $ density of air,
$g = $ the gravitational constant (ρ/g has a value of 0.125 in SI units), and
$C = $ an empirical constant.

The value of C was found to vary with the size distribution of the material involved, and the following values were used: 1.5 for a uniform sand; 1.8 for a naturally graded sand; 2.8 for a sand with a wide range of grain sizes. When sand is driven over a surface containing a large number of hard, immobile elements, Bagnold found that the value of C was about 3.5.

Several problems arise with such an equation. Firstly, it predicts sand flow at velocities below the threshold of movement and therefore the threshold must be independently determined. Secondly, the equation assumes that the mode of transport is both saltation and creep but not suspension, and it may therefore not describe fully the movement of very fine materials and thirdly, the correct value of C must be estimated, and this may well prove difficult, particularly if the equation is extended in its use to predict soil flow. A similar equation, but giving lower values of q for given velocities, was derived by Zingg (1953a) in wind tunnel experiments. In the equation,

$$q = C(d/D)^{3/4} \rho U_*'^3/g \tag{7.16}$$

C has the value 0.83.

Kawamura (1951) developed an expression which took into account the threshold shear velocity (U_{*t}) of the material involved, such that

$$q = (C\rho/g)(U_*' - U_{*t})(U_*' + U_{*t})^2 \tag{7.17}$$

This equation is still not very useful in the field, since the values of C and U_{*t} both have to be found experimentally in a wind tunnel.

Other workers such as Horikawa and Shen (1960), Belly (1964), and Williams (1964), have also measured rates of sand flow in wind tunnels. Their results, and those of research workers already mentioned (see DePloey and Gabriels, Chapter 3), depend on the accuracy and efficiency of the apparatus used to sample the rates of movement of material. A variety of 'sand traps' were compared in this respect by Horikawa and Shen (1960). The traps fall into 'horizontal' and 'vertical' types. The former type consists of a compartmented trough set into the bed of the wind tunnel, parallel to the direction of flow, so that grains bouncing or rolling along the bed are trapped in one of the compartments according to the length of 'hop'. The latter type consists of an opening or series of openings arranged vertically, so as to trap all particles moving past a unit width, irrespective of the height of movement. The horizontal traps were found to require a considerable length before they became efficient, and the vertical traps had inefficiencies due largely to the interference they caused to the air flow; in particular, the build up of 'back pressure' at the mouths of the openings (due to there being little or no exhaust) lead to a deflection of the air flow and presumably its load around the traps. In the light of information from these experiments, Horikawa and Shen designed an extremely efficient vertical trap which included a much improved exhaust system.

7.4.2 Potential erosion

Chepil (1945c) considered that Bagnold's sand flow equation (Equation (7.15)) was adequate in relation to shear velocity, but because for field soils the value of C varied so much over time and space, even within a single field, he felt that a more useful measurement would be potential erosion at a site. This would enable soil loss predictions to be made for individual fields, and the variables in the equation could be manipulated in order to achieve a controlled level of erosion.

An identification and evaluation of the variables in the wind erosion system (discussed above) was undertaken by Chepil and his co-workers over a period of 25 years, culminating in the publication of a wind erosion equation (Woodruff and Siddoway, 1965), similar to the universal soil loss equation for water erosion:

$$E = f(I', K', C', L', V) \qquad (7.18)$$

in which

E = the potential erosion measured in tons per acre per annum,
I' = a soil erodibility index,
K' = a soil ridge roughness factor,
C' = a local climatic factor,
L' = the median unsheltered field length along the prevailing wind erosion direction, and
V = the equivalent vegetation cover measured in pounds per acre.

The precise relationship between all the variables is complex, and calculation of E cannot be represented by a single equation but must be determined by the use of a set of tables and nomographs given in Woodruff and Siddoway (1965).

The credibility of the wind erosion equation as a tool for the planning of erosion controls must depend on the component variables being realistic and the accuracy of the values derived for them (see also Kirkby, Chapter 1). Each of the variables could be the subject of prolonged discussion, but here only the two principal factors, soil erodibility and climate, will be considered.

The most important variable is the soil erodibility index, I', a measure of soil loss from a wide, unsheltered, isolated field with a bare, smooth surface. It is based on climatic conditions in the vicinity of Garden City, Kansas, during the period 1954–56. The other variables in the equation modify the conditions for which I' is calculated, and accordingly alter the estimate of soil loss. The data on which I' are based are taken from several studies. Soil erodibility was found in the wind tunnel to decrease with an increase in the weight percentage of non-erodible clods (>0.84 mm diameter) (Chepil, 1950b) as determined by a dry sieving procedure. However, the boundary layer and scale limitations in the wind tunnel mean that a relationship had to be found between those values and actual soil loss in the field. This was carried out during the period 1954–56 using 69 sample fields in the Garden City area (Chepil, 1960). The depth of soil removed from each field was converted to tons per acre and related to erodibility. Measurements of soil loss of less than half an inch (1.27 cm) were not made due to various practical difficulties, and in such cases qualitative assessments were given in terms of either 'no visual evidence of erosion or plant damage', or 'visual evidence of erosion which is insufficient to kill the plant'. The fact that many of the sample fields (45) did not have a measurable amount of erosion and that there was a considerable spread in the removal/erodibility relationship, would seem to cast some doubt on the accuracy of I'. The erodibility index also makes adjustments for windward slopes less than 152 m in length, the value of I' increasing with both slope and position towards the top of the slope.

As mentioned above, the soil erodibility index was calculated for one time at one location; for all other times and places a climatic factor adjustment must be made. The climatic factor, C' (Chepil, Siddoway, and Armbrust, 1963), is based on the principle of erosion varying directly as the cube of windspeed and inversely as the square of soil moisture (effective precipitation being used as a surrogate measure). The climatic index for any place in a given year (C_1) is given by

$$C_1 = \frac{100}{2.9} \frac{v^3}{(P - E)^2} \tag{7.19}$$

in which

$\bar{v}$ = the mean wind velocity (at 9.1 m) for the year, and

$(P - E)$ = the Thornthwaite (1931) precipitation effectiveness index, itself given by

$$115\left(\sum_{i=1}^{12} (P_i/T_i - 10) \right) \tag{7.20}$$

in which

P_i = the monthly rainfall (in inches), and

T_i = temperature (in °F).

The term $100/2.9$ expresses C_1 as a percentage of the mean annual value of C for Garden City, Kansas. The year in this case refers to the period June 1 to May 31, since it was found that 80 per cent of dust storms occurred in the season from January 1 until May 31, and this was greatly influenced by the prevailing weather conditions in the preceeding months of June 1 until December 31. C' is the mean of all C_1 values for a location.

Subsequently, the annual values of C' for use in the wind erosion equation have been replaced by mean monthly values, in order to take account of the monthly variation in mean wind velocity (Woodruff and Armbrust, 1968). In the monthly calculation, the annual $P - E$ index is retained as the most appropriate measure of the soil moisture condition. The use of either monthly or annual values of C' introduces a statistical element into the wind erosion equation.

In addition to its use in the wind erosion equation, the climatic index has been used on its own as a predictor of the potential severity of erosive conditions. Figures 7.7 and 7.8 show, for both Garden City and Dodge City the number of dust storm days for a period of calendar years and the mean value of C for the previous three years, C_3, (ending May 31 in that calendar year and plotted against that same year). The assumption is made that the severity of erosion is proportional to the number of dust storm days and clearly there is

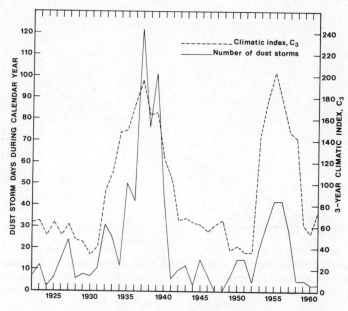

Figure 7.7. Number of dust-storms and the climatic index C_3 for each year of record at Dodge City, Kansas (after Chepil, Siddoway and Armbrust, 1963)

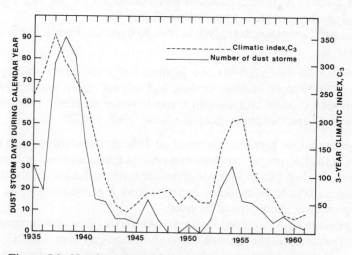

Figure 7.8. Number of dust-storms and the climatic index C_3 for each year of record at Garden City, Kansas (after Chepil, Siddoway, and Armbrust, 1963)

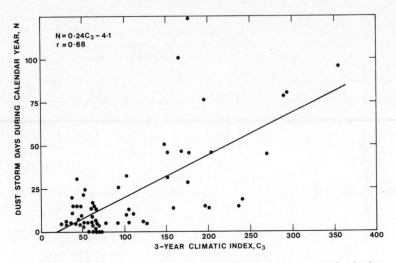

Figure 7.9. Regression of annual number of dust-storms, N, beginning January 1, on the 3-year climatic index, C_3 ending May 31 of the preceding year (combined records for Dodge City and Garden City, Kansas) (after Chepil, Siddoway, and Armbrust, 1963)

a general coincidence between high values of both the number of dust storm days and the climatic index. The data for both towns were therefore combined to produce a single plot of the climatic index against the number of dust storm days, and a linear regression equation was determined for use as a predictor of the likely severity of erosive conditions (Figure 7.9).

The basis of the regression analysis is open to question. Firstly, as the author points out:

'The reason for the relatively low incidence of dust storms during the period 1954–57 may be that farmers had learned more about how to control wind erosion and were in a much better economic position to control it' (Chepil, Siddoway, and Armbrust, 1963, p. 450).

This suggests that the regression is based on different populations, each having their own statistical properties. A second criticism, also outlined in the paper, is that the spread of values does not allow accurate predictions of the number of dust storm days to be made; it merely allows a general indication of the possible severity of conditions to be given.

It was suggested by Chepil, Siddoway, and Armbrust (1963) that severe erosion conditions occurred in the Garden City and Dodge City area when $C_3 > 125$ per cent and N (the number of dust storms predicted from the regression) exceeded 25, at which point special precautions were necessary in addition to those normally taken. This critical value of C_3 can be calculated for other locations, which are likely to have adapted to a similar probability of erosive conditions occurring, using the equation

$$C_{3(crit)} = 125 \frac{C'}{100} \tag{7.21}$$

in which

C' = the mean value of the climatic index for that location.

7.5 CASE STUDY: OBSERVATIONS ON SOIL BLOWING IN THE VALE OF YORK

During the spring and early summer of 1974 and 1975 observations on soil blowing were made in the Vale of York, N. England, in an area of predominantly fine sandy soils near the small towns of Barmby Moor and Market Weighton. The observations included monitoring of rates of wind erosion on individual fields, assessment of relative differences in erosion severity between fields, and examination of the factors influencing the wind-erosion processes.

Rates of erosion were measured using two techniques: firstly by means of a 'catcher' that sampled the passage of material being transported at various heights above the surface and secondly by means of a 'relative soil height measurer' that allowed small changes in the relative height of the soil, at sample points, to be recorded.

The 'catcher' (Figure 7.10) sampled at six heights between 15 cm and 152 cm. Each smapler consisted of a plastic bottle with an inlet and outlet glass tube, mounted on a wind vane which rotated about a central aluminium pole. In the field the inlet always pointed into the wind and the efficiency of the sampler compared very favourably with a modified version of the Horikawa and Shen sampler (Horikawa and Shen, 1960) when tested in a wind tunnel.

Estimates of the total load transported past a point were made by a manual integration of values between 15 cm to 152 cm (including interpolated values between sampled levels) and by estimating the quantity of material moving below 15 cm using the technique used by Williams (1964). The latter estimates were reasonably substantiated by several sets of field measurements taken at levels between 1 cm and 15 cm, in fixed directions. Using an array of catchers, a clear indication of the pattern of variation in the amount of material moved past sample point over a field surface was obtained.

Observations showed that much of the movement was confined to the lowest levels over the surface and most of the load was redistributed within each field, particularly towards its leeward edges, into damp hollows or onto rough surfaces. Little of the material was completely removed from the fields.

Rates of movement varied considerably over very short distances due to rapid changes in erosivity and erodibility. Because of this variation, the traditional method of describing soil loss per unit area became rather meaningless

S. J. Wilson and R. U. Cooke

Figure 7.10

Figure 7.11

Many sampled points recorded a considerable passage of material but with little or no net soil loss. However soil losses of up to 21 tonnes per hectare, averaged over areas of two hectares, were recorded during a single sand storm lasting about seven to eight hours. In practical terms a measure of weight of material passing a unit width per unit of time is more useful, since it is the blasting effect of sand which causes the greatest damage to crops.

In order to clarify the pattern and amount of erosion in a particular field, a system for measuring changes in the relative height of the soil surface at sample points was designed. The relative soil height measurer (Figure 7.11) enabled points on the surface to be located from a movable steel bar suspended about 25 cm above the surface, so that measurements could be made without equipment interfering with the erosion processes. Using a steel rule with a flattened end, changes of 0.05 cm could be measured. The technique proved to be very successful and again indicated a pattern of soil loss or gain.

Generally, soil losses were greatest from certain vulnerable locations and showed a general increase towards the leeward of an eroded field. During a single sand storm depth losses of up to 2.4 cm were recorded at individual points and an average depth loss of 0.62 cm was recorded over 15 points at approximately 100 m downwind from the field boundary. The net soil loss over the same two hectare area sampled by the 'catchers' was about 44 tonnes per hectare.

Whilst the 'catchers' and 'relative soil height measurers' clearly produced different results, the order of magnitude was the same; further testing and refinement of the techniques is required to ensure improved accuracy.

The measurements of soil movement indicated the occurrence of rapid changes in the microtopography of badly affected fields. Evidence of past erosion was given by the banks of deposits in hedges (up to 3 m in height) and by the frequently scoured lower central areas of fields.

The soils which are worst affected in the area are fine sandy soils. Samples from eroded and non-eroded fields showed that when the silt/clay content was less than 10 per cent the soil was easily eroded, between 10 and 15 per cent of the soil was less easily blown and when it exceeded 15 per cent soil was seldom blown.

The critical threshold wind shear velocity required to initiate erosion on several examples of fine sandy fields, sown with sugar beet which either had not yet emerged or was in the small seedling stage, was found to be approximately 30 cm/sec. This was calculated from wind velocity measurements at 29·cm and 124 cm above the surface.

The shelter effect of trees and hedges appeared to be very limited and erosion often started only ten metres or less from the windward edge of the field. However the main effect of the hedges was to prevent saltation from continuing into neighbouring fields, by acting as a trap, thus preventing isolated areas of erosion becoming contiguous during a major sand storm.

Row crops such as sugar beet, carrots, and potatoes were affected by erosion to a greater degree of severity and for a longer period of time during the spring and early summer months than cereal crops. If no remedial action is taken to prevent or restrict erosion, damage to the soil and crops will continue to be severe in certain years.

Further research would seem to be necessary in order to improve existing erosion management techniques and develop new ones. The approaches briefly outlined in this case study were promising and could be improved. The use of aerial photographs of areas affected by erosion, taken immediately after the event, would undoubtedly improve the accuracy of identifying areas at risk and and closer correlation with soil types and perhaps even soil moisture and soil temperature.

7.6 CONCLUSION

The study of agricultural soil erosion by wind has made substantial advances in the last thirty or so years, especially in the context of understanding the details of the environmental factors that influence the nature and rate of soil movement. It is important to emphasize, however, that much of the most important and fundamental work in the United States is only directly applicable to the particular environment of the Great Plains and the soil-management problems encountered there. While, of course, this work is relevant to semiarid lands elsewhere, and indeed to the study of the processes of wind erosion everywhere, the results, and particularly the prediction equations, can only be used with extreme caution in humid environments. In eastern England, for instance, where wind erosion is an increasingly serious problem, the climatic index (C) is of little value because it is based on climatic data that do not comprehend some of the local climatic circumstances that are favourable to wind erosion. There is thus a real need to develop new prediction equations pertinent to the conditions in eastern England and elsewhere. Towards this end the monitoring of wind erosion, as described briefly in the Vale of York, provides a promising way forward.

REFERENCES

Beadnell, H. J. L. (1910). Sand dunes of the Libyan Desert. *Geog. J.*, **35**, 386.

Belly, P.-Y. (1964). *Sand Movement by Wind*. U.S. Army Corps of Engineers (Coastal Engineering Research Center), *Tech. Mem.* 1.

Bryan, K. (1923). Wind erosion near Lees Ferry, Arizona. *Am. J. Sci.*, **6**, 291–307.

Caborn, J. M. (1957). Shelterbelts and microclimate. *Forestry Commission Bull.* **29** (HMSO, Edinburgh).

Chepil, W. S. (1945a). Dynamics of wind erosion. I. The nature of movement of soil by wind. *Soil Sci.*, **60**, 305–320.

Chepil, W. S. (1945b). Dynamics of wind erosion. II. Initiation of soil movement. *Soil Sci.*, **60**, 397–411.

Chepil, W. S. (1945c). Dynamics of wind erosion. III. Transport capacity of the wind.

Chepil, W. S. (1950a). Properties of soil which influence wind erosion. I. The governing principle of surface roughness. *Soil Sci.*, **69**, 149–162.

Chepil, W. S. (1950b). Properties of soil which influence wind erosion. II. Dry aggregate structure as an index of erodibility. *Soil Sci.*, **69**, 403–414.

Chepil, W. S. (1954). Factors that influence clod structure and erodibility of soil by wind. III. Calcium carbonate and decomposed oganic matter. *Soil Sci.*, **77**, 473–480.

Chepil, W. S. (1955a). Factors that influence clod structure and erodibility of soil by wind. IV. Sand, silt, and clay. *Soil Sci.*, **80**, 155–162.

Chepil, W. S. (1955b). Factors that influence clod structure and erodibility of soil by wind. V. Organic matter at various stages of decomposition. *Soil Sci.*, **80**, 413–421.

Chepil, W. S. (1956). Influence of moisture on erodibility of soil by wind. *Proc. Soil Sci. Soc. Am.*, **20**, 288–292.

Chepil, W. S. (1957). Width of field strips to control erosion. *Kansas Agr. Expt. Sta. Tech. Bull.* 92 pp.

Chepil, W. S. (1959a). Equilibrium of soil grains at the threshold of movement by wind. *Proc. Soil Sci. Soc. Am.*, **23**, 422–428.

Chepil, W. S. (1959b). Wind erodibility of farm fields. *J. Soil and Water Conserv.*, **14**, 214–219.

Chepil, W. S. (1960). Conversion of relative field erodibility to annual soil loss by wind. *Proc. Soil Sci. Soc. Am.*, **24**, 143–145.

Chepil, W. S. (1961). The use of spheres to measure lift and drag on wind-eroded soil grains. *Proc. Soil Sci. Soc. Am.*, **25**, 243–245.

Chepil, W. S., and Siddoway, F. H. (1959). Strain gauge anemomenter for analysing various characteristics of wind turbulence. *J. Meteorol.*, **16**, 411–418.

Chepil, W. S., Siddoway, F. H., and Armbrust, D. V. (1963). Climatic index of wind erosion conditions in the Great Plains. *Proc. Soil Sci. Soc. Am.*, **27**, 449–452.

Chepil, W. S., Siddoway, F. H., and Armbrust, D. V. (1964). Wind erodibility of knolly terrain. *J. Soil and Water Conserv.*, **19**, 179–181.

Chepil, W. S., and Woodruff, N. P. (1963). The physics of wind erosion and its control *Adv. in Agron.*, **15**, 211–302.

Cornish, V. (1897). On the formation of sand dunes. *Geog. J.*, **9**, 278–309.

Cornish, V. (1900). On desert sands bordering the Nile Delta. *Geog. J.*, **15**, 1–32.

Cressey, G. B. (1928). The Indiana sand dunes and shore lines of the Lake Michigan Basin. *Geog. Soc. Chicago Bull.*, **8**, 23.

Davies, D. B., and Harrod, M. F. (1970). The process and control of wind erosion. *N.A.A.S. Quat. Rev.*, **88**, 139–150.

Gilbert, G. K. (1895). Lake basins created by wind erosion. *J. Geol.*, **3**, 47–49.

Gilbert, G. K. (1914). Transportation of debris by running water. *US Geol. Surv. Prof. Paper*, **86**.

Hagen, L. J., and Skidmore, E. L. (1971). Turbulent velocity fluctuations and vertical flow as affected by windbreak porosity. *Trans. Am. Soc. Agric. Engrs*, **14**, 464–465.

Horikawa, K., and Shen, H. W. (1960). *Sand Movement by Wind (On the characteristics of sand traps)*, U.S. Beach Erosion Board, Tech. Mem. No 119. 51 pp.

Kádár, L. (1934). A study of the sand sea in the Libyan Desert. *Geog. J.*, **83**, 470–478.

Kalinske, A. A. (1943). The role of turbulence in river hydraulics. *Iowa State Univ. Proc. 2nd Hydraulic Conf. Bull.*, **27**.

Kawamura, R. (1951). Study on sand movement by wind. *Report* of the Institute of Science and Technology, Univ. of Tokyo, **5** (3–4), 95–112. (in Japanese; English review in Horikawa and Shen (1960)).

Keyes, C. R. (1910). Deflation and the relative efficiencies of erosional processes under conditions of aridity. *Bull. Geol. Soc. Am.*, **21**, 565–598.

King, W. J. H. (1916). The nature and formation of sand ripples and dunes. *Geog. J.*, **47**, 189–209.

O'Brien, M. P., and Rindlaub, B. D. (1936). The transportation of sand by wind. *Civil Eng.* (New York), **6**, 325–327.

Olsson-Seffer, P. (1908). Relation of wind to topography of coastal drift sands. *J. Geol.*, **16**, 549–564.

Pollard, E., and Millar, A. (1968). Wind erosion in the East Anglian Fens. *Weather*, **23**, 415–417.

Radley, J., and Simms, C. (1967). Wind erosion in East Yorkshire. *Nature*, Lond., **216**, 20–22.

Radley, J., and Simms, C. (1969). Wind erosion on the Sandlands of East Yorkshire. *Proc. North of England Soils Dis. Group*, **5**, 20–23.

Robinson, D. N. (1968). Soil erosion by wind in Lincolnshire, March 1968. *East Midland Geographer*, **4**, 351–362.

Skidmore, E. L., and Hagen, L. J. (1970). Evapotranspiration and the aerial environment as influenced by windbreaks. *Great Plains Agr. Council Pub.*, **50**, 339–368.

Skidmore, E. L., and Woodruff, N. P. (1968). *Wind Erosion Forces in the United States and Their Use in Predicting Soil Loss.* United States Department of Agriculture, Washington, D.C., *Agricultural Handbook No.* 346, 42 pp.

Sneesby, N. J. (1953). Wind erosion and the value of shelter belts. *Agric.*, **60**, 263–271.

Sneesby, N. J. (1966). Erosion control on the Black Fens. *Agric.*, **73**. 391–394.

Sneesby, N. J. (1968). Shelter against soil erosion. *Agric.*, **75**, 550–551.

Task Committee on Preparation of Sedimentation Manual (1965). Sediment transportation mechanics: wind erosion and transportation, (Progress Report). *J. Hydraulic Div., Proc. Am. Soc. Civil Engrs.*, **91**, 111–112, 267–287.

Thornthwaite, C. W. (1931). Climates of N. America according to a new classification. *Geog. Rev.*, **21**, 633–655.

Udden, J. A. (1894). Erosion, transportation, and sedimentation performed by the atmosphere. *J. Geol.*, **2**, 318–331.

White, C. M. (1940). Equilibrium of grains on bed of stream. *Proc. Roy. Soc. Lond.*, **174A**, 322–334.

Wilkinson, B. *et al.* (1969). Survey of wind erosion on sandy soils in the East Midlands. *Experimental Husbandry*, **18**, 53–59.

Williams, A. M. (1971). Wind protection for soil and crops. *N.A.A.S. Quat. Rev.*, **92**, 148–154.

Williams, G. (1964). Some aspects of the eolian saltation load. *Sedimentation*, 3, 257–287.

Woodruff, N. P., and Armbrust, D. V. (1968). A monthly climatic factor for the wind erosion equation. *J. Soil and Water Conserv.*, **23**. 103–104.

Woodruff, N. P., and Siddoway, F. H. (1965). A wind erosion equation. *Proc. Soil Sci. Soc. Am.*, **29**, 602–608.

Zingg, A. W. (1953a). Wind tunnel studies of the movement of sedimentary material. *Iowa State Univ. Proc. 5th Hydraulic Conf. Bull.*, **34**, 111–135.

Zingg, A. W. (1953b). Speculations on climate as a factor in the wind erosion problem of the Great Plains. *Trans. Kansas Acad. Sci.*, **56**, 371–377.

Bibliography

Astbury, A. K. (1958). *The Black Fens.* Cambridge, The Golden Head Press. 217 pp.

Bagnold, R. A. (1941). *The Physics of Blown Sand and Desert Dunes.* London, Chapman and Hall. 265 pp.

Brunt, D. (1934). *Physical and Dynamical Meteorology.* Cambridge, University Press. 428 pp.

Hillel, D. (1971). *Soil and Water: Physical Principles and Processes.* London, Academic Press. 288 pp.

Jensen, M. (1954). *Shelter Effect.* Copenhagen, The Danish Technical Press. 264 pp.

Soil Erosion
Edited by M. J. Kirkby and R. P. C. Morgan
© 1980 John Wiley and Sons Ltd.

CHAPTER 8

Implications

R. P. C. Morgan

Senior Lecturer in Applied Physical Geography,
National College of Agricultural Engineering, Bedford, UK

8.1 INTRODUCTION

The ultimate objective of research on soil erosion is to solve erosion problems by adopting suitable conservation measures. Although the techniques of controlling erosion are well-documented (Stallings, 1957; FAO, 1965; Hudson, 1971), the basis for their selection and design remains largely empirical, founded on many years of experimentation and practice by agricultural engineers. Research on conservation thus lags behind that on erosion in which, as a result of detailed studies of the mechanics of the processes involved, empiricism is giving way to deterministic modelling. In the light of this development, described fully in the previous chapters, current approaches to soil conservation research are reviewed with the aim of providing pointers for future work. A theoretical approach to soil conservation is adopted, based on the material presented in the earlier chapters. By comparing this theoretical approach with what happens in practice, those topics are identified where further investigation is required if the scope of erosion modelling is to be extended to encompass soil conservation measures. The comparison also provides a perspective for a discussion of the work carried out to date on modelling the effects of soil conservation.

8.2 SOIL LOSS TOLERANCES

Satisfactory erosion control implies obtaining the maximum sustained level of production from a given tract of land whilst maintaining soil loss below a threshold level which theoretically permits the rates of soil erosion and soil formation to balance. Since, in practice, this theoretical state of equilibrium is difficult to either predict or achieve, a less rigorous target is commonly pursued. This is to keep the rate of erosion below the level required to maintain soil fertility in the medium term, say over 20 to 25 years. In this case, allowance

253

is being made for the use of manures and fertilizers to replace those plant nutrients lost by erosion. Even so, there are doubts whether achieving this lesser target adequately maintains the soil resources of many areas, particularly in Great Britain and Europe, where the rate of soil formation is very slow (Kirkby, Chapter 1). Further, even if this productivity target is met, the rate of erosion may still be too high to prevent the pollution of lakes and rivers by sediment and by the plant nutrients and pesticides carried either in the runoff or attached to the sediment particles (Moldenhauer and Onstad, 1975).

8.2.1 Temporal level of resolution

Before specifying the maximum rate of soil loss which may be considered acceptable, it is necessary to define the erosive event against which protection is desired. As indicated by Thornes (Chapter 5), the magnitude of this event is best considered in terms of it exceeding particular thresholds. There has been much research aimed at identifying the thresholds at which significant erosion commences. Critical wind velocities have been established as $4 \, m \, s^{-1}$ and 1 $km \, s^{-1}$ when velocity is measured at heights of 1 m and 15 m above the ground respectively (Chepil, 1945; Woodruff and Zingg, 1952). A critical rainfall intensity of $25 \, mm \, h^{-1}$ has been defined by Hudson (1965) for water erosion based on studies in Zimbabwe. Unfortunately, this last value cannot be taken as universal and is probably only applicable to tropical areas of intense rainfall. In many temperate areas of the world this threshold value of intensity is rarely exceeded and yet significant erosion still occurs. Consequently, lower threshold values of $6 \, mm \, h^{-1}$ and $10 \, mm \, h^{-1}$ have been identified in Germany (Richter and Negendank, 1977) and Great Britain (Morgan, 1977) respectively.

These threshold values are, of course, exceeded too frequently for them to form a basis for conservation design. Any measure aimed solely at protection against these events of low magnitude would be totally inadequate to deal with less frequent but more erosive events. An alternative approach is to provide protection against the most effective event. There is some difficulty in recognizing both the frequency and magnitude of this event (Thornes, Chapter 5), however, because the frequency distribution of erosion events, when weighted by magnitude, is often multimodal. As much erosion may take place in a single occurrence of the 100-year return period event as in 50 occurrences of the 5-year return period event. In deciding against which of these events to design protective measures, further complications arise.

First, the effects of the two types of event may be quite different. Those resulting from the 100-year event may be longer laster and more dramatic causing, for example, headward erosion of gullies and the formation of new channel heads (Nossin, 1964; Thornes, 1976; Anderson and Calver, 1977). Second, events of comparable erosive magnitude may have different return periods for different processes (Rapp, 1960; Iveronova, 1969; Rapp, Axelsson, Berry, and Murray-Rust, 1972). Third, the magnitudes of events of the same frequency may differ. In the eastern Midlands of England a daily rainfall of 37 mm has a five-year return period and is sufficient to cause a major erosion event by local standards (Morgan, 1977) whereas in the Darjeeling Hills, India, a total of 50 mm is received on average on 12 days every year and the erosion resulting from an event with a five-year return period is catastrophic (Starkel, 1976).

Clearly there can be no universal approach towards deciding the size and frequency of the event for which to design erosion control measures. Ultimately, the selection is governed by cost. To provide protection against extreme events with return periods of a hundred years or more is very expensive and, in rural areas, where the danger to human life is relatively low, unrealistic. The compromise solution, adopted in the design of channels to convey excess runoff, is to protect against the event with a ten-year return period. In some tropical areas, where even this event is of too great a magnitude to allow protection at reasonable cost, a five-year return period event is used for design purposes (Chow and Chang, 1968).

No guidelines exist for acceptable rates of soil loss from individual storms. This is because conservation works are rarely designed to deal with single events. Since the amount of erosion depends not only on the energy of the erosive agent but also on the erodibility of the soil and the protection afforded by the plant cover prevailing at the time, design for single events requires predicting both the magnitude of the event with the selected return period and the probability of it coinciding with the soil and plant cover conditions particularly favourable for erosion. Where, because of seasonal rainfall regimes or cropping practices, the plant cover varies during the year, predictions of this kind are not attempted and an alternative approach to design is used aimed at keeping erosion rates below a selected mean annual value.

A rate of 1.1 $kg\,m^{-2}$ is widely accepted as the maximum permissible annual soil loss (Mitchell and Bubenzer, Chapter 2). Although, in the light of the comments made earlier and in comparison with erosion rates measured in many parts of the world (Fournier, 1960), the figure may seem somewhat high, it represents a realistic value to set as a target when viewed against what farmers are likely to achieve. If attained, a new and lower target figure can be set. Lower maximum permissible values are sometimes used. For example, 0.5 $kg\,m^{-2}\,y^{-1}$ is recommended as a target where soils are thin and slow-forming (Smith and Stamey, 1965) and 0.2 $kg\,m^{-2}\,y^{-1}$ where soils are very erodible (Hudson, 1971). Arnoldus (1977) has established a range of values for tolerable mean annual soil loss from 0.2 $kg\,m^{-2}\,y^{-1}$ on soils where the rooting depth is up to 25 cm to 1.1 $kg\,m^{-2}\,y^{-1}$ for soils with over 150 cm rooting depth (Table 8.1). Where erosion rates are naturally very high, however, higher threshold values may have to be adopted. In steep, particularly mountainous, terrain with high rainfall, a figure of 2.5 $kg\,m^{-2}\,y^{-1}$ represents a more realistic target.

Table 8.1. Recommended values for maximum permissible soil loss (kg m^{-2} y^{-1})

Meso-scale (e.g. field level)	
Deep fertile loamy soils; values used in the Mid-West of USA	0.6–1.1[a]
Thin, highly erodible soils	0.2–0.5[b, c]
Very deep loamy soils derived from volcanic deposits, e.g. in Kenya	1.3–1.5[b]
Soil depths: 0–25 cm	0.2[d]
25–50 cm	0.2–0.5[d]
50–100 cm	0.5–0.7[d]
100–150 cm	0.7–0.9[d]
over 150 cm	1.1
Probable realistic value for very erodible areas, e.g. mountains in the tropics	2.5
Macro-scale (e.g. drainage basins)	0.2
Micro-scale (e.g. construction sites)	2.5

After [a] Wischmeier and Smith (1965); [b] Hudson (1971); [c] Smith and Stamey (1965); [d] Arnoldus (1977).

8.2.2 Spatial level of resolution

Erosion problems occur at a range of scales from individual localities such as roadside embankments or cuttings and the outside of river bends to large drainage basins. The factors which control the rate of erosion vary with the size of the area being considered and it is necessary to take account of this scale-dependence when designing conservation measures. McGuinness, Harrold, and Edwards (1971) demonstrated the importance of scale in their study of erosion near Coshocton, Ohio. For areas less than 0.01 km^2 the most important factors influencing sediment yield were found to be rainfall energy and plant cover whereas for a basin of 15,540 km^2 runoff was the most important factor and sediment yield was unrelated to rainfall. Using evidence of studies of sediment concentrations in rivers (Langbein and Schumm, 1958; Fournier, 1960; Rougerie, 1967), measurements of soil loss on hillsides (Roose, 1971) and investi-

Table 8.2. Factors affecting soil loss at different scales

Scale of analysis			Evidence
Macro	Meso	Micro	
Climate	Lithology relief		Sediment yield of rivers (Rougerie, 1967)
Climate	Lithology relief	Micro-climate lithology (soil)	Drainage density (Gregory and Gardiner, 1973; Morgan, 1973)
Climate	Altitude relief		Erosion rates (Fournier, 1960; Langbein and Schumm, 1958; Diaconu, 1969)
Climate		Plant cover micro-climate	Erosion plots (Roose, 1971)

gations of the factors controlling drainage density (Morgan, 1973; Gregory and Gardiner, 1975), the effect of scale on the factors influencing erosion can be determined. The resulting pattern of scale relationships, shown in Table 8.2, has been confirmed by Thornes (1976) in his study of soil erosion in southern Spain.

Although some of the sediment eroded from hillsides, embankments and cuttings finds its way into rivers, much is deposited en route on the lower slopes and in flood plains where it remains in temporary storage. Because of this, erosion rates expressed per unit area are highest for small areas and decrease as the areal unit becomes larger. There is some support for a general relationship in which sediment yield increases with the 0.8 power of drainage area (Strand, 1975; Livesey, 1975) but the exact nature of the relationship is likely to alter locally depending on the proportion of the eroded sediment which is delivered to the streams. Studies of sediment delivery ratios show that this proportion varies from 3 to 90 per cent, decreasing with greater basin area and lower average slope (Maner, 1958; Roehl, 1962; Renfro, 1975).

It would seem logical to take account of these scale effects on erosion when selecting an appropriate value for maximum permissible soil loss but they appear to be ignored in current conservation practice. The widely accepted value of $1.1 \, \text{kg m}^{-2} \, \text{y}^{-1}$ is most appropriate for field size units. For areal units of different size, other values may be more realistic, say $2.5 \, \text{kg m}^{-2} \, \text{y}^{-1}$ for local vulnerable sites and $0.2 \, \text{kg m}^{-2} \, \text{y}^{-1}$ for areas above $10 \, \text{km}^2$ (Table 8.1).

8.3 STRATEGIES FOR SOIL CONSERVATION

The design of strategies for erosion control must be based on the studies of the mechanics of detachment and transport of soil particles by rainsplash, runoff, and wind, described in the earlier chapters. Conservation measures must therefore involve protection of the soil from raindrop impact, increasing the infiltration capacity of the soil to reduce the volume of runoff, improving the aggregate stability of the soil to increase its resistance to erosion, and increasing the roughness of the ground surface to reduce the velocity of runoff and wind (Morgan, 1979). The ways in which these changes in the erosion system are brought about are generally grouped under the headings of mechanical and agronomic techniques and soil management.

Agronomic measures and soil management influence both detachment and transport (Table 8.3) whilst mechanical measures influence the transport processes but have little effect on detachment. The type of measure required clearly depends on whether transport or detachment is the major problem. Agronomic measures are frequently successful on their own and even more effective when combined with good soil management. Mechanical measures, however, are rarely effective without agronomic measures to support them. Further advantages of agronomic over mechanical controls of erosion are

Table 8.3. Soil conservation practices

Practice	Control over					
	Rainsplash		Runoff		Wind	
	D	T	D	T	D	T
Agronomic measures						
Covering soil surface	*	*	*	*	*	*
Increasing surface roughness	−	−	*	*	*	*
Increasing surface depression storage	+	+	*	*	−	−
Increasing infiltration	−	−	+	*	−	−
Soil Management						
Fertilizers, manures	+	+	+	*	+	*
Subsoiling, drainage	−	−	+	*	−	−
Mechanical Measures						
Contouring, ridging	−	+	+	*	+	*
Terraces	−	+	+	*	−	−
Shelterbelts	−	−	−	−	*	*
Waterways	−	−	+	*	−	−

− no control; + moderate control; * strong control (adapted and enlarged from Voetberg, 1970)
D = Detachment, T = Transport.

that they are less costly to introduce, frequently require no special equipment or machinery, need less maintenance and, as shown by Temple (1972a) are more easily incorporated into existing farming systems. Where, because of steep slopes, heavy rain or exposure to strong winds, erosion is severe, no single type of measure is wholly effective and mechanical and agronomic controls must be combined with sound soil management in an integrated soil conservation system.

The most appropriate strategy for conservation depends upon selecting the required level of spatial resolution, identifying the most important factors influencing erosion at that scale, determining a value for maximum acceptable soil loss either on an annual basis or for a single event of a chosen return period, and determining the most suitable techniques for influencing the relevant controlling factors at the given scale. Further discussion is most conveniently presented at four scales: national, regional, field and site.

8.4 NATIONAL CONSERVATION PLANNING

At the national level the aim is to conduct surveys to identify the main regions of erosion risk as a base for formulating policy. The surveys make use of rainfall or erosivity data, as in the studies of Zimbabwe (Stocking and Elwell, 1973) and Malaysia (Morgan, 1974); sediment-discharge rating curves, exemplified by work in Yugoslavia (Jovanović and Vukčević, 1958) and Romania (Diaconu,

1969); or rely on applying empirical formulae to predict soil loss as, for example, in the work of Low (1967) on Peru.

Conservation strategy at this scale is rarely more than a statement of intent to minimize soil loss in the problem areas identified by the surveys and does not involve detailed design work. Instead, attention is focussed on the development of macro-economic models to assess the financial effects of erosion (Nicol, Heady, and Wade, 1976), the provision of the legislation necessary to implement conservation policy (Christy, 1971) and the setting up and funding of institutions to carry out research and provide advisory and extension service. These matters, all vital to the success of any soil conservation scheme, are beyond the scope of this book.

8.5   REGIONAL CONSERVATION PLANNING

The basic procedure of planning conservation work at the regional scale is to carry out some form of land evaluation and classification, select an appropriate landuse system and, where necessary, design conservation measures. The latter are aimed at reducing the energy available for erosion which, as indicated in Table 8.2, at this scale of perception, is largely a function of climate and relief. Agronomic measures are directed at selecting landuses which result in less runoff whilst mechanical measures are employed to break up the lengths of open run for runoff and wind, using terraces and shelterbelts respectively. Where a risk of excess runoff remains, an artifical drainage or waterway system is constructed to dispose of it non-erosively.

8.5.1   Land evaluation

The technique of land evaluation commonly used in soil conservation planning is the land capability classification. Generally, each country, planning unit or conservation service has its own classification scheme but most are adaptations of that developed by the Soil Conservation Service in the United States (Klinge-biel and Montgomery, 1966). The rationale of the classification system is that correct use of the land is the best means of erosion control. The system provides a method of assessing the extent and degree of limitations to the agricultural use of the land, emphasizing particularly erosion risk, soil depth, wetness, and climate. On the basis of the severity of these limiting factors, the land is divided into capability classes, each class representing a particular degree or severity of limitation. In many schemes a simple coding system and a flow chart have been devised to aid assigning land to its appropriate class (Hudson, 1971).

In the United States system, land is assigned to one of eight classes ranging from Class I, land with no or very slight limitations to use, to Class VIII, land with severe limitations which must be left in its natural condition. Only land in the first four classes is suitable for arable farming. Recommended conservation measures go along with each class. In addition to the modifications made to the United States system to take account of conditions in different countries, special-purpose schemes have been devised for rangelands (Renner and Allred, 1962), assessing the suitability of land for irrigation (Bureau of Reclamation, 1953), forestry, recreation and wildlife (McCormack, 1971) and making specific recommendations on the treatment of hilly lands in the tropics (Sheng, 1972a).

Many criticisms have been levelled at the land capability classification (Cruickshank and Armstrong, 1971; Leven *et al.*, 1974; Young, 1976) but, as indicated elsewhere, these generally arise from attempts to use the system for purposes for which it was not intended, usually within a wider context of resource appraisal and environmental management (Morgan, 1979). The main purpose of the classification is to furnish basic data on limitations to the use of the land along with general guidelines on landuse and conservation practice. Since the system does not, however, relate these guidelines to the processes of erosion occurring in an area or to any understanding of how conservation measures actually work, it seems desirable to presage this type of land evaluation with a survey of erosion risk.

Many surveys of erosion tend to be static in character, concerned with mapping features such as rills, gullies and blow-outs, but do not attempt to map the factors influencing soil loss. Static surveys have been used in Zimbabwe (Jones and Keech, 1966) but an effort to make them more dynamic was made by carrying out sequential surveys, using aerial photography taken at different dates, and analysing the trends or changes in erosion (Keech, 1969). Maps showing both erosion features and influencing factors are produced as morpho-conservation maps by the ITC in The Netherlands (Verstappen and van Zuidam, 1968; Verstappen, 1970). By developing further the techniques of special-purpose geomorphological mapping, an erosion survey system has been devised (Williams and Morgan, 1976; Morgan, 1979) aimed at mapping the areas and features of accelerated erosion such as sheetwash, rills, and gullies; the basic landform patterns of crests, streams, and relief; and important aspects of the soils, slopes, landuse, and man-made structures. Maps are produced at 1 : 25,000 scale following two interpretations of panchromatic stereo-photography, separated by a field survey. From the discussions provided by De Ploey and Gabriels (Chapter 3) and Thornes (Chapter 5), it seems logical to extend the scope of the field survey towards identifying the critical soil and slope conditions for accelerated erosion and the localized points of instability where, for example, gullying may be initiated. The aerial photographic interpretation relies on feature recognition by appearance in stereoscopic image and association with particular positions in the landscape. Tonal contrasts and textural patterns are used to a lesser extent although they are important for identifying areas of sheetwash and wind erosion (Bergsma, 1974).

The information on the maps can be assessed qualitatively by identifying the most important erosion processes at work, the areas of high erosion risk and erosion-deposition sequences in downslope, downwind, and downstream directions. Descriptive evaluations of this type have been made using the ITC system by van Genderen (1970) and Rao (1975). Simple numerical assessments of erosion risk can be made by covering the maps with a grid and assigning a score or rating value to the area falling within each grid square in respect of erosion intensity, slope, landuse, soils, and erosivity, and summing the scores (Stocking and Elwell, 1973; De Ploey and Gabriels, Chapter 3).

These techniques of assessing erosion risk are limiting in so far as they do not provide much of a base for understanding the erosion system nor do they indicate what conservation work is required. An alternative approach is to use the information portrayed on the maps to derive empirical equations similar to those discussed by Mitchell and Bubenzer (Chapter 2) for gully head advancement. Unfortunately, none of the equations discussed actually relates to maps of erosion risk and attempts to derive equations from such maps have not been entirely successful (Williams and Morgan, 1976).

The empirical approach, however, should be regarded as a short-term expedient to be replaced eventually by a refined mapping system closely linked to deterministic models of erosion and conservation systems. At present the best that can be achieved is to use the information contained on the maps as a data base for predicting soil loss using one or more of the empirical equations summarized in Table 2.11 (Mitchell and Bubenzer, Chapter 2) and compare the results with assessments of erosion risk using scoring or rating systems, a technique employed in the erosion survey of Moel Famau Country Park, County Clwyd (Baker *et al.*, 1979)

8.5.2 Landuse allocation

The important role of landuse in runoff and erosion control at the regional scale is stressed in all studies of watershed management (Vanoni, 1975). Particular emphasis has been placed in earlier chapters on its effects on soil water storage capacity and rates of transfer of sediment from hillslopes to rivers and through drainage basins in general (Thornes, Chapter 5) and on the extent of the area on the hillsides contributing runoff and sediment, depending on whether overland flow is of the Hortonian or saturation type (Kirkby, Chapter 1). Following the comments made in these chapters it is recommended that another development of erosion surveys should be the identification of erosion-response regions, areas which, by virtue of relatively uniform conditions of slope, soil and landuse, show a similar response to meteorological events. One procedure for identifying such regions in the field has been described by Morgan (1978a) using a principal components analysis of data describing the location, form and soils of slope sites and slope profiles. These units may also provide a suitable base for developing models to route runoff and sediment through drainage basins. Once the units have been recognized for a particular area, land capability analysis can be applied to them and the appropriate level of landuse determined.

None of the systems of classifying land permit the effects of the proposed landuse on soil loss to be assessed and, surprisingly, no conservation texts recognize the need for such assessments on the regional scale. Eventually, assessment may be possible by linking erosion mapping and deterministic modelling but, at present, recourse must be made to empirical models. One approach is to take the sediment yield prediction equation of Kirkby (1976) and, assuming runoff to be the most important factor affected by landuse or agronomic conservation measures at this scale, apply it to allocate landuse according to slope steepness. With the equation

$$Q_s = 170 Q_w^2 \tan S \qquad (8.1)$$

where Q_s is the sediment yield ($cm^3 cm^{-1} y^{-1}$), Q_w is the runoff ($m^3 m^{-1} y^{-1}$) and S is the slope angle (degrees), which is applicable to a standard slope length of 10 m, the following procedure has been developed for use in Peninsular Malaysia (Morgan, 1979).

A value of $0.2 kg m^{-2} y^{-1}$ is selected as an acceptable threshold level of erosion. This is close to the natural rate of erosion under tropical rain forest (Douglas, 1967). With a slope length of 10 m, a square metre discharges across a width of 0.1 m, and assuming a bulk density for the sediment of $1.0 g cm^{-3}$, the threshold value converts to a target figure for Q_s of $20 cm^3 cm^{-1} y^{-1}$. The annual runoff is estimated from the mean annual rainfall using rainfall–runoff ratios. With typical ratios of 0.25 for rain forest, 0.35 for rubber and 0.75 for urban areas (Low and Goh, 1972), the runoff values for Kuala Lumpur, with a rainfall of 2695 mm, are 673, 943, and 2021 mm or 0.67, 0.94, and 2.02 $m^3 m^{-1}$ respectively. Rewriting equation (8.1) to predict S, given Q_s and Q_w, results in the maximum permissible slopes of 14.7° for rain forest, 7.6° for rubber, and 1.7° for urban landuse if the acceptable erosion threshold is not to be exceeded. One implication of this is that on slopes greater than 14.7° the natural rate of erosion in the Kuala Lumpur area exceeds the threshold value and the forest cover must therefore be maintained as a conservation measure. In the case of rubber cultivation, the runoff, and slope values assume mature trees with a dense ground cover. If no ground cover is used the maximum acceptable slope will be much lower. It can be exceeded, however, if ground covers are combined with other conservation practices, for example, bench terracing.

Although this procedure is statistically unsound because the figure of 170 in Equation (8.1) is an empirically-derived constant with no firm physical base, it is no less acceptable than determining maximum permissible slopes using the Universal Soil Loss Equation and, under certain conditions, gives more realistic results (Baker *et al.*, 1979).

8.5.3 Critical slope length

Discussion in earlier chapters has established that rainsplash, overland flow, rill wash, and gullying, in that order, produce increasingly severe erosion. Since splash erosion alone rarely exceeds acceptable rates, conservation work is directed at erosion by overland flow. If this can be controlled, rill and gully erosion will also be prevented. On unvegetated hillsides and on land used for cropping, overland flow tends to be of the Hortonian type (Kirkby, Chapter 6) and it is implicit in Horton's work (1945) that a critical slope length (l_c) can be defined at which overland flow becomes erosive. Provided the effective slope length can be maintained below this critical value, serious soil erosion will not occur. The technique for achieving short slope lengths is to break up the hillside into segments using terraces. Deciding on a suitable spacing for the terraces therefore requires being able to estimate the critical slope length.

For steady state conditions of Hortonian overland flow, the discharge (Q_w) at slope length (l) may be expressed as

$$Q_w = \sigma l \cos S \tag{8.2}$$

where σ is the supply rate, taken as equal to the rainfall intensity minus the infiltration capacity of the soil. From the Maning equation of flow velocity

$$Q_w = \sigma l \cos S = \frac{r^{5/3} \sin^{1/2} S}{n} \tag{8.3}$$

where r is the depth of flow which, for shallow flows of infinite width equates with the hydraulic radius, and n is a roughness coefficient. Since, also from Manning's formula

$$r = \left(\frac{vn}{\sin^{1/2} S} \right)^{3/2} \tag{8.4}$$

then, for a given supply rate (σ) and a critical or maximum safe velocity (v_c) for water flow without erosion,

$$l_c = \frac{v_c^{5/2} n^{3/2}}{\sigma \sin^{3/4} S \cos S} \tag{8.5}$$

A similar equation may be developed using critical tractive force instead of critical velocity.

The main problem with this approach is that it is based on keeping the velocity of flow below that at which erosion is initiated rather than being related to acceptable erosion rates. It is likely to lead to terrace spacings closer than are really necessary and the division of hillsides into units too small to efficiently farm. An alternative approach with soil loss as the starting point can be developed from the model presented by Kirkby (Chapter 6).

From equation (6.9)

$$C = D_0 h \tag{6.9}$$

where C is the transporting capacity, D_0 is the maximum value of detachment capacity and h is the mean travel distance of detached particles, a critical value of h (h_c) can be determined for a given value of C, taken as the maximum acceptable rate of erosion (C_t), so that

$$h_c = \frac{D_0}{C_t} \tag{8.6}$$

Table 8.4. Formulae for terrace spacing

Ramser, 1917	$VI(m) = 0.76S + 0.608$ for erodible soils $VI(m) = 0.1013S + 0.608$ for less erodible soils
United States Soil Conservation Service	$VI(ft) = aS + b$ where a varies from 0.3 in the South to 0.6 in the North and b is 1 or 2 according to the soil
Zimbabwe	$VI(ft) = (S + f)/2$ where f varies from 3 to 6 according to the erodibility of the soil
South Africa	$VI(ft) = (S/a) + b$ where a varies from 1.5 for low rainfall areas to 4 for high rainfall areas and b varies from 1 to 3 according to the soil
Kenya	$VI(m) = 0.3(S + 2)/4$
Israel	$VI(m) = (S/10) + 2$
Algeria/Morocco	$VI(m) = (260\,S)^{-0.3}$ for bench terracing on slopes 10–25 per cent $VI(m) = (64S)^{-0.5}$ for bench terracing on slopes over 25 per cent
Sheng, 1972b	$VI(m) = (S \cdot Wb)/(100 - U)$ for bench terraces in humid tropics, where Wb is the width of the shelf (m) and U is the slope of the riser (expressed as a ratio of horizontal distance to vertical rise and usually taken as 1.0 or 0.75)
China	$VI(m) = Wb/(\cot S - \cot \beta)$ where β is the angle of slope of the riser (normally 70–75°)
New South Wales	$HI(m) = K \cdot S^{-0.5}$ where K varies from 1.0 to 1.4 according to the soil

VI = vertical interval between terraces
HI = horizontal interval between terraces
S = slope (per cent)
After Gichungwa (1970); Hudson (1971); Fournier (1972); Charman (1978).

For steady state overland flow without rills

$$h \propto \frac{(\tan S + \mu \sigma l \tan S/d)}{(\tan \phi + \lambda d_{84}/d)^2} \qquad (6.37)$$

where d is the diameter of the moving grain, d_{84} is the grain size at which 84 per cent of the soil particles is finer, ϕ is the angle of friction, λ is a slope shape factor and μ is an empirical constant. Substituting h_c for h, the critical slope length (l_c) can be determined for a given supply rate (σ), say that with a 10-year return period (σ_{10}) from

$$l_c \propto \frac{(dh_c - \tan S)(\tan \phi + \lambda d_{84}/d)^2}{\mu \sigma_{10} \tan S} \qquad (8.7)$$

In practice, terrace design is not approached in this theoretical way and spacing is based on purely empirical relationships such as those first developed by Ramser (1917) relating slope length between the terraces to slope steepness. A large number of equations have since been developed for different parts of the world, relating spacing to slope, soil erodibility and rainfall (Table 8.4; Hudson, 1971; Sheng, 1972b). In so far as none of the equations include either soil loss or runoff, the present basis for terrace design has somewhat tenuous connections with the erosion processes at work. By developing and testing Equation (8.7) not only are these factors accounted for but, because of the inclusion of a slope shape factor, there is scope for a more judicious spacing and placement of terraces in relation to landform. Greater emphasis may be given, for example, to

Table 8.5. Types of terraces

Diversion terraces	Used to intercept overland flow on a hillside and channel it across slope to a suitable outlet, e.g. grass waterway or soak away to tile drain; built at slight downslope grade from contour.
—Mangum type	Formed by taking soil from both sides of embankment.
—Nichols type	Formed by taking soil from upslope side of embankment only.
—Broad-based type	Bank and channel occupy width of 15 m.
—Narrow-based type	Bank and channel occupy width of 3–4 m.
Retention terraces	Level terraces; used where water must be conserved by storage on the hillside.
Bench terraces	Alternating series of shelves and risers used to cultivate steep slopes. Riser often faced with stones or concrete. Various modifications to permit inward sloping shelves for greater water storage or protection on very steep slopes (Fournier, 1972) or to allow cultivation of tree crops and market-garden crops.

After Hudson (1971); Morgan (1979).

positioning terraces at the point of inflexion on the convexity, frequently a
critical location for the onset of erosion by overland flow and rills (Evans,
Chapter 4).

Three main types of terraces are used for erosion control: diversion, retention
and bench (Table 8.5; Schwab *et al.*, 1966; Hudson, 1971; Fournier, 1972;
Morgan, 1979).

8.5.4   Runoff disposal

Where landuse management alone is insufficient to prevent runoff on hillsides,
the excess water has to be removed without causing erosion. To effect removal
a network of waterways is constructed comprising diversion ditches, grass
waterways, and, where terraces are integrated in the system, terrace channels
(Table 8.6; Figure 8.1: Hudson, 1971).

Table 8.6. Types of waterways

Diversion ditches	Placed upslope of areas where protection is required to intercept water from top of hillside; built across slope at slight grade so as to convey the intercepted runoff to a suitable outlet.
Terrace channels	Placed upslope of terrace bank to collect runoff from inter-terraced area; built across slope at slight grade so as to convey the runoff to a suitable outlet.
Grass waterways	Used as the outlet for diversions and terrace channels; run downslope, at grade of the sloping surface; empty into river system or other outlet; located in natural depressions on hillside.

After Hudson (1971); Morgan (1979).

Most research on runoff disposal techniques has concentrated on the design
of the channels and, except in the case of the advantages of parallel over other
terrace layouts for ease of cultivation (Smith, 1956), few studies have been made
of the overall waterway system. Given that the aim is to avoid too rapid a
delivery of runoff and to reduce flood peaks, it would seem logical to design the
system with regard to the effects of catchment shape and channel morphometry
on its hydrology. To achieve its objectives standard hydrological analysis
(Strahler, 1964) indicates that the network should have the bulk of its area in
the upper catchment, form an elongated drainage basin and have a high bifurcation ratio. Typical layouts with a single grass waterway and diversion ditches
and terrace channels draining laterally into it conform to these requirements.
The subcatchments, comprising the inter-terrace areas, also usually conform
because the recommendations on terrace design (Table 8.7; Hudson, 1971) set

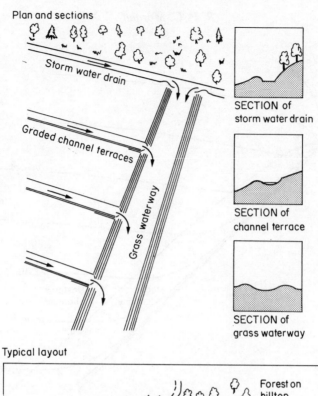

Plan and sections

Storm water drain

Graded channel terraces

Grass waterway

SECTION of
storm water drain

SECTION of
channel terrace

SECTION of
grass waterway

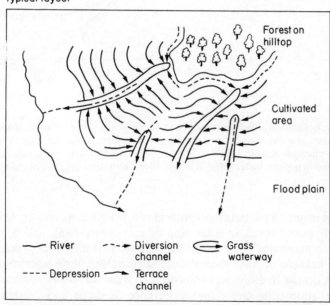

Typical layout

Forest on
hilltop

Cultivated
area

Flood plain

— River   - - - ➤ Diversion
channel   ⊏⊐ Grass
waterway

- - - - Depression   ➤ Terrace
channel

Figure 8.1. The basic components of a waterway system as used
for soil conservation: (a) stormwater drain or diversion channel;
(b) graded terrace channels; (c) grass waterways (after Hudson,
1971; Morgan, 1979); (reproduced by permission of Batsford Ltd,
London)

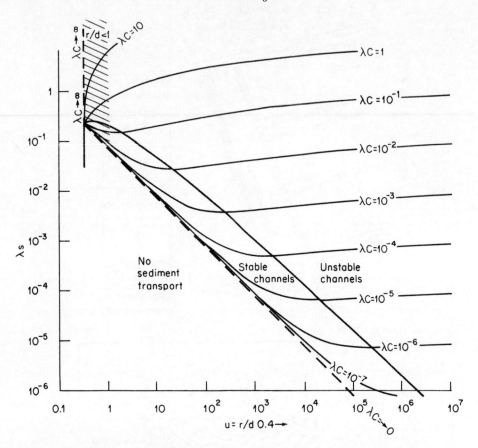

Figure 8.2. Relationship between sediment concentration and channel slope, depth and grain size showing the curve defining the maximum sediment concentration at which the channel remains stable. The shaded area is where roughness elements project from the water (after Kirkby, 1978). Notation is given in the text

a maximum length to the terrace to prevent excess ponding of water in the catchment which would result in breaching of the terrace bank and a minimum length which maintains a catchment of rectangular rather than square shape.

Perhaps because of their theoretically ideal nature, little research has been carried out on the hydrological behaviour of these waterway systems and few soil conservation texts stress the importance of shape and pattern in their design. Research has been limited to demonstrating the general effectiveness of terrace and waterway systems in reducing mean annual runoff and soil loss by comparing catchments where these measures have been installed with those where no conservation is practised. As in the case of the studies of Spomer,

Table 8.7. Design criteria for terraces

Maximum length:	normal	250 m (sandy soils) to 400 m (clay soils)
	absolute	400 m (sandy soils) to 450 m (clay soils)
Maximum grade:	first 100 m	1:1000
	second 100 m	1:500
	third 100 m	1:330
	fourth 100 m	1:250
	where a constant grade is used, 1:250 is recommended	
Ground slopes:	diversion terraces	usable on slopes up to 7°; on steeper slopes the cost of construction is too great and the spacing too close to allow mechanized farming
	retention terraces	recommended only on slopes up to 4.5°
	bench terraces	recommended on slopes of 7 to 30°

After Hudson (1971); Morgan (1979).

Saxton, and Heinemann (1973) in five catchments in southwest Iowa, ranging from 30 to 156 ha in size, the results have limited value because the data from these small catchments cannot be extrapolated to larger drainage basins where, as noted in Section 8.2.2, different controlling factors over runoff and sediment yield may prevail. By paying more attention to the hydrological characteristics of a terrace and waterway system it should be possible not only to improve on present design procedures but also, by applying simulation models to route the peak flow through the system and estimate its attenuation and time lag downstream, to test its efficiency. Simulation models are also required to route sediment through the system and no attempts have yet been made to develop and test these.

The problem of individual channel design within the system focuses on the need for a stable waterway with sufficient capacity to transmit the runoff supplied to it without scour or fill. The form of the channel required can be defined in terms of three equations: a flow resistance equation which expresses the relationship between slope, flow depth, and flow velocity; a sediment discharge equation which describes the hydraulics of sediment transport and usually defines transport capacity; and an equation describing the cross-sectional shape of the channel, usually in terms of a width-depth ratio. Adjustments to the form of the channel to attain stability can be made to its width, depth, and slope. In addition to these three degrees of freedom, natural channels have a fourth which is the ability to develop meanders. However, since straight channels are used in soil conservation work, the latter can be ignored.

The most satisfying approach theoretically to combining and solving the equations governing channel design would appear to be the application of maximum efficiency or minimum work criteria, as developed by Kirkby (1977) in a study of alluvial channels, relating sediment transport to variations in hydraulic geometry. In this study, the maximum efficiency criterion is applied to a single dominant flow rather than to the frequency distribution of all flows which is the normal practice with this approach. This raises the possibility of selecting the peak flow with a ten-year return period, the event customarily adopted in soil conservation design, as the single dominant flow. The following equations are used for flow resistance and sediment transport respectively:

$$Q_w = \left(\frac{2g}{f}\right)^{1/2} wR^{3/2} \tan^{1/2}S \tag{8.8}$$

where f is the Darcy–Weisbach roughness factor, taken as

$$\frac{1}{\sqrt{f}} = 1.77 \ln(R/d_{84}) + 2.0 \tag{8.9}$$

w is the channel width and R is the hydraulic radius; and

$$C = 8(g\Delta d^3)^{1/2} w\left(\frac{R \tan S}{\Delta d} - 0.047\right)^{3/2} \tag{8.10}$$

which is the Meyer–Peter and Müller equation for bedload transport where Δ is the ratio of submerged sediment density to water density, taken as equal to 1.65. These equations are combined to yield the expression:

$$k \cdot \frac{C}{Q_w} = 3.43 \frac{(yk \tan S - 0.077)^{3/2}}{y^{3/2}(k \tan S)^{1/2}(1.77 \ln y + 2.0)} \tag{8.11}$$

where $y = R/d_{84}$ and $k = d_{84}/d$. By plotting $k \tan S$ against y for a family of curves of $k \cdot C/Q_w$ (Figure 8.2), it is found that, except for very high sediment concentrations, there is a minimum slope for each value of $k \cdot C/Q_w$ at which sediment may be carried and that, over the range of values of $k \cdot C/Q_w$, the respective minimum slope values lie on a maximum efficiency or minimum work curve:

$$k \cdot \tan S = \frac{0.0115 \ln y + 0.208}{y} \tag{8.12}$$

Applying this approach to channel design the procedure would be to predict the peak runoff with a ten-year return period, select an acceptable level of sediment discharge, determine the value of $\ln y$ from Equation (8.11), select a design slope (S), determine the appropriate value for y from Equation (8.12) and, for a given value of d_{84}, calculate the design value of R. Using either the Manning equation:

$$v = \frac{R^{2/3} \sin^{1/2}S}{n} \tag{8.13}$$

or the Darcy–Weisbach equation:

$$v^2 = \frac{2g}{f} R \tan S \tag{8.14}$$

the flow velocity (v) is estimated for the design values of S and R. For a channel of rectangular cross-section, the design value of R is equal to the depth of the channel (D) and, from continuity, channel width (w) is calculated from $w = Dv/Q_w$. For a channel of parabolic cross-section, the approximation $D = 1.5 R$ is used to determine D. Again from continuity, $Q_w = v/A$, where A is cross-sectional area of the channel, and $w = A/0.67D$.

In practice, combining the relevant equations in this way is a complex process and obtaining realistic solutions for design purposes is only possible for very limiting conditions, as with the case of designing irrigation canals based on regime theory (Withers and Vipond, 1974). In the above procedure, Equation (8.12) is extremely critical as this provides the solutions based on the minimum work criterion.

A further problem in developing procedures for channel design exists in the selection of suitable equations. Whilst there is considerable agreement that the Manning or the Darcy–Weisbach equations are suitable for flow velocity, a large number of sediment discharge equations exist from which to choose. Vanoni (1975) discusses twelve. In a comparative analysis of five, Cole *et al.* (1973) show that the Engelund–Hansen formula gives the best predictions although none is entirely satisfactory. The Engelund equation is used by Maza Alvarez and Cruickshank Villanueva (1973) for the design of stable channels in alluvium. These authors, however, use their own flow resistance equation in which velocity is related exponentially to flow depth, slope and bed material. The equation is applicable to flow over a sand bed in the lower regime with ripples and dunes on the bed surface and was tested on three rivers in Mexico and found to be satisfactory.

Because of the difficulties outlined above, a simpler but more empirical procedure is adopted in practice for channel design. The peak runoff is estimated for the selected return period and decisions made on the slope of the channel and the waterway lining, normally grass. A value of the Manning coefficient of roughness is determined for the grass cover, usually from tables based mainly on data from experiments in flumes. Taking account of the slope, the proposed grass cover and the erodibility of the soil, and acceptable or maximum permissible value for flow velocity is chosen, again from tables, based on many years of engineering experience (Schwab *et al.*, 1966; Hudson, 1971). The design value for the hydraulic radius (R) is calculated from the Manning Equation (8.13).

An alternative approach, taking account of the fact that whether a particular cross-sectional form of channel is stable or not depends less on the hydraulics of sediment transport and more on the frequency distribution of flows and the stability of the bank material (Kirkby, 1977), is to adopt the concept of critical tractive force as the basis for channel design. The relevant equations, expressing respectively the tractive force on the bed, the tractive force on the bank and the tractive force ratio, are:

$$\tau_0 = \gamma R \sin S \qquad (8.15)$$

$$\tau_0 = 0.75\gamma R \sin S \qquad (8.16)$$

$$\frac{\tau_s}{\tau_c} = \sqrt{1 - \frac{\sin^2 \theta}{\sin^2 \phi}} \qquad (8.17)$$

where τ_0 is the unit tractive force, γ is the unit weight of water, τ_s is the critical tractive force on the bank, τ_c is the critical tractive force on the bed, θ is the angle of the channel side to the horizontal and ϕ is the angle of repose of the bank material.

The design procedure is to determine the angle of repose appropriate to the bed and bank material, select a suitable side slope which is gentler than the angle of repose and, from Equation (8.17), determine the tractive force ratio. The maximum permissible tractive force for the bed and bank material is obtained from tables or graphs, based on laboratory experiments (Withers and Vipond, 1974), and this value is divided by the tractive force ratio to obtain τ_s. Substituting the value of τ_s for τ_0 in Equation (8.16), the value of R is obtained. The value of the maximum permissible tractive force is substituted for τ_0 in equation (8.15) and the value of R again determined. Whichever is the lower of the two determinations of R is used as the design value. The tractive force ratio is only applicable to coarse non-cohesive soils (Withers and Vipond, 1974). For cohesive soils, the cohesive forces are so large in proportion to the gravity forces which cause soil particles to roll down the slope that the gravity component can be neglected. In this case, the tractive force ratio is taken as equal to 1.0 (Schwab et al., 1966).

Using either critical velocity or critical tractive force as the criterion for channel design removes the need for decisions on an acceptable rate of sediment discharge and a suitable sediment transport equation. Although this simplifies the design procedure, it means that current channel design practice is one step divorced from what really matters. Two problems arise from this which frequently have to be resolved. Firstly, the channel dimensions arrived at from applying the tractive force approach differ from those obtained by using critical velocity. Generally, the use of velocity yields a deeper and narrower channel. Secondly, adjustments may be required to the design width and depth to allow for easier construction of the channel (Hudson, 1971). By carrying out further research aimed at relating channel design more closely to the runoff and sediment being carried in the channel, a sounder theoretical base will be provided for making judgements on what constitutes a stable channel and what adjustments to width and depth are permissible.

Grass waterways, as terrace outlets, have become less popular in the United States in recent years. They take up land which could otherwise be used for crops. The farmer has to raise planting and cultivating equipment when crossing them. Maintenance is becoming increasingly costly and difficult as the use of grass-killing herbicides on the adjacent fields increases (Moldenhauer and Onstad, 1977). The junction between the terrace and the grass waterway is the most vulnerable point for erosion in the runoff disposal system and requires regular attention. The method adopted to overcome these problems is to eliminate the waterway, replace it with an underground outlet in the form of a tile drain, and grade the terrace channel to a soak-away depression or storage area which serves as the inlet to the drain. The design procedures for tile drains are beyond the scope of this book but may be found in Schwab *et al.* (1966) and Voetberg (1970). The main requirement is that the runoff stored on the ground surface must be removed within 48 h otherwise damage may occur to the crops. Soil losses from tile-outlet terraces are much reduced compared with those from conventional terrace and waterway systems. Laflen, Johnson, and Reeve (1972) show that less than 5 per cent of the soil delivered to the soak-aways is transported into the drainage system.

8.5.5 Shelterbelts

Shelterbelts serve to break up the length of open wind blow and therefore perform a similar function in wind erosion control to that of terraces in controlling water erosion on hillsides. The theoretical basis for shelterbelt design is, however, weaker than that for terrace design. None of the equations described by Wilson and Cooke (Chapter 7) can be readily adapted to determine a critical length of blow at which the wind becomes erosive and, therefore, to indicate suitable belt spacing.

A reasonable working approach would appear to be to take the following equation for sediment discharge per unit width (Bagnold, 1941):

$$Q_s = c \sqrt{\frac{d}{D}} \cdot \frac{\rho}{g} V_*^3 \tag{7.15}$$

where D is a standard grain diameter of 0.25 mm, d is the mean grain diameter of the soil, ρ is the density of the air, V_* is the drag velocity and c is an empirical constant. The first two terms on the right hand side of the equation may be replaced by a constant B,

$$Q_s = B \frac{\rho}{g} V_*^3 \tag{8.18}$$

and since, from Bagnold (1937),

$$BV_* U = hg \tag{8.19}$$

where U is the horizontal velocity of the grain and h is the mean travel distance of the soil particles, then

$$Q_s = \frac{hg}{V_* U} \cdot \frac{\rho}{g} \cdot V_*^3 \tag{8.20}$$

which, for a critical or maximum acceptable value of sediment discharge

(Q_{s_c}), yields a critical travel distance (h_c):

$$h_c = \frac{Q_{s_c} U}{\rho V_*^2} \qquad (8.21)$$

Because of difficulties in determining U in the above equation and in relating the travel distance of individual particles to shelterbelt spacing, empirical relationships, based largely on data obtained from wind tunnel experiments, are used instead as the basis for shelterbelt design. The data show that, under ideal conditions, the distance protected from erosion in the lee of a shelterbelt is approximately equal to seventeen times the height of the belt, assuming the belt to be perpendicular to the wind direction. Allowing for variations in the direction of the wind, Woodruff and Zingg (1952) have developed the following equation for determining shelterbelt spacing:

$$L = 17H(Vt/V)\cos\alpha \qquad (8.22)$$

where L is the distance apart of the belts, H is the height of the belt, V is actual wind velocity measured at 15 m bove the ground surface, Vt is the threshold wind velocity taken to be 34 km h^{-1} and α is the angle of deviation of the prevailing wind from a line perpendicular to the belt.

Field studies in the Fens (Sneesby, 1953; pers. comm.) indicate that this theoretical relationship between the height of the belt and the distance protected is rarely achieved because of the variable growth and poor maintenance of the trees comprising the belt. A more realistic value for the distance of protection is twelve times the height of the belt. The relationship also varies with the belt height so that, for hedges, the distance protected rises to thirty times the height of the belt. The absolute distance protected from erosion is, of course, greater for trees than for hedges because of their greater height.

The critical factor determining the distance protected is the porosity of the belt which influences the drag exerted by the belt on the wind flow. The effect of the belt is best appreciated in terms of a transfer of momentum from the air flow to the belt and attempts have been made to determine drag coefficients for various types of belt using the momentum balance equation (Hagen and Skidmore, 1971; Skidmore and Hagen, 1977). These show that with slat fence barriers, for example, drag decreases linearly with increasing porosity until the barrier is 40 per cent open and then very sharply as the porosity increases from 40 to 60 per cent. In spite of this decrease in drag with increasing porosity barriers exerting the greatest drag are not the most effective because, after reaching a minimum immediately in the lee of the barrier, the wind speed increases more rapidly leeward of low porosity barriers than is the case with more porous barriers. Thus, behind barriers of low porosity, the area of sheltered ground is much reduced. Since with barriers of high porosity little drag is exerted on the wind flow and little shelter is also provided, there must be an optimum porosity for wind erosion control where the drag reduces the wind speed below the threshold level for erosion for the greatest distance.

Because of difficulties in evaluating drag forces, measurements of wind velocity are used as a surrogate in determining the effectiveness of shelterbelts. Various investigations, summarized in Skidmore and Hagen (1977), show that the greatest overall reductions in wind speed over a distance thirty times the barrier height are achieved with a porosity of 40 per cent. Expressed in simple terms, this level of porosity describes a barrier which is translucent but not transparent. It should be possible to see objects on the other side of the barrier without being able to identify them.

The reductions in wind speed associated with a barrier of 40 per cent porosity may be expressed by the following empirical relationships (Skidmore and Hagen, 1977):

$$V_x/V_0 = 0.85 - 4e^{-0.2H'} + e^{-0.3H'} + 0.0002H'^2 \tag{8.23}$$

leeward of the belt and

$$V_x/V_0 = 0.502 + 0.197x - 0.019x^2 \tag{8.24}$$

windward of the belt, where x represents the distance from the belt measured in units of barrier height, V_0 is the wind velocity in the open field, and $H' = x/\sin \beta$, where β is the acute angle of incident wind. These equations may be used to test the effectiveness of shelterbelts in relation to given threshold wind velocities (Wilson and Cooke, Table 7.4). Taking an open field wind velocity of 44 km h^{-1}, a barrier of 40 per cent porosity will provide protection for a distance of fourteen times the barrier height for soils where the critical velocity for erosion is 30 km h^{-1} but for a distance of only eight times the barrier height for more erodible soils where the critical velocity is 18 km h^{-1}. These distances assume that the wind is perpendicular to the barrier and will be reduced where the wind direction deviates from this.

The effectiveness of a shelterbelt is also dependent upon its length. With short belts the protected area in the lee is triangular in shape but with longer belts it becomes rectangular. Wind velocities are frequently increased at the ends of short barriers and thus longer ones are preferred. The recommended length is about twelve times the height of the belt for winds at right angles to the barrier but, to allow for deviations of up to 45° in wind direction, this figure is increased to twenty-four times the height.

Since between 50 and 72 per cent of the soil eroded by the wind is moved in saltation (Chepil, 1945; Wilson and Cooke, Chapter 7), much of the material is being transported within a metre of the ground surface. To provide a barrier and a filter to this movement, tree belts have to be supplemented by a lower shrub layer. A complete belt may comprise up to two tree and three shrub rows with an overall width of about 9 m. In addition to the tree and shrub belts, other belts are now widely used in wind erosion control including maize, sorghum, tall wheatgrass, sugar cane, and rye.

Shelterbelts may be laid out in parallel, grid, and herringbone patterns. The most effective layout is that which gives maximum protection averaged over all winds with velocities above the threshold for erosion. Where there is a strongly dominant direction of erosive winds, it is standard practice to align the shelterbelts in parallel fashion at right angles to it. Where erosive winds come from several directions the layout which provides maximum protection overall may not include barriers at right angles to any of these directions. Thus maximum protection may be afforded by a scheme in which complete protection against erosion is not obtained against any single wind direction.

Although, in contrast to the situation with waterway and terrace design, no recognized method exists for determining the optimum layout of shelterbelts, a system for evaluating the effectiveness of different layouts is being developed and tested (Skidmore and Hagen, 1977) using Equations (8.23) and (8.24) in conjunction with an expanded version of Equation (7.13):

$$F_x = \sum_{j=1}^{16} \sum_{i=1}^{n} \overline{V}_{xij}^3 f_{ij} \qquad (8.25)$$

where F is the total wind velocity at distance x from the barrier, $\overline{V}$ is the mean wind velocity at distance x in the ith speed group and the jth compass direction, and f is the duration of the wind in the ith speed group and the jth direction. By substituting in turn the open wind velocity and the reduced wind velocity for $\overline{V}_x$, two values for F_x are obtained and the difference between them is a measure of the efficiency of the shelterbelt.

8.6 FIELD CONSERVATION PLANNING

Studies of soil loss from small plots in the field (FAO, 1965; Kellman, 1969; Roose, 1971; Temple, 1972b; Morgan, 1977) show that, in identical conditions of rainfall, slope and soil, different plant covers result in different rates of erosion. Generally, row crops produce more erosion than either grass or tree crops (Table 8.8). Where erosion takes place at accelerated rates it brings about a deterioration in land quality which, at a field level, is expressed in terms of decreased yield, poorer and weaker crops, increased risk of pest infestation and disease and, therefore, in greater costs and reduced income to the farmer. Erosion control at the field scale depends largely on maintaining the protection afforded to the soil by the plant cover and is therefore mainly concerned with agronomic techniques.

The properties of the plant cover which determine its protection-effectiveness are height, ground cover, and density. The importance of cover was demonstrated in the mosquito gauze experiments of Hudson and Jackson (1959). Two identical plots were kept free of weeds and over one plot was suspended a fine wire gauze to break the force of the raindrops, absorb their impact and allow the water to fall to the ground surface from a low height as a fine spray.

Table 8.8. Soil loss under different vegetal covers

Cover	Slope (per cent)	Soil loss $(kg\,m^{-2}\,y^{-1})$	
Primary tropical rain forest	—	0.009	(1)
Secondary tropical rain forest	—	0.013	(1)
	7	0.003	(2)
Temperate mixed woodland	19	0.005	(3)
Dense savanna grass	4	0.005–0.02	(2)
Lucerne	9	0.025	(4)
Blue grass	9	0.007	(4)
Maize (shifting cultivation plot)	—	0.03	(1)
Maize (1st year in maize/oats/ clover rotation)	9	4.4	(4)
Maize (with grass bunds)	—	0.7	(5)
Maize (no conservation)	—	1.2	(5)
Maize (rows up and down slope)	9	9.4	(4)
Maize (on subsoil)	9	12.8	(4)
Coffee (clean weeded)	—	2.2	(5)
	—	1.8–5.5	(5)
Banana	7	1.5	(2)
Banana (trash mulch)	—	0.05	(5)
Manioc	7	9.0	(2)
Sorghum	2	0.3–1.2	(2)
	—	2.6–5.2	(5)
Cotton	2	0.05–1.9	(2)
Ground nuts	2	0.3–1.2	(2)
Crotalaria	7	4.0–5.0	(2)
Upland rice (1st year)	—	0.017	(1)
Upland rice (12th year)	—	0.289	(1)
Bare soil in humid tropics	7	10.0–17.0	(2)
Bare soil in savanna	4	1.8–3.0	(2)
Bare soil in temperate climate	19	1.0	(3)

After (1) Kellman, 1969; (2) Roose, 1971; (3) Morgan, 1977; (4) FAO, 1965; (5) Temple, 1972b.

The mean annual soil loss over six years was 141.3 m³ ha⁻¹ for the bare plot and 1.2 m³ ha⁻¹ for the plot protected by gauze. With taller vegetation the effect of raindrop energy absorption is much reduced. With woodland and tree crops raindrops collect on the leaves of the canopy and coalesce into larger diameter drops. It can be seen from studies of the relationship between terminal velocity and height of fall (Gunn and Kinzer, 1949), that, on falling from the canopy, these drops attain 80 to 90 per cent of their terminal velocity before striking the ground surface.

It is generally accepted that vegetation should cover at least 70 per cent of the ground surface to give adequate protection against erosion (Fournier, 1972; Elwell and Stocking, 1976; Evans, Chapter 4). With decreasing plant cover erosion begins to increase rapidly when this threshold percentage is reached and even more rapidly when the cover becomes less than 35 per cent. As pointed out by Thornes (Chapter 5) and indicated by the laboratory studies of De Ploey, Savat, and Moeyersons (1976), a plant cover may not always bring about the expected reductions in erosion, however, particularly on steep slopes or where the vegetation is tussocky. Although the mean velocity of overland flow is reduced on vegetated surfaces, many eddies develop on steep slopes downstream of grass blades and vortex erosion may occur upslope of tufts of vegetation. The combined effect may be sufficient to match the erosive power of overland flow on an unvegetated surface of the same slope.

Increased erosion is likely to be most marked where the density of the plant cover is insufficient. The importance of density is emphasized by Hudson (1971) who cites a reduction in soil loss in Zimbabwe from 1.23 to 0.07 kg m^{-2} y^{-1} as a result of increasing the planting density of maize from 2.5 to 3.7 plants m^{-2} and using a trash mulch.

8.6.1 Plant selection

The selection of strategies for erosion control based on agronomic techniques is reliant on data on soil loss from erosion plots (De Ploey and Gabriels, Chapter 3) which indicate the reductions in erosion that may be achieved with different plant covers and different treatments. These data form the basis for determining values for the C-factor in the Universal Soil Loss Equation (Mitchell and Bubenzer, Chapter 2). Although these plot experiments indicate which are the most effective treatments, they do not give more than a crude understanding of why particular treatments work. It is therefore difficult to extrapolate the results to new conditions. Thus, in spite of the vast amount of data on soil loss under a wide range of plants available from experimental stations in the United States, similar plot studies are necessarily being carried out elsewhere in the world (Roose, 1971; Lal, 1976) to furnish data for different crops and different conditions. Rather than continuing to repeat this type of work indefinitely, it would seem to be more valuable to use the information already available to build up models of the effects of plant covers on erosion.

The following effects need to be considered:

(a) reduction in the impact of falling raindrops as a result of rainfall interception;
(b) increased resistance to air and water flow through greater roughness;
(c) deposition of sediment brought about by a reduction in flow velocity;
(d) increased infiltration capacity resulting from root growth opening up the soil;
(e) additional structural stability following an increase in the organic content of the soil; and
(f) lower runoff on vegetated as against bare soil resulting from greater interception, infiltration, and evapotranspiration.

In terms of the modelling equations presented by Kirkby (Chapter 6) and Wilson and Cooke (Chapter 7), (a), (d), and (f) affect the supply rate of runoff (σ) and (a) also influences the effective rainfall intensity (i) at the ground surface; (b) and (c) affect the constants μ in Equation (6.37), n in Equation (6.49) and z' in Equation (7.1) all of which incorporate roughness parameters; and (e) influences the resistance of the soil to erosion, a factor inadequately accounted for by the grain size term (d). In the context of water erosion (a), (d), and (f) affect the hydrological controls over soil loss and (b) and (c) affect the hydraulic controls.

The research base for modelling the effects of rainfall interception is very weak. Most studies of interception have been carried out under forest and these show considerable variations in the proportion of rain that is intercepted depending on the rainfall intensity, the morphology of the tree canopy, whether the trees are broad- or needle-leaved and, in the case of deciduous trees, the time of year. As examples (Barry and Chorley, 1976; Jackson, 1977), German pine forests intercept as much as 94 per cent of rainfall at low intensities but only 15 per cent at high intensities. The equivalent figures for tropical rain forests are 85 per cent and 12 per cent. In German beech forests, 43 per cent of the summer rainfall is intercepted but only 23 per cent of the winter rainfall. Average figures suggest that about 20 to 30 per cent of the mean annual rainfall is intercepted by a forest. When considering the difficulties of measuring interception (Jackson, 1975), it is not surprising that so few studies have been made. For information on interception by agricultural crops it is still necessary to refer to the pioneer work of Wollny (1890). Since almost no work has been done on the changes in the drop-size distribution and energy of that part of the intercepted rainfall which reaches the ground surface by drip from the vegetation, it is possible at present to make no more than inspired guesses as to the effects of different plant covers on the amount and intensity of the rain.

A much sounder basis exists for modelling the combined effects of infiltration and evapotranspiration on runoff within an overall water balance model. Two approaches, infiltration models, and soil moisture storage models, are used and sometimes successfully combined. These models are outlined by Kirkby (Chapter 6) following a theoretical discussion by Thornes (Chapter 5) and developed more fully in Kirkby (1978). The influence of a plant cover is incorporated in the models through its control over the rate at which evapotranspiration takes place and the soil moisture store becomes depleted between storms. This rate is normally expressed by the potential evapotranspiration (E_0) but, with a plant cover, the E_0 value is modified by an E_t/E_0 ratio, where E_t represents the actual evapotranspiration of the plant. E_t/E_0 ratios vary with the type of plant, the stage of plant growth, the development of the leaves and the density of the plant cover. Values are available for a wide range of plants including agricultural crops (Jackson, 1977; Nieuwolt, 1977).

The simplest approach to modelling the hydraulic effects of a plant cover is

to concentrate on the roughness imparted to the flow of water and air by vegetation. This approach allows the possibility of determining the level of roughness
required to maintain soil loss below a selected maximum permissible value for
a given runoff. The selection of an appropriate plant cover is then a matter of
matching various types of plant with roughness values.

Applying this approach to erosion by overland flow involves determining the
required value of h for a selected transport capacity (C) from Equation (6.9)
and, substituting this value in Equation (6.37) to obtain the value for $\mu \cdot Q_w \cdot S/d$.
From Equations (6.35), (6.36) and (6.37)

$$\mu \cdot Q_w \cdot S/d \simeq \frac{(f/2g)^{1/3} \cdot (Q_w \cdot S)^{2/3}}{\Delta \cdot d} = \frac{r \cdot S}{\Delta \cdot d} \tag{8.26}$$

By substituting the second of these expressions for the first in Equation (6.37),
the appropriate value of the Darcy-Weisbach coefficient f can be predicted
for the required value of h. Unfortunately this result is of little practical value
at the present time because there is no body of data available on typical f values
for plant covers. Engineers have preferred to express roughness effects in terms
of Manning's n, in spite of the disadvantage that this is an empirical rather than
a dimensionless coefficient and varies with the depth of flow. Nevertheless for
immediate practical use, it may be more convenient to replace $\mu \cdot Q_w \cdot S/d$
by the right hand term in Equation (8.26), rewriting the latter by expressing
r in terms of the Manning equation for flow velocity. From this modified equation, the appropriate value of n for the required value of h may be predicted.

Although this approach is potentially more useful, it is not without problems.
Most of the theory on the hydraulics of overland flow is based on adaptations
of theories developed for channel flow, supported by laboratory experiments,
and, as pointed out by Thornes (Chapter 5), the properties of overland flow
as observed in the field may not conform to those expected from theory. The
field studies of Emmett (1970; 1978), Pearce (1976), and Morgan (in press)
indicate that overland flow has very low Reynolds and Froude numbers but,
rather than being considered as laminar flow, is best described as disturbed
flow because of the effect of raindrop impact. These studies suggest that the flow
should be described by the modified version of the Manning equation (Savat,
1977):

$$Q_w = \frac{\sin^{0.95} S r^{2.7}}{n} \tag{8.27}$$

thereby allowing for the greater depths of flow experienced in the field.

A further problem concerns the validity of the available data on n values for
different plant covers (Ree, 1958; Petryk and Bosmajian, 1975; Ree, Wimberley,
and Crow, 1977; Table 8.9) for overland flow. Most of these values have been
determined for channel flow and there is some evidence to suggest that they are
an order of magnitude too low for overland flow. For example, the value of

Table 8.9 Values for Manning's *n*

Short grass (5–15 cm)	0.030–0.060	(1)
Medium grass (15–25 cm)	0.030–0.085	(1)
Long grass (25–60 cm)	0.040–0.150	(1)
Ploughed, harrowed field	0.049	(2)
Cultivated field	0.049	(2)
Cereals	0.075	(2)
Mowed meadow	0.075	(2)
Meadow with high grass	0.149	(2)
Wheat	0.100–0.125	(3)
Sorghum	0.040–0.110	(3)
Dense grass cover	0.21–0.62	(4)
Fair grass cover	0.31–0.51	(4)
Poor grass cover	0.25–0.28	(4)
Bare soil to poor grass cover	0.20–1.0	(5)
	0.35	(5)
	0.20–1.7	(7)

After (1) Hudson, 1971; (2) Voetberg, 1970; (3) Petryk and Bosmajian, 1973; (4) Ree, Wimberley, and Crow, 1977; (5) Emmett, 1970; (6) Pearce, 1976; (7) Morgan, in press.

0.02 is commonly used for bare soil (Hudson, 1971) but field studies of overland flow over bare or partially vegetated soil surfaces yield values of 0.2 to 1.0 (Emmett, 1970), 0.35 (Pearce, 1976) and 0.2 to 1.7 (Morgan, in press).

The effects of a plant cover on rill erosion may be modelled using Equation (6.49). This expresses the detachment capacity of rill flow (D_R) as

$$D_R \propto (Q_w \cdot S)^m \qquad (8.28)$$

where the value of exponent *m* is determined empirically. This equation is basically a restatement of the relationships between soil loss and slope length, a surrogate for runoff, and slope steepness obtained by Zingg (1940) and discussed earlier by Evans (Chapter 4) and Thornes (Chapter 5). Kirkby (Chapter 6) quotes values for *m* ranging from 1.7 to 3.5, the lower values being associated with a denser plant cover. Although the value of *m* also varies with the intensity of the rainfall and the steepness of slope (D'Souza and Morgan, 1976) and the erodibility of the soil (Gabriels, Pauwels, and De Boodt, 1975), sufficient studies have been carried out to enable reasonable judgements to be made on the appropriate value of *m* for a wide range of conditions (Table 8.10).

For modelling the effect of a plant cover on wind erosion, the equation for wind velocity (Bagnold, 1937; Equation (7.1)) can be adapted to predict the required value for roughness coefficient (z') to maintain wind velocity (V^*), as measured at height (z), below a selected critical value V_c^*:

$$z' = \frac{z}{V_z/\exp 5.75\, V_c^*} \qquad (8.29)$$

Table 8.10. Values for m in the relationship between rate of soil loss (Q_s) and slope (S) in the form $E \propto S^m$

m	Sources	Remarks
1.4	Zingg, 1940; Musgrave, 1947; Kirkby, 1969	Data from field plots up to 15 per cent slope in USA
2.0	Hudson and Jackson, 1959	High intensity rain in Zimbabwe; field plots
2.0	Kirkby, 1969	Rills; cohesionless soils
1.6	Horváth and Erödi, 1962	0–5 per cent slopes
0.7	Horváth and Erödi, 1962	6–12 per cent slopes
0.4	Horváth and Erödi, 1962	Slopes over 12 per cent
0.5	D'Souza and Morgan, 1976	Convex slopes, 3.5–14 per cent
0.4	D'Souza and Morgan, 1976	Straight slopes, 3.5–14 per cent
0.1	D'Souza and Morgan, 1976	Concave slopes, 3.5–14 per cent
−3.8	Heusch, 1970	12–65 per cent slopes: field plots in Morocco
0.6	Gabriels, Pauwels, and De Boodt, 1975	Soil particles 0.05 mm
1.7	Gabriels, Pauwels, and De Boodt, 1975	Soil particles 1.0 mm
1.9	Morgan, 1977	10–19 per cent slopes, sandy soils, in lowland England
1.1	Lal, 1976	Bare fallow, Nigeria: 1–15 per cent
1.1	Lal, 1976	Maize, Nigeria; 1–15 per cent
0.5	Lal, 1976	Mulch, Nigeria; 1–15 per cent

Table 8.11. Values of z' (cm)

Urban area	750
Wheat	1.3
Stubble	0.74
Sugar beet	0.4
Grass	0.3
Short (mowed) grass	0.2
Smooth water	0.018
Harrowed land	0.082

After Voetberg (1970); Cooke and Doornkamp (1974)

Selection of the appropriate plant cover or ground surface condition for the predicted level of roughness is hampered by lack of data on typical values of z' (Table 8.11).

In current practice, plant selection is made without recourse to any of the modelling procedures described above and is based largely on intuition or, at best, on data on soil losses associated with different crops derived from plot studies. Selection is made in the context of either plant succession, as in rotation;

or with several plants being grown at the same time, as with intercropping; or the use of ground covers, as under plantation crops. Where erosion plot data are available, cropping patterns can be designed which permit the growing of essential food or cash crops with the minimum of erosion. For example, taking data from Table 8.8 for upland areas in the Philippines, it can be calculated that a shifting cultivation practice of first-year maize, second-year upland rice, followed by ten years of secondary forest growth, will give a mean annual soil loss of 15 g m^{-2} which is close to that obtained by not using the land for agriculture at all but maintaining it under primary rain forest. Where no information on typical erosion rates exists, use is sometimes made of the Universal Soil Loss Equation to predict the value of C, the cropping factor, required to reduce erosion to an acceptable level. The problem of matching this value to various crops and vegetation covers is resolved either by taking typical C values from data tabulated in Wischmeier and Smith (1978), summarized by Mitchell and Bubenzer (1978, Chapter 2), or by estimating values according to the procedure devised by Wischmeier (1975) based on conditions of canopy, ground cover and previous landuse.

The design of strategies using agronomic techniques for erosion control is covered fully in various texts on soil erosion (Fournier, 1972; Morgan, 1979) and soil conservation (Stallings, 1957; Hudson, 1971). In this design work factors other than the protection-effectiveness of different plants must be considered. Since the protective value depends on a strong, healthy growth, how well, and how quickly a protection crop becomes established are important as are any adverse effects this crop may have on the growth and yield of the main crop. Plant selection is therefore a matter of matching as closely as possible crop requirements with the environmental conditions. A further factor of importance and one which may influence the success of any conservation scheme is whether the protection crop has any economic value.

8.6.2 Plant simulation

Where climatic and soil conditions are such that a satisfactory plant cover cannot be established at the time of year when erosion risk is greatest, a mulch is used to simulate a vegetation cover. The material used varies from straw, maize stalks, banana leaves, and palm fronds, depending on what is locally available, laid on the soil surface, to the stubble of a previous crop or a meadow residue with roots extending into the soil. The studies relating soil loss to bare ground indicate that about 70 per cent of the soil surface must be covered and, for straw, an application rate of 0.5 kg m^{-2} is sufficient to achieve this. A sparser covering affords inadequate protection and a denser one may affect the emergence of the main crop. The latter consideration does not apply to established tree crops and, under these, a denser covering may successfully suppress weed growth.

A mulch reduces the surface area exposed to raindrop impact and decreases flow velocity by imparting roughness. The shear stress of air and water flow is distributed amongst the mulch elements instead of acting entirely on the soil particles. On decomposition a mulch adds organic matter to the soil.

That a mulch effectively reduces soil loss has been shown in both field (Borst and Woodburn, 1942; Lal, 1976) and laboratory (Lattanzi, Meyer, and Baumgardner, 1974) studies. Attempts to model the effects of mulching on erosion have concentrated not on changes in the values of μ, m, and z' in the equations described above but on changes in the flow velocity. Foster and Meyer (1975) have established the following relationship between the cube of the velocity ratio and a straw mulch rate:

$$\left(\frac{V_m}{V_0}\right)^3 = \exp\{-3.0[1.0 - \exp(-2.15M^{0.65})]\} \qquad (8.30)$$

where V is the flow velocity for conditions with mulch (m) and without mulch (o) and M is the mulch rate ($t\,ac^{-1}$). From this equation the mulch rate can be predicted which will reduce a given velocity without mulch (V_0) to a maximum permissible velocity (V_m). The mulch rate can be related to the area of bare or exposed ground (E), using the expression

$$E = e^{-1.27M} \qquad (8.31)$$

Whilst tests of these equations have been encouraging (Foster and Meyer, 1975), they apply only to straw mulches and similar equations need to be derived for other types of mulch.

A similar approach has been used in wind tunnel experiments by Lyles and Allison (1976) to test the effectiveness of standing crop residues and the following relationship has been derived:

$$\frac{V^*}{V_t^*} = 1.638 + 17.044\frac{NA_s}{A_t} - 0.177\frac{L_y}{L_x} + (1.0236)^C - 1 \qquad (8.32)$$

where V^* is the drag velocity for a given open wind velocity, V_t^* is the critical drag velocity for the soil particles in question, N/A_t is the number of stalks in area $A_t(cm^2)$, A_s is the average silhouette area or projected area facing the flow of a single stalk (cm^2), L_y is the distance between stalks in a line perpendicular to the wind direction (cm), L_x is the distance between stalks in the wind direction (cm) and C is the percentage of dry soil aggregates greater than 1 cm in diameter. Further work is required to test this equation under field conditions.

The development of equations based on critical velocity must be seen as an early stage in the modelling process which will place design procedures for mulching on a similar level to those currently adopted in waterway design. More detailed modelling requires an approach similar to that outlined above for assessing the effects of plant covers. The only contribution to this end made so far is the empirical determination of values for m (Lal, 1976; Table 8.10).

8.6.3 Mechanical controls

The two most important mechanical measures used to control erosion at a field scale are contour strip cropping and contour bunds. By farming land in alternate strips, deviating from the contour by not more than 2 per cent, effective erosion control against runoff can usually be achieved on slopes up to 8.5°. The crops providing good and poor protection are alternated so that soil eroded from one strip is filtered out and deposited in the next strip downslope. A similar technique is employed to control wind erosion by aligning the strips at right angles to the wind. The width of the strips is varied with the erodibility of the soil. Although Chepil (1960) has produced a complex empirical technique for determining strip widths for wind erosion control, based on wind tunnel experiments, few studies have been made of the mechanics of the erosion-control process. Design widths are based on practical experience and reference is made to tables of recommended widths (FAO, 1965; Cooke and Doornkamp, 1974; Table 8.12).

Table 8.12. Recommended strip widths

Water erosion (Soils with fairly high water intake)	
2–5 per cent slope	30 m
6–9 per cent	25 m
10–14 per cent	20 m
15–20 per cent	15 m
Wind erosion (Strips perpendicular to wind direction)	
Sandy soil	6 m
Loamy sand	7 m
Sandy loam	30 m
Loam	75 m
Silt loam	85 m
Clay loam	105 m

After FAO (1965); Chepil and Woodruff (1963)

Contour bunds are earth banks, 1.5 to 2 m wide, thrown across the slope with a maximum permissible deviation of 10 per cent from the contour, to form buffer strips at 10 to 20 m intervals. Although little reduction in runoff occurs, there is considerable reduction in soil loss, as demonstrated by Roose (1966) in West Africa. The bunds are usually planted with grasses or trees to reduce splash erosion.

8.6.4 Soil management

Sound soil management is critical to the success of soil conservation schemes because only by maintaining the fertility and structure of the soil will a good plant cover be obtained. Where the structure of the soil breaks down and erosion ensues, vital plant nutrients and organic matter are removed from the soil in runoff and attached to eroded particles. Whilst the nutrients may be replaced by using mineral fertilizers, organic material cannot be replenished in this way. Yet it is the organic content which contributes to the aggregate stability of the soil through chemical bonding of clay and humus compounds. Soils with less than 3.5 per cent organic content or 2 per cent organic carbon are considered erodible (Evans, Chapter 4).

Whilst regularly applying organic matter to the soil is the most practical way of maintaining structural stability, actually increasing the organic content of an erodible soil to make it more resistant is a lengthy and difficult process. Studies at Samaru, Nigeria (Jones, 1971) show that ploughing in the residue of a maize crop, at about 0.5 to 1 $kg\,m^{-2}$, results in an addition of only 0.004 to 0.017 per cent to the organic carbon in the top 20 cm of soil. Maintaining the organic content of the soil at a satisfactory level using farmyard manure requires an application rate of 1 $kg\,m^{-2}$ a rate at which it is almost impossible to maintain supplies over a large area. Because of variations in the rate at which different manures decompose and beneficially affect soil structure (Fournier, 1972) and the rapid rate at which the organic content of a soil declines some four years after application (Chepil, 1955), regular use of organic matter in the farming system is essential. In the long term the most satisfactory method of building up the organic content is through grass leys. A three year ley increases the organic matter by an amount equal to that obtained by applying farmyard manure at 1.2 $kg\,m^{-2}\,y^{-1}$.

In modelling the effects of these structural changes in the soil, it is necessary to vary the value of the grain size parameter (d) in the modelling equations. This cannot be achieved simply, for example, by considering aggregate size distribution rather than particle size distribution because aggregates of clay minerals do not behave in the same way as sand particles of equivalent size because of their lighter weight. The development of a satisfactory modelling procedure probably awaits the results of more fundamental research into the mechanics of erosion of cohesive soils (Paaswell, 1973; Arulanandan and Heinzen, 1978).

The other important aspect of soil management concerns tillage practice. The purpose of tillage is to break up the soil to provide a fine seed bed in which the crops may grow. Conventional tillage practice, comprising the sequence of ploughing, one or more harrowings and planting, has a wide degree of applicability and it is only where the soil is so weak that it pulverizes or where compaction at plough depth creates a hard pan that problems arise. A wide range of special tillage operations is available to reduce soil loss in problem areas (Schwab *et al.*, 1966). These may be classified according to the aims of reducing pulverization pan break-up, and increasing depression storage.

The pulverizing effect of conventional tillage is counteracted by reducing the number of operations on the land. This is achieved by cultivating only the small strips of land required for seed beds and leaving wide untilled zones (strip-zone tillage), carrying out tillage whilst a mulch is retained on the ground (mulch tillage), or completing as many activities as possible in the one pass (minimum tillage) as with plough-plant operations. Break-up of the hard pan is achieved by sub-soiling. Depression storage is increased on arable land by forming ridges, or, where even higher storage is required, connecting the ridges with cross ties to form rectangular depressions, a technique known as tied ridging. Ridging also imparts roughness to wind and water flow and ridges of 4 to 7 cm in height are sometimes used to reduce wind erosion (Woodruff and Siddoway, 1965).

Selection of the most appropriate practice must be related carefully to the soil. For example, increasing depression storage by tied ridging may, on clay soils, induce waterlogging followed by mass movement (Gray and Brenner, 1970). Subsoiling followed by drainage may be inappropriate on sandy soils especially where the drainage system removes water so rapidly that the crops are starved of moisture. Recent studies have emphasized the importance of only carrying out cultivation when soil conditions, particularly soil moisture content, are right. Although this work is aimed at avoiding structural breakdown of the soil by compaction and smearing (Davies, 1974; Spoor, 1976), there is no reason to doubt that restricting cultivation to those times when the soil's resistance to the shear forces of tines is greatest would be a valuable erosion control measure.

Surprisingly, few soil conservation texts stress the importance of organic matter and tillage methods as basic techniques of erosion control and further investigations into the role that they might play in conserving soil would undoubtedly prove worthwhile. Linking tillage practices with erosion modelling requires:

(a) assessing their effects on soil structure, for example by analysing the soil mechanics of soil-tine force interactions (Spoor, 1976), and making adjustments to the grain size parameter (d) similar to those outlined above for modelling structural changes;

(b) examining the effects of pan break-up and increased depression storage on the supply rate of runoff (σ);

(c) studying the effects of compaction of the soil as a result of the passage of farm machinery on the supply rate of runoff (σ) (Davies, Eagle, and Finney, 1972; Spoor and Muckle, 1974; Martin, 1979); and

(d) researching the effect of surface roughness on the modelling parameters μ, n and z'.

8.7 SITE MANAGEMENT

This section is concerned with soil loss in special, often localised, problem areas where erosion rates are extremely high and which may be major sources of sediment production in the region as a whole. Although erosion control measures are again based largely on a combination of agronomic and mechanical techniques, an additional approach employed at this scale is to tackle the problem through soil erodibility (Table 8.2). Attempts are made to increase the resistance of the soil to erosion either by applying various adhesive sprays and soil conditioners or by covering the soil with an artificial surface such as concrete.

8.7.1 Soil conditioners

Temporary soil stability, lasting from two weeks to six months, can be achieved by applying various oil or rubber based emulsions. These are poly-functional polymers which develop chemical bonds with the clay minerals in the soil and result in the formation of soil aggregates which, in turn, increases the size and improves the stability of the pore spaces and raises the infiltration rate. The greater aggregate stability also makes the soil more resistant to wind erosion.

Although soil stabilizers have been used for many years on newly-seeded areas on golf courses, road cuttings, embankments and sand dunes, research into their efficiency has not produced a base for modelling their effects. Work has concentrated instead on investigating, for the numerous conditioners now commercially available, appropriate application rates, the length of time for which stability is provided, effects on plant emergence and the cost of application per unit area. Wind tunnel tests (Lyles *et al.*, 1969; Armbrust and Dickerson, 1971; Armbrust, 1977) show that only six out of 34 materials tested met the requirements of acceptable cost, no adverse effect on plant growth or emergence, reduction of erosion for up to two months and easy application. Further, effective temporary control over erosion is only achieved with spray-on materials if the whole of the soil surface is covered. This compares with the 70 per cent coverage required for erosion control using mulches and plant covers.

Tests on the use of soil conditioners for water erosion control (Gabriels *et al.*, 1977) show that certain bitumen treatments are only effective for a few storms and that asphalt and latex emulsions, particularly when applied with low dilution rates, may only seal the surface and result in increased runoff and erosion. The latter problem can be overcome by incorporating the emulsion in the top 10 to 20 cm of the soil but what is critical then is the size of the aggregates that are produced. Because the emulsion causes the aggregates to repel water, rainfall must either infiltrate the soil through the pore spaces or form runoff. If the aggregates are too small, infiltration rates are low and increased runoff is produced. For most emulsions the aggregates must be at least 2 mm in size and ideally greater than 5 mm (Pla, 1977) but experiments with poly-acrylamide conditioners suggest that high infiltration rates may be achieved regardless of the size distribution of the aggregates. These emulsions are, however, expensive to use.

8.7.2 Stream bank erosion

The problems of scouring of the channel sides by flowing water and slumping of the bank as moisture conditions change with the rise and fall of the water level are tackled in three ways. First, an attempt is made to reduce the flow velocity to an acceptable level, particularly at the point where the bank is most vulnerable. Stakes are placed in the channel bed and small jetties and groynes built out from the banks to induce roughness, thereby decreasing velocity, as well as deflecting the flow away from the bank itself. The latter effect may only transfer the problem further downstream, however.

The second approach is to mould or shape the bank into a stable form using the design procedures adopted for waterways based on critical tractive force (Equations (8.15), (8.16), and (8.17); Section 8.5.4). Where the bank cannot be shaped in any way to give a stable slope, the third approach of bank strengthening is employed. Some erosion control may be obtained by planting grasses which are resistant to periodic flooding but where this is insufficient, an artificial facing is provided in the form of rock-filled wire cages or gabions, or concrete. These structures must be securely anchored to avoid collapse.

8.7.3 Roadside erosion

The main problems associated with roads are posed by the excess runoff generated from the road itself because water cannot penetrate its hard and compacted surface, and bank stability of cuttings and embankments.

In erodible areas it is customary to place roads on ridges (Hudson, 1971) as this reduces the catchment area from which runoff may derive. It also results in less expense on bridges and culverts to deal with the crossing of streams. Even so, the runoff problem cannot be entirely solved in this way because, wherever the ridge or divide slopes into cols and along spurs, erosion by overland flow and gullying can occur (Yair, 1972). Side ditches, usually grass-lined, with regularly spaced outlets are the standard method of runoff control. The ditch is designed so that the permissible velocity for runoff is not exceeded (Turner, 1963).

Various techniques are available to improve bank stability. These range from benching, used to reduce the overall slope angle of the bank, establishment of herbaceous plants, construction of diversion ditches and spillways to dispose of runoff, to strengthening of the bank by applying soil conditioners, facing it with cement or laying erosion nets (Barnes, 1973; Turelle, 1973; Dunn, 1975).

8.8 MULTI-SCALE PROBLEMS

This review has emphasized the importance of identifying the most appropriate scale of operation for designing soil conservation strategies. When it comes to gullying and sand dune migration, however, multi-scale strategies are required. With these most severe forms of soil erosion the controlling processes operate at both regional and local scales and it is for this reason that effective erosion control is difficult to implement and the costs of conservation and the risk of failure are both extremely high.

8.8.1   Gully erosion control

The relevant scales of operation for dealing with gully erosion are the watershed, where runoff must be controlled, and the channel, where a stable form must be provided. Attempts to control gullying frequently fail, firstly, because, to reduce costs, only partial and inadequate solutions are implemented, and secondly, because wrong measures are sometimes adopted owing to incorrect identification of the problem. For instance, gullies may develop by surface erosion through the enlargement and coalescence of small depressions and knicks on a hillside or they may form by subsurface erosion and the collapse of pipes and tunnels (Thornes, Chapter 5). Measures to deal with the first type of development will not help and may even aggravate the problem of subsurface erosion. A further difficulty is in establishing the main source of the sediment which may be the adjacent hillsides, the gully head or the channel bed and banks further downstream. The importance of determining the source is emphasized by the studies of Piest, Bradford, and Spomer (1975) who show that measures to reduce runoff may be effective in reducing channel erosion but are unlikely to have much effect on erosion at the gully head because the latter can take place in quite moderate runoff events.

Control measures at the watershed scale concentrate on reducing runoff by landuse management. To deal with surface runoff, the gullies and surrounding headwaters are treated with mulches, grasses, trees, and combinations thereof to increase roughness, decrease flow velocity and induce infiltration. The treatment is usually supported by building temporary dams in each channel to trap sediment and thereby reduce channel depth and slope (Miller, Woodburn, and Turner, 1962). For subsurface erosion, the aim is to obtain a more uniform pattern of infiltration and soil water movement than occurs in tunnelled areas. Thus, mechanical measures in the form of diversion ditches and terraces which are often used to control surface runoff are inappropriate because, by causing ponding, they afford a concentrated supply of water which may open up a new pipe system or increase erosion in an existing one. Control of subsurface erosion is best achieved by establishing a healthy grass cover (Colclough, 1965) or, where conditions are too marginal for improved pasture, a mixture of scrub grasses and trees. If tunnelling has reached an advanced stage, the tunnels must be broken up by ripping prior to sowing of the grass. Where tunnelling is associated with specific soil problems such as a high sodium content, chemical treatments may be used. Spraying of soluble salts on the surface causes the clay particles to flocculate and form resistant bonds. This treatment is too expensive to employ over large areas and gives only temporary control.

Conservation measures in the channel may include reshaping the gully to a new stable design, particularly where it may serve as an outlet for terrace channels. Usually such landforming operations are impracticable and simple mechanical controls are employed instead, using earth and brush dams, piling, bricks, boulders, or planks to form barriers across the channel, inducing sedimentation and changing the slope both in the main gully and the tributaries (Thornes, Chapter 5). These temporary structures are designed with a spillway to cope with overtopping during high flows. The required spacing for the dams is obtained from the equation (Heede and Mufich, 1973):

$$SP = \frac{HE}{K \cdot \tan S \cdot \cos S} \qquad (8.33)$$

where SP is the distance between the dams, S is the channel slope, HE is the effective dam height and K is a constant which equals 0.3 for $\tan S \leqslant 0.2$, and 0.5 for $\tan S > 0.2$. A numerical procedure has been devised (Heede and Mufich, 1973) for dam design which takes account of costs. Application of this to conditions in the Rocky Mountains shows that wire-bound and loose-rock dams are most economical at a height of 0.45 m, single fence dams at 0.75 m and double fence dams at 1.7 m.

These temporary dams have a high risk of failure. They must be carefully constructed, properly anchored and regularly inspected and maintained (Heede, 1976). On large gullies, permanent concrete structures are sometimes installed to control the overfall of water at the gully head. These structures are designed to cope with the peak runoff with a ten-year return period or greater and to dissipate the energy of the flow in a manner which protects both the structure and the area downstream. The main components of the structure are an inlet, a conduit and an outlet. The various forms of these components are described in Schwab *et al* (1966).

8.8.2 Dune stabilization

The techniques of stabilizing moving sand dunes along sea and lake shores and on desert margins are based mainly on experience, much of it gained in the late eighteenth century in Les Landes, France. The use of shelterbelts is hampered by the difficulty of establishing trees and shrubs in shifting sands. The broad scale agronomic approach to conservation is supplemented, therefore, by localized treatments on individual dunes. Stabilization begins by breaking up any hard-pan by deep ploughing. A wicker fence is then built on the windward side of the dune aligned at right angles to the wind to serve as a sediment trap. Sand-binding grasses, for example, marram grass, and trees are planted in the lee of the fence to complete the barrier. As sand piles up against the barrier, a second line of grasses and trees is established to the leeward and this sequence is continued until sand deposition has formed a relatively stable slope. Once this has occurred, large areas leeward of these barriers can be planted with trees and shrubs (Wolf, 1929).

Best results are obtained where already established tree and shrub stock, grown in nurseries, is planted. Sowing seeds invariably ends in failure. Even with nursery stock, however, additional help is needed in the early stages to get plants established and this may be provided by using soil conditioners. A styrene-butadiene latex stabilizer was successfully used in this way on dune sands at Scolt Head Island on the Norfolk coast (Haas and Steers, 1964).

The stabilization of dunes with a plant cover brings about changes in the properties of the sandy soil. Within eight years of successfully establishing the fast-growing *Tamarix aphylla* on the sands surrounding the Al Hasa oasis in Saudi Arabia, the depth of an organic horizon had increased from zero to 1 cm and the calcium carbonate content had decreased from about 30 per cent to 15 per cent (Stevens, 1974).

8.9   MODELLING SOIL CONSERVATION

Although recent years have seen a considerable advance in the development of models of soil erosion, particularly with the move away from empiricism towards modelling based on the mechanics of the processes involved, no such trend has occurred in the research on soil conservation. Yet there would appear to be several advantages in modelling the effects of conservation measures on the erosion processes. First, the value of many of the soil erosion models would be enhanced by making them applicable to a wide range of plant cover conditions. This would enable soil losses under various crops to be predicted. Second, the effectiveness of conservation measures could be assessed before implementing them. Third, by providing a method of comparing the effects of different conservation practices, the advice given to farmers, recreation planners, and land managers could be improved. Where a particular conservation scheme is unacceptable for economic or social reasons (Temple, 1972a), alternative strategies could be investigated.

At present the soil conservationist is able to identify and map the erosion features of an area, classify the land according to its erosion risk and, based on a wide range of experience, built up by agricultural engineers over many years, suggest control measures which will probably work. The design of the conservation strategy and its detailed components such as terraces and waterways remains empirically based, sometimes relating to data on soil loss under different conditions, more often to the use of the Universal Soil Loss Equation, but occasionally is merely intuitive. The result is considerable variation in approach.

The procedures for waterway design emphasize protection against an individual erosion event whereas agronomic measures are aimed at minimizing mean annual soil loss. Waterway design is based on a well-established flow velocity equation whereas the selection of agronomic measures is unrelated to any consideration of the mechanics of erosion.

If the erosion models being currently developed are to achieve their maximum potential value they must be capable of incorporating the effects of conservation practices. Before conservation measures can be included, however, their influence on the mechanics of erosion must be investigated. In this context the experiments of Foster and Meyer (1975) on mulching, Lyles and Allison (1976) on standing crop residues and Skidmore and Hagen (1977) on shelterbelts indicate the type of study which is required. Ultimately the scope of this work must be extended from its concern with flow velocities to encompass sediment removal.

Two approaches may be suggested for including soil conservation within an erosion model. One is to define and parameterize a separate sub-system for conservation inputs. The other is to take account of conservation work within the existing sub-systems of the model. The latter would appear to be the more realistic approach because it will enable the effects of conservation to be considered directly. Further, as indicated above with reference to developments of the model presented by Kirkby (Chapter 6), some, albeit limited, data which could be used are already available on the interception of rainfall by different plant covers, the depression storage characteristics of some mechanical conservation measures, and the effects of roughness on flow velocity. One difficulty which will be encountered with this approach is where a conservation measure influences more than one sub-system in an erosion model. Allowance must then be made for interactive effects.

The inclusion of conservation techniques in erosion models will focus attention on aspects hitherto largely neglected in conservation work. A much greater appreciation of scale, both temporal and spatial, will be obtained with a consequent improvement to the design of strategies for soil conservation. Much conservation work is presently aimed at the field scale but by giving greater emphasis to conservation at the watershed and site levels, more thought will need to be given to what is an appropriate maximum permissible value for soil loss. A greater understanding will also result of the way in which measures appropriate to different scales interlink. The importance of this scale-linkage has been stressed with reference to gully erosion control and dune stabilization.

In addition to these benefits to conservation research, the linking of conservation with erosion models will enhance the development of the latter. With the requirement of designing conservation measures to deal with erosion events with a given return period, methods will be needed, most probably based on stochastic techniques, to predict mean annual soil loss from frequency distributions of meteorological events and to take account of the probability of these events coinciding with conditions of poor plant cover. These developments mean that whilst the period of experimentation to determine which conservation measures work is virtually over, basic research into all the conservation techniques is required to provide the foundation for modelling. This is the challenge for the future.

REFERENCES

Anderson, M. G., and Calver, A. (1977). On the persistence of landscape features formed by a large flood. *Trans. Inst. Br. Geogr. New Series*, **2**(2), 243–254.

Armbrust, D. V. (1977). A review of mulches to control wind erosion. *Trans. Am. Soc. Agric. Engrs*, **20**, 904–905, 910.

Armbrust, D. V., and Dickerson, J. D. (1971). Temporary wind erosion control: cost and effectiveness of 34 commercial materials. *J. Soil and Water Conserv.*, **26**, 154–157.

Arnoldus, H. M. J. (1977). Predicting soil losses due to sheet and rill erosion. *FAO Conservation Guide*, Vol. 1, pp. 99–124.

Arulanandan, K., and Heinzen, R. T. (1978). Factors influencing erosion in dispersive clays and methods of identification. *Int. Assoc. Scient. Hydrol. Pub.*, **122**, 75–81.

Bagnold, R. A. (1937). The transport of sand by wind. *Geogr. J.*, **89**, 409–438.

Bagnold, R. A. (1941). *The Physics of Blown Sand and Desert Dunes*, Methuen, London. 265 pp.

Baker, C. F., Morgan, R. P. C., Brown, I. W., Hawkes, D. E., and Ratcliffe, J. B. (1979). Soil erosion survey of the Moel Famau Country Park. *Clwyd County Council Planning Department, Country Park Research Report No. 2, Nat. Coll. Agr. Engng Silsoe, Occasional Paper*, No. 7.

Barnes, R. C., Jr. (1973). Erosion control structures. In: *Soil Erosion: Causes and Mechanisms; Prevention and Control*, Highway Research Board, Special Report No. 135, Washington, D.C.

Barry, R. G., and Chorley, R. J. (1976). *Atmosphere, Weather, and Climate*, Methuen, London, 3rd edn. 432 pp.

Bergsma, E. (1974). Soil erosion sequences on aerial photographs. *ITC Journal*, **1974–3**, 342–376.

Borst, H. L., and Woodburn, R. (1942). The effect of mulching and methods of cultivation on runoff and erosion from Muskingum silt loam, *Agric. Engr.*, **23**, 19–22.

Bureau of Reclamation (1953). *Bureau of Reclamation Manual. Volume V. Irrigated Land use. Part 2: Land Classification*, Bureau of Reclamation, US Department of the Interior, Denver, Colorado. 130 pp.

Charman, P. E. V. (1978). Soils of New South Wales: their characterization, classification and conservation. *NSW Soil Conservation Service, Technical Handbook No. 1.* 236 pp.

Chepil, W. S. (1945). Dynamics of wind erosion. III. Transport capacity of the wind. *Soil Sci.*, **60**, 475–480.

Chepil, W. S. (1955). Factors that influence clod structure and erodibility of soil by wind. V. Organic matter at various stages of decomposition. *Soil Sci.*, **80**, 413–421.

Chepil, W. S. (1960). Conversion of relative field erodibility to annual soil loss by wind. *Proc. Soil Sci. Soc. Am.*, **24**, 143–145.

Chepil, W. S., and Woodruff, N. P. (1963). The physics of wind erosion and its control. *Adv. in Agron.*, **15**, 211–302.

Chow, S. K., and Chang, K. K. (1968). Drainage in the Republic of Singapore. *J. Instn Engrs (Singapore)*, 17–23.

Christy, L. C. (1971). Legislative principles of soil conservation. *FAO Soils Bulletin No. 15.* 68 pp.

Colclough, J. D. (1965). Soil conservation and soil erosion control in Tasmania: tunnel erosion. *Tasmanian J. Agric.*, **36**, 7–12.

Cole, C. V., Tarapore, Z. S., Kanhere, V. N., and Dixit, J. G. (1973). Sediment discharge formulae: a comparative analysis. In: *International Symposium on River Mechanics, Volume 1, Sediment Transportation*, International Association for Hydraulic Research, Bangkok, pp. 547–560.

Cooke, R. U., and Doornkamp, J. C. (1974). *Geomorphology in Environmental Management*, Oxford University Press, London. 413 pp.

Cruickshank, J. G., and Armstrong, W. J. (1971). Soil and agricultural land classification in County Londonderry. *Trans. Inst. Br. Geogr.*, **53**, 79–94.

Davies, D. B. (1974). Soil structure and crop production. In: *Soil Type and Land Capability* (D. Mackney, Ed.), Soil Survey, Technical Monograph No. 4, Harpenden, pp. 117–124.

Davies, D. B., Eagle, D. J., and Finney, J. B. (1972). *Soil Management*, Farming Press, Ipswich, 254 pp.

De Ploey, J., Savat, J., and Moeyersons, J. (1976). The differential impact of some soil loss factors on flow, runoff creep and rainwash. *Earth Surface Processes*, **1**, 151–161.

Diaconu, C. (1969). Résultats de l'étude de l'écoulement des alluvions en suspension des rivieres de la Roumanie. *Bull. Assoc. Scient. Hydrol.*, **14**, 51–89.

Douglas, I. (1967). Natural and man-made erosion in the humid tropics of Australia, Malaysia, and Singapore. *Int. Assoc. Scient. Hydrol. Pub.*, **75**, 17–30.

D'Souza, V. P. C., and Morgan, R. P. C. (1976). A laboratory study of the effect of slope steepness and curvature on soil erosion. *J. Agric. Engng Res.*, **21**, 21–31.

Dunn, C. S. (1975). Control of erosion on highways. Paper presented to *PTRC Summer Annual Meeting*, University Warwick, 7–11 July (1975).

Elwell, H. A., and Stocking, M. A. (1976). Vegetal cover to estimate soil erosion hazard in Rhodesia. *Geoderma*, **15**, 61–70.

Emmett, W. W. (1970). The hydraulics of overland flow on hillslopes. *US Geol. Surv. Prof. Paper*, **662–A**. 68 pp.

Emmett, W. W. (1978). Overland flow. In: *Hillslope Hydrology* (M. J. Kirkby, Ed.), Wiley, Chichester, pp. 145–176.

Fang, Z., Liu, B., Zhou, P., Ren, L., Liu, Q., and Zhang, H. (1980). Terraces in the loess plateau of China. In *Soil Conservation* (Ed. R. P. C. Morgan), Wiley, Chichester (in press).

F.A.O. (1965). *Soil Erosion by Water*, Rome. 284 pp.

Foster, G. R., and Meyer, L. D. (1975). Mathematical simulation of upland erosion by fundamental erosion mechanics. In: *Present and Prospective Technology for Predicting Sediment Yields and Sources*—Proceedings of Sediment-Yield Workshop, United States Department of Agriculture Sedimentation Laboratory, Oxford, Mississippi, November 1972 (Agricultural Research Service Report ARS-S-40), United States Department of Agriculture, Washington, D.C., pp. 190–207.

Fournier, F. (1960). *Climat et érosion: la relation entre l'érosion du sol par l'eau et les précipitations atmosphériques*, Presses Universitaires de France, Paris. 201 pp.

Fournier, F. (1972). *Soil Conservation*. Nature and Environment Series, Council of Europe. 194 pp.

Gabriels, D., Maene, L., Lenvain, J., and De Boodt, M. (1977). Possibilities of using soil conditioners for soil erosion control. In: *Soil Conservation and Management in the Humid Tropics* (D. J. Greenland and R. Lal, Eds), Wiley, Chichester, pp. 99–108.

Gabriels, D., Pauwels, J. M., and De Boodt, M. (1975). The slope gradient as it affects the amount and size distribution of soil loss material from runoff on silt loam aggregates. *Med. Fac. Landbouww. Rijksuniv. Gent.*, **40**. 1333–1338.

Gichungwa, J. K. (1970). *Soil Conservation in Central Province (Kenya)*, Ministry of Agriculture, Nairobi.

Gray, D. H., and Brenner, P. R. (1970). The hydrology and stability of cutover slopes. In: *Proceedings, Symposium on Interdisciplinary Aspects of Watershed Management*, Montana State University, Bozeman, ASCE, pp. 295–326.

Gregory, K. J. and Gardiner, V. (1975). Drainage density and climate. *Z. für Geomorph.*, **19**, 287–298.

Gunn, R., and Kinzer, G. D. (1949). The terminal velocity of fall for water droplets. *J. Meteorol.*, **6**, 243–248.

Haas, J. A., and Steers, J. A. (1964). An aid to stabilization of sand dunes: experiments at Scolt Head Island. *Geogr. J.*, **130**, 265–267.

Hagen, L. J., and Skidmore, E. L. (1971). Windbreak drag as influenced by porosity. *Trans. Am. Soc. Agric. Engrs*, **14**, 464–465.

Heede, B. H. (1976). Gully development and control: the status of our knowledge. *USDA Forest Service, Research Paper RM-169*, Rocky Mountain Forest and Range Experiment Station, Fort Collins, Colorado.

Heede, B. H., and Mufich, J. G. (1973). Functional relationships and a computer program for structural gully control. *J. Environ. Management*, **1**, 321–344.

Heusch, B. (1970). L'érosion du Pré-Rif. Une étude quantitative de l'érosion hydraulique dans les collines marneuses du Pré-Rif occidental. *Ann. Rech. Fores. Maroc*, **12**, 9–176.

Horton, R. E. (1945). Erosional development of streams and their drainage basins: a hydrophysical approach to quantitative morphology. *Bull. Geol. Soc. Am.*, **56**, 275–370.

Horváth, V., and Erödi, B. (1962). Determination of natural slope category limits by functional identity of erosion intensity. *Int. Assoc. Scient. Hydrol. Pub.*, **59**, 131–143.

Hudson, N. W. (1965). *The Influence of Rainfall on the Mechanics of Soil Erosion with Particular Reference to Southern Rhodesia*. M.Sc. Thesis, University of Cape Town.

Hudson, N. W. (1971). *Soil Conservation*, Batsford, London. 320 pp.

Hudson, N. W., and Jackson, D. C. (1959). Results achieved in the measurement of erosion and runoff in Southern Rhodesia. *Proc. Third Inter-African Soils Conference*, Dalaba, pp. 575–583.

Iveronova, M. I. (1969). Opyt kolichestvennogo analiza protsessov sovremmenoi denudatsii. *Izvestiya AN SSR, Ser. Geograf.* **1969–2**, 13–24. Transl.: An attempt at the quantitative analysis of contemporary denudation process. Russian Translating Programme, NLL Boston Spa, RTS 7436, 1972.

Jackson, I. J. (1975). Relationships between rainfall parameters and interception by tropical forest. *J. Hydrol.*, **24**, 215–238.

Jackson, I. J. (1977). *Climate, Water, and Agriculture in the Tropics*, Longman, London. 248 pp.

Jones, M. J. (1971). The maintenance of soil organic matter under continuous cultivation at Samaru, Nigeria. *J. Agric. Sci.*, **77**, 473–482.

Jones, R. G. B., and Keech, M. A. (1966). Identifying and assessing problem areas in soil erosion surveys using aerial photographs. *Photogrammetric Record*, **5**(27), 189–197.

Jovanović, S., and Vukčević, M. (1958). Suspended sediment regimen on some watercourses in Yugoslavia and analysis of erosion processes. *Int. Assoc. Scient. Hydrol. Pub.*, **43**. 337–359.

Keech, M. A. (1969). Mondaro Tribal Trust Land. Determination of trend using air photo analysis. *Rhod. Agric. J.*, **66**, 3–10.

Kellman, M. C. (1969). Some environmental components of shifting cultivation in upland Mindanao. *J. Trop. Geog.*, **28**, 40–56.

Kirkby, M. J. (1969). Erosion by water on hillslopes. In: *Water, Earth, and Man* (R. J. Chorley, Ed.), Methuen, London, pp 229–238.

Kirkby, M. J. (1976). Hydrological slope models: the influence of climate. In: *Geomorphology and Climate* (E. Derbyshire, Ed.), Wiley, London, pp. 247–267.

Kirkby, M. J. (1977). Maximum sediment efficiency as a criterion for alluvial channels. In: *River Channel Changes* (K. J. Gregory, Ed.), Wiley, Chichester, pp. 429–442.

Kirkby, M. J. (1978). Implications for sediment transport. In: *Hillslope Hydrology* (M. J. Kirkby, Ed.), Wiley, Chichester, pp. 325–363.

Klingebiel, A. A., and Montgomery, P. H. (1966). *Land Capability Classification.* Soil Conservation Service Agricultural Handbook No. 210, United States Department of Agriculture, Washington, D.C. 21 pp.

Laflen, J. M., Johnson, H. P., and Reeve, R. C. (1972). Soil loss from tile-outlet terraces. *J. Soil and Water Conserv.*, **27**, 74–77.

Lal, R. (1976). Soil erosion problems on an alfisol in western Nigeria and their control. *IITA Monograph No. 1.* 160 pp.

Langbein, W. B., and Schumm, S. A. (1958). Yield of sediment in relation to mean annual precipitation. *Trans. Am. Geophys. Union*, **39**, 1076–1084.

Lattanzi, A. R., Meyer, L. D., and Baumgardner, M. F. (1974). Influence of mulch rate and slope steepness on interrill erosion. *Proc. Soil Sci. Soc. Am.*, **38**, 946–950.

Leven, A. A., Meurisse, R. T., Carleton, J. O., and Williams, J. A. (1974). Land response units. *Proc. Soil Sci. Soc. Am.*, **38**, 140–143.

Livesey, R. H. (1975). Corps of Engineers methods for predicting sediment yields. In: *Present and Prospective Technology for Predicting Sediment Yields and Sources*— Proceedings of Sediment-Yield Workshop, United States Department of Agriculture Sedimentation Laboratory, Oxford, Mississippi, November 1972 (Agricultural Research Service Report ARS-S-40), United States Department of Agriculture, Washington, D.C., pp. 16–32.

Low, F. K. (1967). Estimating potential erosion in developing countries. *J. Soil and Water Conserv.*, **22**, 147–148.

Low, K. S., and Goh, K. C. (1972). The water balance of five catchments in Selangor, West Malaysia. *J. Trop. Geog.*, **35**, 60–66.

Lyles, L., and Allison, B. E. (1976). Wind erosion: the protective role of simulated standing stubble. *Trans. Am. Soc. Agric. Engrs*, **19**, 61–64.

Lyles, L., Armbrust, D. V., Dickerson, J. D., and Woodruff, N. P. (1969). Spray-on adhesives for temporary wind erosion control. *J. Soil and Water Conserv.*, **24**, 190–193.

McCormack, R. J. (1971). The Canada Land Use inventory: a basis for land use planning. *J. Soil and Water Conserv.*, **26**, 141–146.

McGuinness, J. L., Harrold, L. L., and Edwards, W. M. (1971). Relation of rainfall energy and streamflow to sediment yield from small and large watersheds. *J. Soil and Water Conserv.*, **26**, 233–235.

Martin, L. (1979). Accelerated soil erosion from tractor wheelings: a case study in mid-Bedfordshire. *Proceedings, Seminar on Agricultural soil erosion in temperate non-Mediterranean climates*, University Louis Pasteur, Strasbourg, Semptember 1978. 157–162.

Maner, S. B. (1958). Factors affecting sediment delivery rates in the Red Hills physiographic area. *Trans. Am. Geophys. Union* **39**, 669–675.

Maza Alvarez, J. A., and Cruickshank Villanueva, C. (1973). Stable channels in alluvium. In: *International Symposium on River Mechanics, Volume 1. Sediment Transportation*, Bangkok, International Association for Hydraulic Research, pp. 715–722.

Miller, C. R., Woodburn, R., and Turner, H. R. (1962). Upland gully sediment production. *Int. Ass. Scient. Hydrol. Pub.* **59**, 83–104.

Moldenhauer, W. C., and Onstad, C. A. (1975). Achieving specified soil loss levels. *J. Soil and Water Conserv.*, **30**, 166–168.

Moldenhauer, W. C., and Onstad, C. A. (1977). Engineering practices to control erosion. In: *Soil Conservation and Management in the Humid Tropics* (D. J. Greenland and R. Lal, Eds), Wiley, Chichester, pp. 87–92.

Morgan, R. P. C. (1973). The influence of scale in climatic geomorphology: a case study of drainage density in West Malaysia. *Geogr. Ann.* **55–A**, 107–115.

Morgan, R. P. C. (1974). Estimating regional variations in soil erosion hazard in Peninsular Malaysia. *Malay. Nat. J.*, **28**, 94–106.

Morgan, R. P. C. (1977). Soil erosion in the United Kingdom: field studies in the Silsoe area, 1973–75. *Nat. Coll. Agr. Engng. Silsoe, Occasional Paper No. 4*, 41 pp.

Morgan, R. P. C. (1978). Identifying hillslope response units: a case study of a small catchment in Peninsular Malaysia. *J. Trop. Geog.*, **46**, 37–46.

Morgan, R. P. C. (1979). *Topics in Applied Geography: Soil Erosion*, Longman, London. 114 pp.

Morgan, R. P. C. (in press). Field studies of sediment transport by overland flow. *Earth Surface Pressures.*

Musgrave, G. W. (1947). The quantitative evaluation of factors in water erosion: a first approximation. *J. Soil and Water Conserv.*, **2**, 133–138.

Nicol, K. J., Heady, E. O., and Wade, J. C. (1976). National and interregional models of land use, water allocation, soil loss control, and agricultural policy. *Trans. Am. Soc. Agric. Engrs*, **19**, 90–96.

Nieuwolt, S. (1977). *Tropical Climatology*, Wiley, London. 207 pp.

Nossin, J. J. (1964). Geomorphology of the surroundings of Kuantan (eastern Malaya). *Geol. en Mijnb.*, **43**, 157–182.

Paaswell, R. E. (1973). Causes and mechanisms of cohesive soil erosion: the state of the art. In: *Soil Erosion: Causes and Mechanisms; Prevention and Control*, Highway Research Board, Special Report No. 135, Washington, D.C., pp. 52–74.

Pearce, A. J. (1976). Magnitude and frequency of erosion by Hortonian overland flow. *J. Geol.*, **84**, 65–80.

Petryk, S., and Bosmajian, G. III. (1975). Analysis of flow through vegetation. *J. Hydraul. Div.*, *Am. Soc. Civil Engrs*, **101**, 871–884.

Piest, R. F., Bradford, J. M., and Spomer, R. G. (1975). Mechanisms of erosion and sediment movement from gullies. In: *Present and Prospective Technology for Predicting Sediment Yields and Sources*—Proceedings of Sediment-Yield Workshop, United States Department of Agriculture Sedimentation Laboratory, Oxford, Mississippi, November 1972 (Agricultural Research Service Report ARS-S-40), United States Department of Agriculture, Washington, D.C., pp. 162–176.

Pla, I. (1977). Aggregate size and erosion control on sloping land treated with hydrophobic bitumen emulsion. In: *Soil Conservation and Management in the Humia Tropics* (D. J. Greenland and R. Lal, Eds), Wiley, Chichester, pp. 109–115.

Ramser, C. E. (1917). *Prevention of the erosion of farm lands by terracing*, Bulletin No. 512, United States Department of Agriculture, Washington, D.C.

Rao, D. P. (1975). Applied geomorphological mapping for erosion surveys: the example of the Oliva basin, Calabria. *ITC Journal*, **1975–3**, 341–350.

Rapp, A. (1960). Recent development of mountain slopes in Kärkevagge and surroundings, northern Scandinavia. *Geogr. Ann.*, **42**, 65–200.

Rapp, A., Axelsson, V., Berry, L., and Murray-Rust, D. H. (1972). Soil erosion and sediment transport in the Morogoro River catchment, Tanzania. *Geogr. Ann.*, **54-A**, 125–155.

Ree, W. O. (1958). Retardation coefficients for row crops in diversion terraces. *Trans. Am. Soc. Agric. Engrs*, **1**, 78–80.

Ree, W. O., Wimberley, F. L., and Crow, F. R. (1977). Manning *n* and the overland flow equation. *Trans. Am. Soc. Agric. Engrs*, **20**, 89–95.

Renfro, G. W. (1975). Use of erosion equations and sediment-delivery ratios for predicting sediment yield. In: *Present and Prospective Technology for Predicting Sediment Yields and Sources*—Proceedings of Sediment-Yield Workshop, United States Department of Agriculture Sedimentation Laboratory, Oxford, Mississippi, November 1972 (Agricultural Research Service Report ARS-S-40), United States Department of Agriculture, Washington, D.C., pp. 33–45.

Renner, F. G., and Allred, B. W. (1962). *Classifying Rangeland for Conservation*, Soil Conservation Service Agricultural Handbook No. 235, United States Department of Agriculture, Washington, D.C.

Richter, G., and Negendank, J. F. W. (1977). Soil erosion processes and their measurement in the German area of the Moselle river. *Earth Surface Proc.*, **2**, 261–278.

Roehl, J. W. (1962). Sediment source areas, delivery ratios and influencing morphological factors. *Int. Assoc. Scient. Hydrol. Pub.*, **59**, 202–213.

Roose, E. J. (1966). *Etude de la méthode des bandes d'arrêt pour la conservation de l'eau et des sols*. ORSTOM, Adiopodoumé, Ivory Coast. 23 pp.

Roose, E. J. (1971). *Influence des modifications du milieu naturel sur l'érosion: le bilan hydrique et chimique suite à la mise en culture sous climat tropical*. ORSTOM, Adiopodoumé, Ivory Coast. 22 pp.

Rougerie, G. (1967). Facteurs climatiques et facteurs édaphiques dans la differentiation des teneurs des eaux de drainage en substances dissoutes. In: *L' évolution des versants* (P. Macar, Ed.), University of Liége, pp. 259–270.

Savat, J. (1977). The hydraulics of sheet flow on a smooth surface and the effect of simulated rainfall. *Earth Surface Proc.*, **2**, 125–140.

Schwab, G. O., Frevert, R. K., Edminster, T. W., and Barnes, K. K. (1966). *Soil and Water Conservation Engineering*, Wiley, New York. 683 pp.

Sheng, T. C. (1972a). A treatment-oriented land capability classification scheme for hilly marginal lands in the humid tropics. *J. Scient. Research Council, Jamaica*, **3**, 93–112.

Sheng, T. C. (1972b). Bench terracing. *J. Scient. Research Council, Jamaica*, **3**, 113–127.

Skidmore, E. L., and Hagen, L. J. (1977). Reducing wind erosion with barriers. *Trans. Am. Soc. Agric. Engrs.*, **20**, 911–915.

Smith, D. D. (1956). Time study of parallel terraces. *Agric. Engng*, **37**, 342–345.

Smith, R. M., and Stamey, W. L. (1965). Determining the range of tolerable erosion. *Soil Sci.*, **100**, 414–424.

Sneesby, N. J. (1953). Wind erosion and the value of shelterbelts. *Agriculture*, **60**, 263–271.

Spomer, R. G., Saxton, K. E., and Heinemann, H. G. (1973). Water yield and erosion response to land management. *J. Soil and Water Conserv.*, **28**, 168–171.

Spoor. G. (1975). Fundamental aspects of cultivations. In: *Soil Physical Conditions and Crop Production*. MAFF Technical Bulletin No. 29, pp. 128–144.

Spoor, G., and Muckle, T. B. (1974). Influence of soil type and slope on tractor and implement performance. In: *Soil Type and Land Capability* (D. Mackney, Ed.), Soil Survey Technical Monograph No. 4, pp. 125–134.

Stallings, J. H. (1957). *Soil Conservation*, Prentice-Hall, Englewood Cliffs, N. J., 575 pp.

Starkel, L. (1976). The role of extreme (catastrophic) meteorological events in contemporary evolution of slopes. In: *Geomorphology and Climate* (E. Derbyshire, Ed.), Wiley, London, pp. 203–246.

Stevens, J. H. (1974). Stabilization of aeolian sands in Saudi Arabia's Al Hasa oasis. *J. Soil and Water Conserv.*, **29**, 129–133.

Stocking, M. A., and Elwell, H. A. (1973). Soil erosion hazard in Rhodesia. *Rhod. agric. J.*, **70**, 94–101.

Strahler, A. N. (1964). Quantitative geomorphology of drainage basins and channel networks. In: *Handbook of Applied Hydrology* (V. T. Chow, Ed.), McGraw-Hill, New York, pp. 4-39-4-76.

Strand, R. I. (1975). Bureau of Reclamation procedures for predicting sediment yield. In: *Present and Prospective Technology for Predicting Sediment Yields and Sources,* —Proceedings of Sediment-Yield Workshop, United States Department of Agriculture Sedimentation Laboratory, Oxford, Mississippi, November 1972 (Agricultural Research Service Report ARS-S-40), United States Department of Agriculture, Washington, D.C., pp. 10–15.

Temple, P. H. (1972a). Soil and water conservation policies in the Uluguru Mountains, Tanzania. *Geogr. Ann.* **54**–A, 110–123.

Temple, P. H. (1972b). Measurements of runoff and soil erosion at an erosion plot scale with particular reference to Tanzania. *Geogr. Ann.,* **54**–A, 203–220.

Thornes, J. B. (1976). Semi-arid erosion systems: case studies from Spain. *London School of Economics, Geogr. Papers, No. 7.* 79 pp.

Turelle, J. W. (1973). Factors involved in the use of herbaceous plants for erosion control on roadways. In: *Soil Erosion: Causes and Mechanisms; Prevention and Control.* Highway Research Board, Special Report No. 135, Washington, D.C., pp. 99–104.

Turner, A. K. (1963). The control of roadside erosion. *Road Research Laboratory Overseas Bulletin No. 17.*

van Genderen, J. L. (1970). *The Morphodynamics of the Crati River Basin, Calabria,* ITC Publications, Series B. No. 56, Delft.

Vanoni, V. A. (1976). *Sedimentation Engineering,* American Society of Civil Engineers, New York.

Verstappen, H. Th. (1970). Introduction to the ITC system of geomorphological survey. *Geogr. Tijds.,* **4**, 85–91.

Verstappen, H. Th., and van Zuidam, R. A. (1968). *ITC System of Geomorphological Survey,* ITC Publications, Delft.

Voetberg, K. S. (1970). *Erosion on Agricultural Lands,* Agricultural University, Wageningen.

Williams, A. R., and Morgan, R. P. C. (1976). Geomorphological mapping applied to soil erosion evaluation. *J. Soil and Water Conserv.,* **31**, 164–168.

Wischmeier, W. H. (1975). Estimating the soil loss equation's cover and management factor for undisturbed areas. In: *Present and Prospective Technology for Predicting Sediment Yields and Sources*—Proceedings of Sediment-Yield Workshop, United States Department of Agriculture Sedimentation Laboratory, Oxford, Mississippi, November 1972 (Agricultural Research Service Report ARS-S-40), United States Department of Agriculture, Washington, D.C., pp. 118–124.

Wischmeier, W. H., and Smith D. D. (1978). *Predicting Rainfall Erosion Losses.* Agricultural Research Service Handbook No. 282, United States Department of Agriculture, Washington, D.C.

Withers, B., and Vipond, S. (1974). *Irrigation: Design and Practice,* Batsford, London, 306 pp.

Wolf, J. J. (1929). The dunes of Sabloney near Arcachon. *Geogr. J.,* **73**, 453.

Wollny, E. (1890). Untersuchungen über die Beeinflussung der Fruchtbarkeit der Ackerkrume durch die Thatigkeit der Regenwurmer. *Forsch. Geb. Agric. Phys.,* **13**, 381–395.

Woodruff, N. P., and Siddoway, F. H. (1965). A wind erosion equation. *Proc. Soil Sci. Soc. Am.,* **29**, 602–608.

Woodruff, N. P., and Zingg, A. W. (1952). Wind-tunnel studies of fundamental problems related to windbreaks. *U.S. Soil Conserv. Serv. Pub., SCS-TP-112.*

Yaır, A. (1972). Observations sur les effets d'un ruissellement dirigé selon la pente des interfluves dans une région semi-aride d'Israël. *Rev. Géogr. Phys. Géol. Dyn.*, **14** 537–548.

Young, A. (1976). *Tropical Soils and Soil Survey*, Cambridge University Press. 468 pp.

Zingg, A. W. (1940). Degree and length of land slope as it affects soil loss in runoff. *Agric. Engng*, **21**, 59–64.

Soil Erosion
Edited by M. J. Kirkby and R. P. C. Morgan
© 1980 John Wiley and Sons Ltd.

CHAPTER 9

Conclusion

R. P. C. Morgan

Senior Lecturer in Applied Physical Geography,
National College of Agricultural Engineering, Bedford, UK

Since the establishment of the Soil Conservation Service in the United States in 1935, the achievements of researchers in soil erosion and conservation have been considerable. Although the Universal Soil Loss Equation has been criticized by many workers because of its weak scientific base and its strange mixture of units, it has been developed as an effective working tool for erosion prediction and conservation design (Mitchell and Bubenzer, Chapter 2). Field trials have been set up comparing erosion under different crops and land treatments and these have been so successful that it is now possible to recommend measures which will probably work to control erosion almost anywhere in the world. At its simplest level, erosion control is no longer a technical or engineering problem but an economic and social one.

Yet, even with the impetus of the Soil Conservation Service and the Department of Agriculture, conservation measures are not widely adopted in the United States (Jacobson, 1969). Very few farmers in Britain practise conservation even in areas where erosion is a problem and hardly any attempt has been made to research suitable measures for British agriculture (Morgan, 1980). If soil conservation is not adopted in the developed world, it is scarcely surprising that it is not widely practised in the third world where erosion is a great cause of concern. The need for soil conservation was emphasized by Bennett (1939), Jacks and Whyte (1949) and many others in the mid-1930s to early 1950s but so little has been achieved that more recent reports (FAO, 1965; Fournier, 1972) are virtually indistinguishable in factual content and sentiment from the earlier studies. The need for soil conservation is as great today as it ever was.

What, therefore, is the way forward? Clearly, much research is required on the economics of soil conservation and the reasons why soil conservation practices are not taken up. Economists and sociologists must become more involved with this work because conservation schemes which are economically and socially attractive will not be developed by agricultural engineers, soil scientists

and geomorphologists alone. Although much of this development is outside
the scope of this book, it provides the framework for examining the future role
of the geomorphologist and the agricultural engineer which is of concern here.
The basic advisory and research tasks of these workers, related to the imple-
mentation of soil conservation measures, are likely to remain unchanged but,
given the need to concentrate on measures which are more acceptable, it will
be necessary to examine the relative merits of different conservation strategies
more closely. The techniques used in designing conservation systems will also
need to be refined to provide a greater range of solutions to a particular problem
and allow more flexibility in selecting the most appropriate strategy.

One line of approach, hinted at by Mitchell and Bubenzer (Chapter 2) is to
modify the Universal Soil Loss Equation. In the long-term, this is likely to
prove limiting, however, and an alternative approach, advocated throughout
this book, is to return to basic principles of the mechanics of erosion and from
these build up a model of the erosion system with a sound physical base within
which the effects of conservation measures can be incorporated. The techniques
to collect the data required with a properly executed experimental design have
been described by De Ploey and Gabriels (Chapter 3). The existing knowledge
of erosion processes and their mechanics have been reviewed by Evans (Chapter
4) and Thornes (Chapter 5) for water and Cooke and Wilson (Chapter 7) for
wind. The basis for model building has been outlined by Kirkby (Chapter 6)
where a suitable framework for a model of water erosion on hillslopes is
presented. The difficulties of including conservation practices in erosion models
has been indicated by Morgan (Chapter 8). From these individual contributions,
it is possible to list certain priorities for research.

First, the whole question of what constitutes a suitable value for acceptable
soil loss needs to be reappraised. Values in current use are probably too stringent
to be met by many farmers but not rigorous enough to prevent pollution of
rivers and reservoirs and permit sustained use of the land in the long-term. At
present, no satisfactory basis exists for determining appropriate tolerance
values.

Second, efforts must be made to develop research on water and wind erosion
in parallel. In recent years water and wind have been viewed as separate processes of erosion, conditioned by different controlling factors. In reality both
processes are related to the movement of fluids and their ability of detach and
transport soil particles and have many control measures in common, particularly
those focussing on plant covers and surface roughness. Although differences
between the processes exist and these must be recognized, there appear to be
sufficient similarities to make some joint studies worthwhile to provide a
common approach to erosion modelling.

Third, erosion modelling is most likely to be successful if it proceeds from
both a hydrological and a hydraulic base. The hydrological component is
required to predict how much runoff arises from a given amount of rainfall.

An analogous component for wind erosion would predict the surface or drag velocity of wind for a wind speed in free air. The hydraulic component is required to predict the sediment yield for given runoff and wind conditions. This latter component must take account of the availability of sediment for transport and must therefore incorporate the detachment phase of erosion as well as the transporting capacity of the eroding agents. With this in mind, it seems more appropriate to develop a model which expresses soil loss in terms of sediment concentration rather than attempting to relate gross runoff or wind characteristics to sediment yield. Such a model would be applicable to both detachment-limited and transport-limited conditions of erosion. It would also enable the effects of conservation measures to be assessed in terms of their direct effects on soil loss. This, in turn, would allow conservation design work to be related to selected soil loss tolerances rather than to critical values of the tractive force or the velocity of the eroding agent, as is the case with present design procedures.

Fourth, within the confines of the model, more research is required on the erodibility of soils. Although the relevant mechanical properties of soils are well-recognized, work on the effects of chemical properties on erodibility is still developing and must be encouraged. This work will increase our knowledge of the role of soil conditioners and organic matter in controlling erosion. Variations in erodibility with season and with tillage practice also need to be investigated to determine how problem soils can be cultivated with minimal damage to their structure.

Fifth, the effects of surface roughness on erosion require further investigation, particularly with regard to changes in infiltration and erodibility and to the interaction with agronomic measures of erosion control and tillage practice. At present there is no standard procedure for measuring or monitoring roughness in the field even though changing surface roughness is the basis of many widely-used conservation techniques.

Sixth, attention must be given to the applicability of erosion models in both time and space. A model must give reasonable predictions of soil loss over events ranging from the frequent, low magnitude occurrence through moderate events to extreme conditions. It must also predict spatial variations in runoff excess and sediment supply, especially in relation to local points of instability in the landscape and slope shape.

Seventh, erosion models must be applicable at a range of scales from a small slope segment to a large watershed. Models of soil erosion on hillslopes which take account of rill and inter-rill processes must be extended to encompass gully erosion and link in with separate models of stream channel systems. If a large watershed is considered as a series of smaller units, an erosion model may be applied to each unit in turn and a suitable routing procedure developed to carry runoff and sediment from one unit to another. The routing procedure may also be used to test the effectiveness of waterway and terrace systems.

The development of models needs to be placed in the wider perspective of practical conservation work. The latter begins with erosion survey and better links between survey procedures and models for predicting soil loss and assessing the effectiveness of conservation measures must be established. Also, conservation design is only a part of the overall planning of landuse and any erosion model is likely to form a sub-model of a much larger model which looks at the planning process. As an example, an erosion model could be the soil loss predictor in an optimization model designed to plan landuse in a way which would minimise soil loss. One approach to this would be to develop a linear programming model similar to that described by Potter, Morgan, and Noble (1978) for runoff management. Optimization models have also been developed to evaluate the economics of soil conservation on a national scale (Nicol, Heady, and Wade, 1976) and an erosion model could suitably replace the Universal Soil Loss Equation in these.

With all these potentially fruitful areas for research and development, the challenge facing workers in soil erosion and conservation today may be different from that which faced those who set up the Soil Conservation Service in the 1930s but it is no less demanding. It is a challenge which must be met.

REFERENCES

Bennett, H. H. (1939). *Soil Conservation*, McGraw-Hill, New York.

FAO (1965). *Soil Erosion by Water. Some Measures for its Control on Cultivated Lands*, Rome. 284 pp.

Fournier, F. (1972). *Soil Conservation*, Nature and Environment Series, Council of Europe.

Jacks, G. V., and Whyte, R. O. (1949). *The Rape of the Earth. A World Survey of Soil Erosion*, Faber, London.

Jacobson, P. (1969). Soil erosion control practices in perspective. *J. Soil and Water Conserv.*, **24**, 123–126.

Morgan, R. P. C. (1980). Soil erosion in Britain: is it an enigma? *Progress in Physical Geography*, **4**, 24–27.

Nicol, K. J., Heady, E. O., and Wade, J. C. (1976). National and interregional models of land use, water allocation, soil loss control, and agricultural policy. *Trans. Am. Soc. Agric. Engrs*, **19**, 90–96.

Potter, M. F., Morgan, R. P. C., and Noble, D. H. (1978). The application of linear programming to run-off management. *J. Environ. Management*, **6**, 43–55.

Subject Index

acceptable rates of erosion (*see also* Soil loss tolerance), 5, 7, 15, 262, 263, 304
acceptable sediment discharge, 272, 273
aerial photography, 49, 69, 71, 72, 74, 83, 89, 92, 96, 248, 260
afforestation, 173
agronomic conservation measures, 257–258, 259, 262, 276, 278, 283, 291, 292, 305
alluvial deposits, 83
alluvial fans, 154, 156, 165
alluvial plains, 82, 84
angle of repose (*see also* Slope stability, Threshold slopes), 272
animal activity in soil, 133, 146, 151, 170, 171, 202
antecedent conditions, 121, 133, 169
arroyo problem, 172
avalanching of sand, 103, 229

badlands, 71, 74, 164
Bagnold's formula for sand flow, 97, 239, 240, 273
bank erosion (*see also* Channel erosion), 289
barchans, 96, 97
basin-scale controls, 162–166
bed load, 55, 82, 155, 156, 190
bed-load transport equation, 49, 270
bed-load trap, 81
Bernoulli effect, 151, 152
bifurcation ratio, 53, 266
Boltzmann equation, 144
boundary conditions, 67, 184
braiding, 157
burning, *see* Fire

catchment shape, 266–268
catena, 162
cation-exchange process, 143, 144, 154

channel erosion, 48–51, 163, 290
channel flow, 141–143
check dams, 157, 165, 290–291
Chezy roughness coefficient, 133
clay content, 7, 116, 120, 124
climatic change, 172, 174, 175
climatic index for wind erosion, 220, 241–245
colluvial deposits, 83
colluvial fans, 71
colluvial slopes, 82
continuity equation, 135, 186, 206, 212
contour bunds, 285
contour ploughing, 15
contouring, 45
contributing area, 166
convective rainfall, 171
critical tractive force (*see also* Tractive shear stress), 263, 271–272, 305
critical velocity, *see* Maximum safe velocity
crop cover factor, 21, 43
crop management factor (*C*-factor in Universal Soil Loss Equation), 9, 26, 38–44, 171, 278, 283
crop residues, 38, 42, 45, 293
crop rotation, 19, 20, 26, 38, 42, 45, 282–283
crust development, 114, 120, 147–148, 198, 232
cultivations, *see* Tillage
cyclonic rainfall, 171

Darcy's law, 139
Darcy–Weisbach roughness coefficient, 133, 134, 137, 191, 204, 270, 280
debris flows, 153, 156
deflation, 101
deposition, 84, 156, 157, 160, 165, 189, 191, 209, 212, 230–231

307